Topics in Current Chemistry Collections

Aims and Scope

The series *Topics in Current Chemistry Collections* presents critical reviews from the journal *Topics in Current Chemistry* organized in topical volumes. The scope of coverage is all areas of chemical science including the interfaces with related disciplines such as biology, medicine and materials science.

The goal of each thematic volume is to give the non-specialist reader, whether in academia or industry, a comprehensive insight into an area where new research is emerging which is of interest to a larger scientific audience.

Each review within the volume critically surveys one aspect of that topic and places it within the context of the volume as a whole. The most significant developments of the last 5 to 10 years are presented using selected examples to illustrate the principles discussed. The coverage is not intended to be an exhaustive summary of the field or include large quantities of data, but should rather be conceptual, concentrating on the methodological thinking that will allow the non-specialist reader to understand the information presented.

Contributions also offer an outlook on potential future developments in the field.

More information about this series at http://www.springer.com/series/14181

Tiago Buckup • Jérémie Léonard
Editors

Multidimensional Time-Resolved Spectroscopy

With contributions from

Johanna Brazard • Tiago Buckup • Thomas P. Cheshire
Marco Garavelli • Zhenkun Guo • Harold Y. Hwang
Jan Philip Kraack • Jérémie Léonard • Xian Li • Jian Lu
Margherita Maiuri • Brian P. Molesky • Andrew M. Moran
Shaul Mukamel • Keith A. Nelson • Artur Nenov
Benjamin K. Ofori-Okai • Ivan Rivalta • Javier Segarra-Martí
Yaqing Zhang

Springer

Editors
Tiago Buckup
Physikalisch-Chemisches Institut
Universität Heidelberg
Heidelberg, Germany

Jérémie Léonard
Université de Strasbourg
CNRS Institut de Physique et
Chimie des Matériaux de Strasbourg
UMR 7504, and Labex NIE
67034 Strasbourg, France

Partly previously published in Top Curr Chem (Z) Volume 375 (2017); Top Curr Chem (Z) Volume 376 (2018).

ISSN 2367-4067
Topics in Current Chemistry Collections
ISBN 978-3-030-02477-2

Library of Congress Control Number: 2018961135

This Springer imprint is published by the registered company Springer Nature Switzerland AG
The registered company address is: Gewerbestrasse 11, 6330 Cham, Switzerland

Contents

Preface

Constant development of new spectroscopy methods promise unchallengeable breakthroughs in unraveling natural phenomena and in the development of new materials, a crucial aspect in several fields of our technological society. Novel methods based on light interaction with matter have been a prime tool to investigate optical, electronic, structural, as well magnetic properties of matter. Particularly in time resolved spectroscopy, methods involving multiple, i.e. non-linear, light-matter interactions have been rapidly developing to become a very valuable spectroscopy tool in the investigation of new phenomena.

This topical collection focuses on recent developments in so-called multidimensional time-resolved spectroscopy methods. Over the last decade, multidimensional methods have undergone tremendous progress and extended over a very broad spectral range from the THz to the ultraviolet, allowing for investigation of all kinds of (bio)molecules, aggregates, solid compounds and composite systems, only to name a few examples. In spite of that, non-specialists are still facing several obstacles in understanding what kind of information multidimensional spectroscopy contains. One of the major obstacles is the tsunami of experimental techniques variants and terminologies used. In this sense, this topical collection pursues three goals. (i) The first one is to offer a very general overview of all the techniques in one single place. This, we believe, is a unique opportunity for newcomers as well as for users of a specific technique to learn and understand how complementary the content information among techniques can be in spite of differing spectral ranges or other technical features. (ii) The second goal is to review what these techniques can do for scientists of akin areas. This is richly illustrated in each one of the contributions of this Topics in Current Chemistry by discussing several recent experimental and numerical examples. (iii) The last objective is to discuss what are the open experimental and computational questions and the challenges still facing the community.

This collection has been organized to largely reflect these questions. The opening chapter by Kraack and Buckup is strongly recommended to anyone new to the field, since it addresses general points like terminology and classification of multidimensional techniques, which are used in the rest of the collection. It also deals with questions about why or when to use multidimensional spectroscopy. The following contributions focus on the different types of techniques, which were grouped in regard to the spectral region. The second and third contributions by Maiuri and Brazard and Segarra-Marti et al. review the experimental and theoretical developments of 2D Electronic Spectroscopy, respectively. Coupling between electronic states is the central topic of these two contributions. The contribution "Ultrafast structural molecular dynamics investigated with 2D infrared spectroscopy methods" by Kraack reviews in detail multidimensional techniques variants in the infrared spectral region. The topic of the chapter "Multidimensional Vibrational Coherence Spectroscopy" is Stimulated Raman spectroscopies and the mapping of Raman active vibrations in different electronic states. In the contribution "Two-Dimensional Resonance Raman Signatures of Vibronic Coherence Transfer in Chemical Reactions" by Guo et al, the latter topic is expanded to elucidate the coherence transfer process in small molecules. Finally, the contribution by Lu et al reviews the multidimensional techniques in the terahertz spectral region and mixtures of these techniques with other variants.

We would like to thank all authors for their contributions and exceptional effort. The Editors also specially thank the referees for their outstanding comments and insights in the review process of this Topics in Current Chemistry.

Tiago Buckup and Jérémie Léonard

Dr. Tiago Buckup
Physikalisch-Chemisches Institut, Universität Heidelberg, Heidelberg, Germany

Dr. Jérémie Léonard
Université de Strasbourg, CNRS Institut de Physique et Chimie des Matériaux de Strasbourg UMR 7504, and Labex NIE, 67034 Strasbourg, France

Topics in Current Chemistry (2018) 376:28
https://doi.org/10.1007/s41061-018-0206-3

REVIEW

Introduction to State-of-the-Art Multidimensional Time-Resolved Spectroscopy Methods

Jan Philip Kraack[1] · Tiago Buckup[2]

Received: 27 February 2018 / Accepted: 13 June 2018 / Published online: 25 June 2018
© Springer International Publishing AG, part of Springer Nature 2018

Abstract

The field of multidimensional laser spectroscopy comprises a variety of highly developed state-of-the-art methods, which exhibit broad prospects for applications in several areas of natural, material, and even medical sciences. This collection summarizes the main achievements from this area and gives basic introductory insight into what is currently possible with such methods. In the present introductory contribution, we briefly outline the general concept behind multidimensional laser spectroscopy, for instance by highlighting the often-employed analogy between multidimensional laser spectroscopy and NMR methods. Our initial introduction is followed by an overview of the most important and widely used multidimensional spectroscopies' classification. Special emphasis is placed on how the contributing spectral region defines a natural way of grouping the techniques in terms of their information content. On this basis, we introduce the most important graphical ways in which multidimensional data is generally visualized. This is done by comparing specifically temporal and spectra axes that make up each single multidimensional data plot. Several central experimental methods that are common to the various techniques reviewed in this collection are addressed in the perspective of recent developments and their impact on the field. These methods include, for example, heterodyne/homodyne detection, fast scanning, spatial light modulation, and sparse sampling methods. Importantly, we address the central and fundamental questions where multidimensional ultrafast spectroscopy can be used to help understanding chemical dynamics and intermolecular interactions. Finally, we briefly pinpoint what we believe are the main open questions and what will be the future directions for technical developments and promotion of scientific understanding that multidimensional spectroscopy can provide for chemistry, physics, and life sciences.

Keywords Ultrafast laser spectroscopy · Multidimensional spectroscopy · Fourier-transform spectroscopy · Photon echo · Coherence spectroscopy · Molecular interactions · Excited states · Coupling

Chapter 1 was originally published as Kraack, J. P. & Buckup, T. Topics in Current Chemistry (2018) 376: 28. https://doi.org/10.1007/s41061-018-0206-3.

Extended author information available on the last page of the article

1 Introduction

Among all kinds of spectroscopy methods, time-resolved laser spectroscopy has been a major player in the elucidation and discovery of matter properties. A vast range of physical processes and photochemical reactions from the temporal range of a few nanoseconds down to a few tens of femtoseconds (10^{-15} s) have been successfully elucidated for very small molecular systems in the gas phase [1] up to complex photo-activated processes in biology [2] and even single molecules [3]. Due to its versatility and broad spectral application, time-resolved laser spectroscopy is currently a present and necessary partner in the characterization and development of new materials for technologically relevant applications like those found in organic voltaics and solar light harvesting [4–9], as well as in data storage [10, 11] and medicine [12–16]. Very few experimental techniques have such an ample impact on so many different areas of science and technology.

In its most fundamental form, time-resolved laser spectroscopy is based on the detection of a signal after the excitation with a pulsed laser [17, 18]. The so-called pump-probe method describes this general approach, where pump-induced changes in the material properties (such as absorption) are measured by a second probe pulse, and are followed in dependence of the time delay between the two pulses. In spite of its generality and widespread applicability and broad information content, time-resolved laser spectroscopy still faces many challenges. In its classical form, time-resolved laser spectroscopy is often unable to unambiguously assign the origin of features of the optical signal, requiring complex global fitting and data analysis approaches [19, 20]. Similarly, classic time-resolved laser spectroscopy is also experimentally unable, in certain cases, to map and disentangle how molecular structures change in time, how different molecules interact, or how, within a certain molecule [21] or aggregate, electronic [22], vibrational and rotational states are coupled. This is notoriously aggravated for large, polyatomic molecules and for even larger (bio-)chemical systems [23], where different electronic states or molecular species can contribute to the signal.

Several experimental approaches based on classical time-resolved laser spectroscopy methods, such as pump-probe spectroscopy, have been developed to overcome such challenges. One of them is based on the tailoring of the electromagnetic properties of the pump pulse. The electric fields of the coherent laser light are described by the spectral phase and amplitude, and the vectorial nature of the field is best conceptually visualized by the field polarization. All these properties can today be experimentally controlled with highest precision and this control is exploited to selectively induce or enhance specific signals of specific molecules and/or suppress undesired ones [24–27]. There is an ever-growing number of applications with these "control knobs" in this active research field of laser spectroscopy, which is consequently coined quantum control spectroscopy. This method has been used as a highly specialized tool to study, address selectively, and enhance specific transient spectroscopy features [28–33].

Being the topic of this collection, the most common way to address some of the ambiguities of the time-resolved laser spectroscopy is based on the use of the so-called multidimensional methods. There have been several implementations of multidimensional spectroscopy methods, and the label "multidimensional" has been loosely applied to describe different types of techniques in which two axes (frequency–frequency or frequency–time) are used to display the data in a "2D plot". Multidimensional spectroscopy has been carried out, for example, by exploiting the dynamic hole-burning effect in the spectral domain, i.e., by actively controlling the (fairly narrowband) central wavelength of the pump spectrum and correlating it with the corresponding changes in a (broadband) probe spectrum. This method works well from the infrared [34] spectral range that addresses vibrational transitions up to the UV region where electronic transitions with high photon energy are considered [35]. The most wide-spread implementation of multidimensional laser techniques nowadays is, however, based on pulse sequences of laser light with temporal durations of only a few femtoseconds. In this case, the signal is at least partly acquired in the time domain using variations of various time delays, as we will explain further below. The spectral information is then obtained by performing Fourier transformations of the time-domain contributions [36–38].

In a very general way, the basic principle of multidimensional spectroscopy relies on the possibility to manipulate the formation of a so-called photon echo signal from the sample [39]. This optical photon echo has a close relation to the Hahn spin-echo [40], which has been known to chemists from the field of NMR spectroscopy for decades. To generate such an echo response, multiple light–matter interactions are exploited to generate a signal. In NMR spectroscopy, the signal stems from the sample's macroscopic magnetization after radio-frequency excitation, and the sample is held in an external magnetic field that is inevitable to only make the otherwise energetically degenerate spin states spectroscopically "visible". In contrast to that, in optical spectroscopy, the signal source is an electric polarization that originates from light–matter interactions between several laser pulses and the sample's transition dipoles. Note that in this case no other external fields are necessary due to the naturally occurring energetic splitting of optical transitions based on electronic, vibrational, or rotational quantum numbers. Despite these very fundamental differences in the investigated systems, the experimental concept is transferable and the analogy between NMR and optical spectroscopy has often been drawn, for instance by using Bloch vector representations to visualize the interaction sequences [41]. In such a description, which will be repeatedly used in this collection in the context of various methods, a first pulse in a general photon echo experiment induces an oscillating electric polarization in the sample. This polarization dephases over time due to time-dependent transition frequencies in the sample. The time dependence of the transition frequencies stem from interactions of the molecules with their environment. Dephasing causes a so-called free-induction decay of the polarization, the evolution of which is mapped out in time by a second laser pulse. That second interaction induces a new polarization in the sample, which is also subjected to relaxation processes, i.e., energy relaxation or again dephasing, dependent on the underlying interactions. Finally, a third light–matter interaction is used to generate another polarization, which radiates again an oscillatory signal field off the sample. Dependent on the exact choice of experimental parameters, i.e., combinations of frequencies and wave-vectors of the preceding light–matter interactions, a photon echo may now form as a signal response, or not. The choice can easily experimentally be made by clever choices of the angles between the different laser beams in combination with associated temporal delays. The basis of the echo formation is associated with a "rephasing" of the Bloch vectors, but only if the evolution of the free induction decay (FID) in the first and third evolution period occurs with different signs of frequencies. If during these two time periods the frequencies exhibit the same sign, no echo will form, and dephasing simply proceeds (non-rephasing) (Fig. 1). Due to the fact that, at least as a theoretical construct, an echo could have formed at an evolution time before the third interaction actually took place, the latter case is sometimes referred to as the "virtual echo" [42, 43].

2 Classification of Multidimensional Techniques

Multidimensional time-resolved spectroscopy can be generally classified regarding the electromagnetic properties of the excitation pulses. The first natural property that is controllable experimentally is the spectral region of the employed light fields (Fig. 2). Multidimensional time-resolved experiments have been successfully demonstrated in a very broad spectral range from the terahertz [44] up to the UV [35, 45, 46]. The exact spectral location of the involved pulses is an important parameter in controlling the type of information that can be harnessed with a certain method. On the one hand, experiments involving high photon energies (e.g., UV–Vis), which are resonant with molecular electronic transitions, are sensitive to interactions between electronic states. On the other hand, experiments involving low-energy photons, like those in the mid-IR or THz range, are mainly sensitive to vibrational transitions in the electronic ground state of molecules. However, these methods can report on electronic dynamics as well, if for example condensed matter samples are investigated that exhibit highly delocalized electronic states such as solids (metals, semiconductors, quantum wells, etc.). In this context, an important property often encountered in time-resolved signals is the laser-induced generation of a coherent superposition of quantum states of the sample. Such a superposition

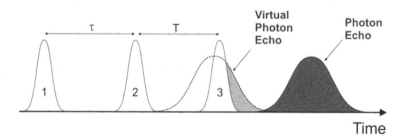

Fig. 1 Scheme of the formation of a photon echo (*black shaded*) and a "virtual" photon echo (*traced line*), i.e., when no photon-echo is formed. The virtual echo has in principle the same shape as the photon echo, but its maximum occurs before the last interaction with the third pulse takes place. Only the tail of the virtual photon echo (*gray shaded*) is detected in an experiment

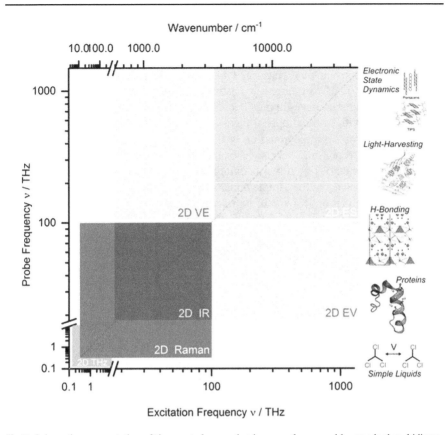

Fig. 2 Schematic representation of the spectral range that is currently covered by standard multidimensional optical techniques together with sketches of typical molecules that are investigated with these methods. *THz* terahertz, *IR* infrared, *ES* electronic spectroscopy, *EV* electronic/vibrational, *VE* vibrational/electronic (spectroscopy). Note the different scales used for the THz and Raman/IR/electronic regions. Molecular schemes, *from top to bottom*: Pentacene molecules adapted from Ref. [134] with the permission of Springer Nature. FMO complex structure adapted from Ref. [135] with permission from Nature Physics. H-bonding structure adapted with permission from Ref. [136]. Copyright 2016 American Chemical Society. Protein structure adapted from Ref. [137] under a creative commons attribution 3.0 unported licence

can be either electronic, vibronic, vibrational or rotational in nature, dependent on the spectral region of investigation, but also dependent on the bandwidth of excitation. Induction of coherence can require for example one short, or two (at least approximately) simultaneous excitations of the sample with multiple frequencies. Coherences between vibrational states can be directly induced in this way in the electronic excited- as well as ground-state, e.g., by a spectrally broadband excitation pulse via a stimulated Raman-type process [47, 48]. As the bandwidth of excitation spectrum provided by the laser light and the time-dependence of the associated electric field are related by a Fourier transformation [49], the possibility to observe such coherences is strongly dependent on the temporal duration of the laser pulses. This effect has motivated an everlasting race in the field of light source development for generating shorter and even shorter pulses with a currently supposed record of about 40 attoseconds (10^{-18} s) [50].

Another way of classifying multidimensional time-resolved techniques is related to the ability to disentangle rephasing (optical echo formation) or non-rephasing (no echo) processes [51]. Most currently existing multidimensional spectroscopy methods are so-called "four-wave-mixing" techniques. In such methods, three light–matter interactions with external optical fields (input) take place and the fourth interaction (output) stems from the emission of the signal light (see Fig. 1). As discussed above, the term "rephasing" originates from the formation of an echo signal due to the different interactions with the incoming pulses, and its existence and dynamics are inherent properties of the sample under study. It is important to stress once again that whether a given optical signal is based on a rephasing or non-rephasing process can be experimentally controlled by the chosen geometry of the incoming and detected beam as well as the temporal ordering of the involved pulses [42]. The use of these experimental parameters is carefully discussed in the following contributions for each experimental technique.

3 2D Plots

As briefly mentioned in the initial section, there are many ways of displaying the time-resolved data that are acquired in multidimensional spectroscopy methods. In most cases, the representation involves the spectral domain using optical frequency representations. Very rare is the direct depiction of the time delay itself between two laser interactions as an axis to display the data. Only few examples exist in the literature, for which the optical period of the light used to generate a signal is so long that it is comparable to dephasing times [52]. The 2D plots associated with multidimensional techniques often rather exhibit data that is processed via a Fourier transformation of time-resolved signals. The reason to use the spectral representation is simply that the signals are intuitively much easier to interpret as compared to the time domain counterparts (just like one would normally analyze a linear absorption signal in the spectral domain rather than in the time domain). The way the transformation is performed depends not only on the experimental details of each technique but also on the nature of the sought molecular property, or dynamics of interest. In essence, most 2D plots share the core idea of correlating an excitation with a detection frequency (Fig. 3a, b, d). However, there exist important special cases in which one axis is represented by a delay (e.g., TR-IVS, Fig. 3c).

Two spectral axes (with frequency (THz) or energy units (eV, cm^{-1})) are often employed in 2D electronic, infrared, and terahertz spectroscopy methods (Fig. 3a, b). The two axes in these cases are obtained in a very similar way. The detection frequency axis is generally obtained by recording the signal spectrum directly with a spectrometer. This can be done either by measuring the spectral intensity, or by measuring the electric signal field by an experimental trick, i.e., through interference with a known local oscillator (heterodyne detection). The second axis is named the excitation axis and is often either obtained from direct spectral scanning of the frequency of the excitation pulse (for example, with a tunable spectral filter), or by a

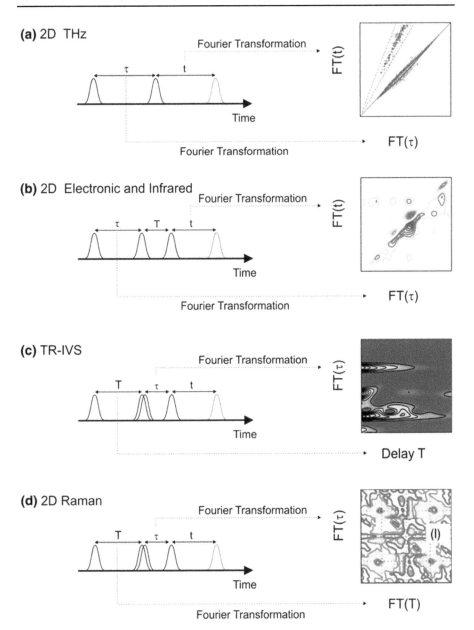

Fig. 3 Schematic 2D plots and pulse representations in different multidimensional time-resolved techniques. **a** 2D THz, **b** 2D electronic and infrared spectroscopy, **c** time-resolved impulsive vibrational spectroscopy, **d** 2D Raman spectroscopy. The selected pulse sequence schemes (*left*) represent the most common implementations of each multidimensional spectroscopy technique. Each method, however, has several different variants, for example, involving different number of individual pulses, beams, and geometries. Copyrights for 2D Maps: **a** Adapted from Ref. [138] with permission of National Academy of Sciences. **b** Adapted from Ref. [139] with permission of Springer Nature. **d** Adapted from Ref. [140], with permission of AIP Publishing

Fourier transformation of the detected signal in dependence of the delay between the first two laser interactions.

2D Raman and other Raman-based multidimensional techniques (Fig. 3c, d) use a slightly different approach. The reason for this variation is a fundamental difference between an absorptive UV–Vis, IR, or THz interaction and a scattering-based (e.g., Raman) excitation process. Time-resolved impulsive vibrational spectroscopy methods for instance (like time-resolved impulsive vibrational spectroscopy (TR-IVS) [53, 54] or pump degenerate four-wave-mixing (pump-DFWM) [55, 56], Fig. 3c) plot the amplitude of active Raman modes in dependence of time. Here, the Raman-frequency axis is obtained by a Fourier-transformation of an oscillatory signal in dependence of the probe interaction delay. The second axis is a purely temporal axis, i.e., a delay between the first pair of excitations, which is most often not Fourier-transformed [57]. 2D Raman goes one step beyond that and performs even a second Fourier transformation on the initial temporal axis. This analysis leads to a frequency–frequency correlation 2D plot with Raman frequencies on the two axes. These two axes allow correlating the induced Raman frequencies in different time intervals, in a similar fashion as in other 2D spectroscopy methods such as 2D IR. Importantly, TR-IVS and 2D Raman methods are of higher-order nonlinearity (in fact fifth-order) than predominately absorptive UV–Vis/IR/THz methods (third-order). This implies that these methods allow and require scanning additional temporal delays. As a consequence, the information content may be significantly different in these two types of methods. To overcome this difference, third-order methods are often combined with pre-excitation pulses or with other perturbations not involved in the axis of the 2D plot. The additional delays that emerge from the different perturbations allow generating an "evolution" of 2D plot [58–60]. In spite of this additional time axis, each individual 2D plot is still generated in the same way and is based on the same idea of correlating quantities as described above.

Finally, we note an additionally important aspect of 2D spectroscopy. This is the inherence of the associated lower-dimensional time-resolved spectroscopy signals from the respective multidimensional plot. In simple words, for example, the excitation axis can always be projected on the detection axis, thereby reducing the data to one-dimensional difference spectra (the projection slice theorem) [41]. In some methods, however, it is not very obvious to recognize this effect, but for other methods the interpretation can be straightforward, which is often even exploited as an internal reference to validate the analysis and interpretability of the data.

4 Experimental Aspects of Multidimensional Time-Resolved Spectroscopy

Multidimensional time-resolved spectroscopy (TRS) signals, like all optical signals, can be detected in different ways. There are two major categories of detection schemes, which are shared by all spectroscopy methods in the next contributions, namely homodyne and heterodyne detection (Fig. 4). These two methods of detecting the optical signal are so vital to multidimensional techniques that the differences as well the advantages and drawbacks between homodyne and

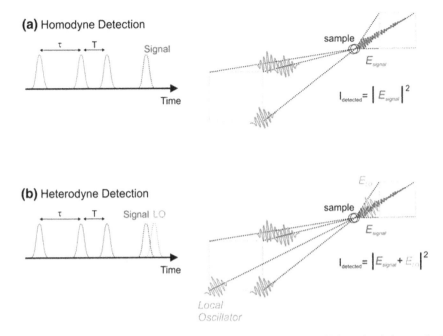

Fig. 4 Detection scheme of **a** homodyne and **b** heterodyne detection in multidimensional time-resolved spectroscopy. *Orange* and *red pulses* refer to the excitation pulses and the signal light, respectively. The local oscillator, which is only present in heterodyne detection, is depicted in *blue*

heterodyne detection have been thoroughly discussed in many reports. This is not repeated here and we refer the readers to Refs. [41, 51] for more details. As a bottom-line, however, we would like to briefly stress that the homodyne variant is the by far experimentally easier and cheaper method, but also contains limited obtainable information content as compared to heterodyne detection.

To explain this difference briefly, a standard optical signal acquired by a detector such as a photodiode or a photomultiplier is an integration in time over the square modulus of the signals electric field generated from a sample. This is the most direct way of detecting optical signals and it is used in several fields in optical science, including time-resolved spectroscopy. In spite of its very easy experimental implementation and widespread relevance, this way of measuring the intensity of an optical signal irradiated from a sample is termed homodyne detection, and it does not deliver the full information on the optical field. Since the electric field is detected and squared, the phase of the electric field is completely lost. This phase can be, however, retrieved in heterodyne detection, i.e., when the signal light is spectrally and spatially overlapped with an additional and well-characterized external electric field, usually called local oscillator (LO). The detector in this case will see the sum of both electric fields, and will allow obtaining the full phase information of the emitted electric field via the characterization of an interference term between the LO and the signal. Heterodyne detection in the field of multidimensional spectroscopy has been pioneered by Joffre's, Fleming's's, and Miller's groups [61–63].

Another central aspect in the state-of-the-art methods of time-resolved multi-dimensional spectroscopy is the scanning of the time delay between laser pulses. Mechanically moving stages takes time and thus 2D spectroscopy methods involving three or more beams take intrinsically longer for acquiring the complete dataset than techniques with a single delay to scan. It is important to note in this context that in order to allow for averaging of signals, interferometric phase stability between the beams is generally required in 2D spectroscopy methods, leading to challenging experimental constraints when delays are actively scanned. However, interferometric phase stability is often required mainly between given pulse pairs, e.g., between the first two pulses, or between the signal and the local oscillator. Phase stability to achieve experimentally becomes the more challenging the shorter the wavelength of the applied laser light is. Hence, the technically most demanding experiments are spectrally located in the UV region, but advances in technical instrumentation have made it possible to obtain phase stability as good as ca $\lambda/187$ over half an hour at ca. 310 nm [64]. Such drastic experimental requirements have led to the development of several novel experimental approaches regarding the achievement of phase stability. The desired phase stability can generally be achieved in several ways. For instance, the use of diffractive optics [61, 65–67] helps to generate phase–coherent pairs of laser beams at a given geometry. In general, the beams generated that way are then delayed with optical wedges [68, 69] rather than delay stages to delay the associated pulses, which allows an advantageous way to maintain phase stability. Alternatively, one can use for example acousto-optic modulators (AOM), liquid crystal masks, and "liquid–crystals on silicon" (LCOS) shapers to generate phase–coherent pulse pairs [70–75]. Both variants allow for the generation of simple beam geometries and pulse delaying with intrinsic phase stability and therefore reducing time for experiments. In addition to these aspects, optical phase modulators allow for a facile and precise generation of desired phases for independent pulses, thereby making it possible to "switch" phases on demand, which is used for example to suppress or even eliminate signal contributions upon averaging (phase cycling) [76].

The experimental time required to scan several delays can be further optimized by exploiting on-the-fly scanning techniques, often called rapid-scan or fast scan [77, 78]. Instead of scanning the delay "step-by-step" between two pulses, and for each delay step averaging a certain number of laser pulses, rapid scan techniques measure the signal only one time at a single time delay (i.e., a single pulse), after which it moves to the next delay [79]. This method of data acquisition offers several advantages. The first one being that a sequence of delays can be potentially measured much faster than by halting at each delay to acquire the signal. It requires, however, the ability to measure the delay instantaneously. This is offered off-the-shelf in several commercial delay stages but has also been implemented by exploiting interferometric detection of continuous wave reference beams [80], or by using spatial light modulators for generating pulse delay. Significantly reducing the time that it takes to acquire a certain measurement, the main advantage of rapid-scan techniques is the ability to filter out the low-frequency noise typical of ultrashort laser sources.

In a similar context of reducing measurement time, sophisticated sampling techniques have been developed (Fig. 5). These methods make use of the possibility to drastically reduce for example the number of data points required to construct an

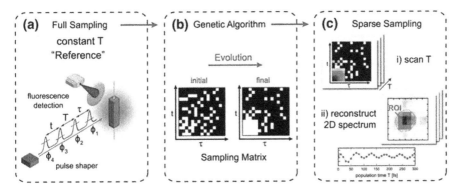

Fig. 5 Acquisition strategy using an optimized sampling approach and pulse shaper. **a** Acquisition of a reference data set at a constant population time T with a full sampling of (τ, t)-parameter space. **b** Optimization of the reduced sampling matrix using the reference data set and a genetic algorithm. **c** Acquisition of 2D spectra at all other population times with the optimized sampling matrix. Figure adapted from Ref. [141], under a Creative Commons Attribution (CC BY) license

interferogram in the time-domain. So-called under-sampling [81, 82] or sparse sampling [83] also requires a high degree of phase-stability as well as an accurate readout of the delay. However, reliable measurements of this sort are technically feasible and nowadays routinely used in laser labs around the world [79, 84–86].

Finally, the time for measurements can be simply and drastically reduced by increasing the repetition rate of the light source [87]. Commercially available amplified laser systems in useful frequency ranges have long been restricted to sub-10 kHz repetition rates, i.e., delivering light bursts only about every 100 μs. Recently, technological breakthroughs in developments of new laser variants have made it possible to shorten the shot-to-shot delay down to less than 10 μs (> 100 kHz). In this context, it is important to consider systems that work at a stability that is comparable to the highly developed few kHz systems, but it has been shown that an increase of repetition rate can indeed lead to a proportionally reduced measurement time [88]. Developments on other routes currently aim at increasing the repetition rate even further (~MHz) [89]. Given the fairly high laser pulse energies that are often used in the experiments (several nano to micro Joules per pulse), however, these developments are all but straightforward and require a careful choice of the employed optics to prevent effects of heat generation and sample damage as well as sufficiently rapid replenishment of the investigated sample volume.

5 What Can a Chemist Learn from Multidimensional Time-Resolved Spectroscopy?

This is the question with which ultrafast spectroscopists are most often challenged when talking to a colleague from a different chemistry discipline. To come up with an answer to this question, it is useful to think about conceptual analogues to the optical methods. The technique that is most straightforward to understand for a chemist is possibly NMR spectroscopy. Fortunately, this is also the method with which chemists are most intimately familiar in terms of signal interpretation. There exist many analogies between multidimensional optical spectroscopy and nuclear magnetic resonances [90–92]. For instance, quantized molecular states (either vibrational, electronic, or other) are the analogues of the spin configurations, albeit without a need to make them spectroscopically visible by the application of electric or magnetic external fields. Also in optical spectroscopy, the properties and dynamics of the molecular states are obtained by following the (laser-induced) FID in time. As will be repeatedly demonstrated in the following contributions, spreading of the optical signals in two dimensions allows for example the precise determination of spectral linewidths and their dissection into homogeneous and heterogeneous contributions. The disentanglement of heterogeneity in the molecular spectra will be one of the central topics in three contributions of this collection, namely the 2D electronic, 2D infrared, and 2D Raman spectroscopies. Heterogeneous contribution to the signal is widely known to be associated with the microscopic dephasing (fluctuations of energy levels). Such information cannot easily be determined from linear or one-dimensional spectroscopy methods like transient absorption.

Also in close analogy to, e.g., widely known multidimensional COSY and NOESY experiments in NMR spectroscopy, there exists tremendous interest in optical multidimensional spectroscopy regarding the observation of so-called cross peaks between different resonances. Such cross peaks may be observed for all types of different resonances and can have manifold origin that need to be, however, evaluated on a case by case basis. For example, static 2D spectra can report on the existence of electrostatic (through space), mechanic (through bond) or electronic coupling in a sample system [93]. In case of optical measurements, the relative intensity between diagonal and cross peak signals is very sensitive to structural properties of the investigated sample, i.e., to relative distances and angles, which determine the coupling strengths. Careful determination of relative intensities under different experimental conditions can then be used in conjunction with theoretical models of different levels to determine precisely molecular structure [33, 93, 94], or even the structure of molecular aggregates [95]. The contribution on 2D IR spectroscopy [96] depicts how multidimensional techniques using infrared light can be applied to recover structural information from vibrational coupling. Cross peaks and couplings are also the topic of other contributions. Intra- and intermolecular coupling between electronic states is reviewed in the 2D electronic spectroscopy contribution. The contribution on 2D THz spectroscopy deals with low-frequency couplings, e.g., between intramolecular vibrational modes in the liquid phase. Similarly important is the calculation of such couplings. Recent development of numerical methods to calculate the complete and accurate 2D electronic response of chromophores including inter-chromophoric coupling is presented in the last contribution of this collection.

Thinking moreover about molecular dynamic effects in NMR spectroscopy, one needs to consider the time dependence of both diagonal as well as cross peaks. In multidimensional NMR spectroscopy, the diagonal peaks are generally stationary, since the possibility of a "population time axis" is often not necessary and not exploited. This is clearly different in the optical counterparts, where extensive use is made out of the fact that pump-induced difference signals are detected and the sample systems undergoes different forms of relaxation. Next to the type of trivial information on energy relaxation for which no multidimensional methods would be needed, significant information is revealed by, e.g., the temporal evolution of infrared and Raman vibrational frequencies and of the multidimensional line shapes of the signals [97–99]. Frequency shifts during the evolution of a photochemical reaction can be traced back to how chemical bondings are forming and breaking in the femtosecond and picosecond time scale. The multidimensional vibrational coherence spectroscopy, but also the 2D infrared and 2D Raman contributions, show for photoreactions involving small (e.g., photodissociation of triiodide, hydrogen bonding) as well as large molecular system (e.g., isomerization of stilbene) how reaction mechanisms can be developed from such experimental frequency shifts. The second significant information provided by multidimensional spectroscopy, i.e., the evolution of lineshapes, is originated due to the changes in the environmental interactions through structural fluctuations on the timescale of femtoseconds to picoseconds and longer after excitation. Such effects are known to cause "spectral diffusion", which is highly relevant for vibrational spectroscopy methods, but to a lesser extent than electronic counterparts [100]. Detailed insight into the origin of this type of evolution is then generally obtained from a comparison of experimental signals and simulations based on molecular dynamics program packages. Even more important, however, are the temporal dynamics of the cross peaks, which may exhibit kinetic (incoherent) as well as oscillatory (coherent) contributions. In this regard, incoherent dynamics can generally originate from energy transfer dynamics in the weak coupling limit between a donor and acceptor state, just like the effect often observed in Förster resonant energy transfer spectroscopy [101, 102]. The dynamics of such energy transfer is a good source for determination of structural properties of the sample, since the transfer rates are highly sensitive to molecular orientation and distances [103, 104]. Another source of incoherent evolution of cross peaks is represented by chemical exchange, i.e., the inter-conversion between different molecular states that are spectroscopically distinguishable. Good examples in this context are the breaking and new formation of chemical bonds like those encountered during the isomerization of double bonds in organic molecules, or the loss/rearrangement of ligands in metal complexes [105–107]. A crucial requirement for these incoherent effects to be observed in multidimensional optical signals is the matching of the timescales for the relevant processes with the energy relaxation dynamics of the pump-induced perturbations. If energy relaxation is too fast compared to energy transfer or chemical exchange, there is simply not enough time for an appreciable transfer or a reaction to occur before the difference signals have ceased. Different strategies have been devised (e.g., "triggered exchange spectroscopy") [90] to overcome that limitation of ultrafast spectroscopy, some of which are discussed in detail in the contribution on infrared multidimensional spectroscopy [96]. Regarding the second point of coherent contributions, the strong coupling regime, for example, can result in oscillatory cross peak intensities, depending on the modulus of the coupling strength relative to the spectral separation of interacting resonances [41]. Next to this, coherent oscillatory contributions in electronic multidimensional spectra may originate from both coherent coupling of electronic states as well as from vibrational wave packet evolution in ground- or excited electronic states. Much effort has recently been undertaken in order to dissect these contributions and an accurate interpretation is often not trivial [100, 108–111]. The role and interpretation of vibronic coupling is a major topic in the 2D electronic spectroscopy contribution.

6 How is this Collection Organized

There are several ways to classify the different kinds of spectroscopies. As discussed previously, the spectral range is a useful way of doing this because it leads to a separation of the most important physicochemical processes involved. In this regard, four contributions of this volume focus mainly on different aspects of vibrational dynamics and interactions, while two contributions deal with predominately with electronic interactions. The reader will notice that though, some physicochemical processes like, e.g., the role of vibrational coupling will eventually appear in more than one contribution, but in different contexts or in different electronic states.

The interaction of very low photon energies (see Fig. 2) with small molecules and other simple physical systems is addressed in the contribution titled "Two-dimensional spectroscopy at terahertz frequencies" by Lu et al. Multidimensional terahertz spectroscopy is used to understand examples like gas-phase molecular rotations, spin precessions in magnetic systems, as well as liquid molecular dynamics and interactions. The contribution also features how terahertz methods can be combined with optical excitation and detection through non-resonant Raman scattering, leading to hybrid methods like 2D THz-Raman techniques.

The next photon energy range is covered by the contribution "Ultrafast Structural Molecular Dynamics Investigated with 2D Infrared Spectroscopies" by Kraack. Several examples of physicochemical processes like vibrational coupling in molecules, spectral diffusion dynamics, chemical exchange of chemical bond formation and breaking, intra- and intermolecular energy transfer are discussed in the light of various kinds of 2D infrared spectroscopy methods. The contribution also covers in detail several state-of-the-art variants of 2D infrared spectroscopy involving, e.g., electrochemistry, diffraction-limited microscopy, transient applications of 2D IR methods, as well as hybrid methods like vibrational-electronic methods are presented.

The role of Raman vibrations in the electronically excited state dynamics is the focus of the contribution "Multidimensional Vibrational Coherence Spectroscopy" by Buckup and Leonard. In particular, the examples portray mapping of structural dynamics along a photoreaction, identification of transient molecular species, and chemical/structural heterogeneity of complex molecular systems in condensed phase.

Another example of how structural dynamics along a photoreaction can be monitored by multidimensional spectroscopies is discussed in the contribution "Two-Dimensional Resonance Raman Signatures of Vibronic Coherence Transfer in Chemical Reactions" by Guo et al. In particular, a variation of a 2D Raman spectroscopy technique, labeled 2D resonance Raman, is used to highlight correlated distributions of reactant and product geometries and structural heterogeneity in an ensemble for a prototype photodissociation reaction of triiodide.

Finally, the last two contributions deal with two aspects of 2D electronic spectroscopy. The contribution titled "Electronic Couplings in (Bio-) Chemical Processes" by Maiuri and Brazard presents the experimental aspect of 2D electronic spectroscopy techniques and the role of these techniques to elucidate ground-state heterogeneity, excitation energy transfer mechanisms, photo-induced coherent oscillations associated with couplings. This experimental approach is extended by the theoretical contribution "Towards Accurate Simulations of Two-Dimensional Electronic Spectroscopy" by Segarra-Martì et al. It details how 2D spectroscopy can be used to characterize the ground-state conformational state of small biological relevant molecules by tracking inter-chromophoric electronic couplings.

7 Conclusions and Outlook

Although the development of techniques, tools, and methods of multidimensional time-resolved spectroscopy have come a long way in the last two decades, researchers have still not managed to establish them as routine spectroscopy tools for chemical analytics. Commercial off-the-shelf setups like those commonly found for transient absorption are currently only available for some 2D infrared and electronic spectroscopy methods from only a handful of companies, while other multidimensional time-resolved spectroscopy methods still use highly specialized optical setups and require skilled users to operate them.

Several of the required technologies that allow multidimensional time-resolved spectroscopy methods to be established as routine spectroscopy tools are already available and have been demonstrated, as discussed above. Fast and easy data acquisition via spatial light modulators, for example, is in this regard an enabling technology, which has the capacity to generate giant leaps in performance and user-friendliness. Other required technologies are still not completely and routinely available for a very broad spectral range from the THz to the UV region, like the turn-key generation, manipulation and characterization of ultrashort, spectrally broadband and tunable pulses, and will perhaps still remain in this status for some time to be restricted to specialized optical labs.

Another point that has prevented the widespread distribution of ultrafast multidimensional spectroscopy in chemistry-related disciplines is strongly connected to the detailed interpretation of the data. It has become clear from a variety of reports, that very often there does not exist a single, unambiguous, and "smoking-gun" interpretation of the signals. In contrast, the vast range of possible contributing molecular dynamics, pathways, as well as possible intermolecular interactions introduce a considerable complexity to the signals and requires sophisticated data treatment, separation of various contributions by multidimensional fitting, model simulations with molecular dynamics based on different force fields and parameters, or often even theory support from high-level and costly quantum chemistry protocols. These points make clear that an accurate interpretation of the signals generally requires detailed theoretical knowledge already by the experimentalists themselves, along with long experience in the different fields of experimental and theoretical ultrafast spectroscopy. As this fact is no technical issue, it will ultimately persist as challenging to researchers in establishing ultrafast spectroscopy as a standard analytical tool.

Despite these still-existing challenges, it is remarkable to recognize that multidimensional laser spectroscopy has evolved into and is still strongly evolving as a highly active field of research that reflects an enormous amount of curiosity and inventive genius of many of the contributing researchers. From the dozens of contributing labs that are active both in development of new methods as well as in testing of possible boundaries of existing ones, there still exists a continuous output of novel scientific and technological insight. For instance, a highly active branch of research is dedicated to combining methods from different frequency ranges. This way, sophisticated combinations of UV–Vis/IR methods have been developed, and it can be anticipated that other mixtures (i.e., UV–Vis/THz) will be demonstrated shortly in the near future. As a side note, even a combination of UV–Vis and NMR has been presented [112]. Moreover, it is also important to keep in mind that most of the so-far-demonstrated methods exclusively rely on the detection of optical signals derived from laser light [113–115]. However, other approaches can be envisioned and are currently being tested such as detection of photocurrents [116–118]. Other labs have even succeeded in combining ultrafast multidimensional optical spectroscopy with the advantages that detection of ultrafast photoelectrons provide, which feature a severely shorter de-Broglie wavelength and, therefore, allow for extreme spatial resolution down to the nanometer-length scale. Such ultrafast nanoscopy [119] is not yet fully explored in its capabilities, but holds great promise regarding possible insight into highly spatially confined processes in chemistry and material science. In a very similar direction, significant efforts are currently being undertaken to advance ultrafast multidimensional spectroscopy to the fewest possible contributing emitters, i.e., ideally single molecules [120]. Such methods might for instance allow studying in detail and in a multidimensional way, how the environment around single molecules influences the spectro-temporal characteristics of different entities.

Multidimensional spectroscopy has been restricted so far to conventional light sources such as table top lasers. It will be interesting and important to see to what extent it will become possible to transfer the corresponding methods and approaches to even shorter wavelengths such as vacuum UV or X-ray ranges [121–123]. These frequency ranges in combination with coherent light emission will, with sufficient photon fluxes, possibly only be available in the near future from large-scale facilities such as the various free-electron lasers that have been set up in the last decades and new ones being currently assembled start operating. There exist current efforts to also develop tabletop variants for these frequency ranges, but these sources are far from being standard. In an attractive manner, such short wavelengths allow the implementation of even attosecond pulses, making available the shortest currently reachable timescales that are relevant mostly for electronic motions. Many different novel phenomena will be possible to be investigated with these methods such as electronic spectroscopy from tightly bound core-near electrons, multi-photon ionization, charge migration after ionization, Auger-processes, auto-ionization, or Coulomb explosions. This way, one might be able to access the ultimately fastest timescales that are relevant for chemistry, physics, and material science, possibly witnessing the impact of electronic movements and non-Born–Oppenheimer dynamics on chemical reactions. It is possible that such new methods exhibit the potential to completely change our understanding of the function and properties of matter as well as provide a clear and portrayable picture of light–matter interaction in terms of absorption of a photon or scattering. Taking into account, however, technical requirements for these methods that need to be fulfilled (e.g., phase stability discussed above), such aspects will become increasingly demanding for shorter wavelengths, given that the optical wavelength is orders of magnitude shorter than currently implemented. Still, we finally trust that this will be eventually possible to advance the understanding of chemistry and physics of, e.g., novel and advanced functional materials as well as fundamental biological matter. We hope that the readers of this volume will overall gain an understanding of the concepts of multidimensional ultrafast spectroscopy, its historical evolution, as well as its great potential that has been developed from this very active research field. As such, this collection outlines how the fields' progress has supported the understanding of several bio-physicochemical processes in recent years.

8 Further Reading

The theoretical concepts, experimental techniques, and simulation methods discussed in this collection of Topics in Current Chemistry can be further explored in several more technical texts. A reader interested in a more basic understanding of ultrafast time-resolved spectroscopy and nonlinear optics is referred to the textbooks of Rulliére [49] and Diels [124]. Though focusing on the techniques of 2D infrared spectroscopy, the book by Hamm and Zanni [41] also presents very well all the basic and advanced aspects of (multidimensional) time-resolved spectroscopy and is strongly recommended.

The technology and concepts behind 2D terahertz spectroscopy have been reviewed recently in a few different contexts. For instance, the fundamentals of time-resolved THz have been covered in Ref. [125], while Ref. [126] provides insights into more advanced concepts such as control of matter with THz laser pulses. Contrasting with that, the literature on 2D infrared spectroscopy is already very extensive, and many excellent overviews have been reported before [106, 127, 128], since this method is in fact often thought of as the predecessor of all other variants. However, it is important to keep in mind that many reviews that appeared over the last years largely focus on special aspects of the spectroscopic understanding at the time of publication. In this regard, the contributions in the context of the present book aim at providing a detailed picture of the latest developments together, an up-to-date outlook for advanced methods and an overall placement of its capabilities for furthering the understanding of chemical physics.

The techniques exploited in the two contributions involving pure Raman methods (multidimensional vibrational coherence spectroscopies and 2D resonant Raman), have been discussed in more technical detail in several sources. A detailed comparison of multidimensional Raman methods has been done in Ref. [54], while a broad technical review of several stimulated Raman spectroscopies variants has been recently presented in Ref. [129]. Other variants of 2D Raman spectroscopy based on

Raman echo effects and challenges of this art of spectroscopy have been discussed in Ref. [130].

There are several interesting aspects, experimental and theoretical, of 2D electronic spectroscopy, which are not addressed in detail in this collection of topics in current chemistry. For example, the combination of pulse shaping and coherent control methods in 2D electronic spectroscopy is a very modern development, as described above, which is reported in Refs. [131, 132]. Finally, the vast number of different experimental implementations of 2D electronic coherent methods can be further read in Ref. [133].

References

1. Dantus M, Rosker MJ, Zewail AH (1988) Femtosecond real-time probing of reactions. 2. The dissociation reaction of ICN. J Chem Phys 89(10):6128–6140. https://doi.org/10.1063/1.455428
2. Sundstrom V (2008) Femtobiology. Annu Rev Phys Chem 59:53–77. https://doi.org/10.1146/annurev.physchem.59.032607.093615
3. Brinks D, Hildner R, van Dijk E, Stefani FD, Nieder JB, Hernando J, van Hulst NF (2014) Ultrafast dynamics of single molecules. Chem Soc Rev 43(8):2476–2491. https://doi.org/10.1039/c3cs60269a
4. Lou YB, Chen XB, Samia AC, Burda C (2003) Femtosecond spectroscopic investigation of the carrier lifetimes in digenite quantum dots and discrimination of the electron and hole dynamics via ultrafast interfacial electron transfer. J Phys Chem B 107(45):12431–12437. https://doi.org/10.1021/jp035618k
5. Robel I, Subramanian V, Kuno M, Kamat PV (2006) Quantum dot solar cells. Harvesting light energy with CdSe nanocrystals molecularly linked to mesoscopic TiO_2 films. J Am Chem Soc 128(7):2385–2393. https://doi.org/10.1021/ja056494n
6. Listorti A, O'Regan B, Durrant JR (2011) Electron transfer dynamics in dye-sensitized solar cells. Chem Mater 23(15):3381–3399. https://doi.org/10.1021/cm200651e
7. Gelinas S, Rao A, Kumar A, Smith SL, Chin AW, Clark J, van der Poll TS, Bazan GC, Friend RH (2014) Ultrafast long-range charge separation in organic semiconductor photovoltaic diodes. Science 343(6170):512–516. https://doi.org/10.1126/science.1246249
8. Falke SM, Rozzi CA, Brida D, Maiuri M, Amato M, Sommer E, De Sio A, Rubio A, Cerullo G, Molinari E, Lienau C (2014) Coherent ultrafast charge transfer in an organic photovoltaic blend. Science 344(6187):1001–1005. https://doi.org/10.1126/science.1249771
9. Jakowetz AC, Bohm ML, Zhang JB, Sadhanala A, Huettner S, Bakulin AA, Rao A, Friend RH (2016) What controls the rate of ultrafast charge transfer and charge separation efficiency in organic photovoltaic blends. J Am Chem Soc 138(36):11672–11679. https://doi.org/10.1021/jacs.6b05131
10. Kirilyuk A, Kimel AV, Rasing T (2010) Ultrafast optical manipulation of magnetic order. Rev Mod Phys 82(3):2731–2784. https://doi.org/10.1103/RevModPhys.82.2731
11. Stamm C, Kachel T, Pontius N, Mitzner R, Quast T, Holldack K, Khan S, Lupulescu C, Aziz EF, Wietstruk M, Durr HA, Eberhardt W (2007) Femtosecond modification of electron localization and transfer of angular momentum in nickel. Nat Mater 6(10):740–743. https://doi.org/10.1038/nmat1985
12. Reichardt C, Guo C, Crespo-Hernandez CE (2011) Excited-state dynamics in 6-thioguanosine from the femtosecond to microsecond time scale. J Phys Chem B 115(12):3263–3270. https://doi.org/10.1021/jp112018u
13. Teuchner K, Ehlert J, Freyer W, Leupold D, Altmeyer P, Stucker M, Hoffmann K (2000) Fluorescence studies of melanin by stepwise two-photon femtosecond laser excitation. J Fluoresc 10(3):275–281. https://doi.org/10.1023/a:1009453228102
14. Ye T, Hong L, Garguilo J, Pawlak A, Edwards GS, Nemanich RJ, Sarna T, Simon JD (2006) Photoionization thresholds of melanins obtained from free electron laser-photoelectron emission microscopy, femtosecond transient absorption spectroscopy and electron paramagnetic resonance

measurements of oxygen photoconsumption. Photochem Photobiol 82(3):733–737. https://doi.
org/10.1562/2006-01-02-ra-762

15. Shim S-H, Gupta R, Ling YL, Strasfeld DB, Raleigh DP, Zanni MT (2009) Two-dimensional IR spectroscopy and isotope labeling defines the pathway of amyloid formation with residue-specific resolution. Proc Natl Acad Sci USA 106(16):6614–6619. https://doi.org/10.1073/pnas.0805957106

16. Alfano RR, Demos SG, Galland P, Gayen SK, Guo Y, Ho PP, Liang X, Liu F, Wang L, Wang QZ, Wang WB (1998) Time-resolved and nonlinear optical imaging for medical applications. Ann N Y Acad Sci 838(1):14–28. https://doi.org/10.1111/j.1749-6632.1998.tb08184.x

17. Tashiro H, Yajima T (1974) Picosecond absorption spectroscopy of excited-states of dye molecules. Chem Phys Lett 25(4):582–586. https://doi.org/10.1016/0009-2614(74)85373-x

18. Shapiro SL, Auston DH (1977) Ultrashort light pulses: picosecond techniques and applications. Topics in applied physics v 18. Springer, Berlin

19. van Stokkum IHM, Larsen DS, van Grondelle R (2004) Global and target analysis of time-resolved spectra. Biochim Biophys Acta Bioenerg 1657(2–3):82–104. https://doi.org/10.1016/j.bbabi o.2004.04.011

20. Ruckebusch C, Sliwa M, Pernot P, de Juan A, Tauler R (2012) Comprehensive data analysis of femtosecond transient absorption spectra: a review. J Photochem Photobiol C Photochem Rev 13(1):1–27. https://doi.org/10.1016/j.jphotochemrev.2011.10.002

21. Cho MH, Vaswani HM, Brixner T, Stenger J, Fleming GR (2005) Exciton analysis in 2D electronic spectroscopy. J Phys Chem B 109(21):10542–10556

22. Ginsberg NS, Cheng YC, Fleming GR (2009) Two-dimensional electronic spectroscopy of molecular aggregates. Acc Chem Res 42(9):1352–1363. https://doi.org/10.1021/ar9001075

23. Ruetzel S, Diekmann M, Nuernberger P, Walter C, Engels B, Brixner T (2014) Multidimensional spectroscopy of photoreactivity. Proc Natl Acad Sci 111(13):4764–4769. https://doi.org/10.1073/pnas.1323792111

24. Brixner T, Gerber G (2001) Femtosecond polarization pulse shaping. Opt Lett 26(8):557–559. https://doi.org/10.1364/ol.26.000557

25. Brixner T, Krampert G, Pfeifer T, Selle R, Gerber G, Wollenhaupt M, Graefe O, Horn C, Liese D, Baumert T (2004) Quantum control by ultrafast polarization shaping. Phys Rev Lett 92(20):208301. https://doi.org/10.1103/PhysRevLett.92.208301

26. Strasfeld DB, Middleton CT, Zanni MT (2009) Mode selectivity with polarization shaping in the mid-IR. New J Phys. https://doi.org/10.1088/1367-2630/11/10/105046

27. Weiner AM (2011) Ultrafast optical pulse shaping: a tutorial review. Opt Commun 284(15):3669–3692. https://doi.org/10.1016/j.optcom.2011.03.084

28. Buckup T, Lebold T, Weigel A, Wohlleben W, Motzkus M (2006) Singlet versus triplet dynamics of beta-carotene studied by quantum control spectroscopy. J Photochem Photobiol A 180(3):314–321

29. Tseng CH, Weinacht TC, Rhoades AE, Murray M, Pearson BJ (2011) Using shaped ultrafast laser pulses to detect enzyme binding. Opt Express 19(24):24638–24646. https://doi.org/10.1364/oe.19.024638

30. Mohring J, Buckup T, Motzkus M (2012) A quantum control spectroscopy approach by direct UV femtosecond pulse shaping. IEEE J Sel Top Quantum Electron 18(1):449–459. https://doi.org/10.1109/jstqe.2011.2138684

31. Consani C, Ruetzel S, Nuernberger P, Brixner T (2014) Quantum control spectroscopy of competing reaction pathways in a molecular switch. J Phys Chem A 118(48):11364–11372. https://doi.org/10.1021/jp509382m

32. Nuernberger P, Ruetzel S, Brixner T (2015) Multidimensional electronic spectroscopy of photochemical reactions. Angew Chem Int Ed 54(39):11368–11386. https://doi.org/10.1002/anie.20150 2974

33. Zanni MT, Ge NH, Kim YS, Hochstrasser RM (2001) Two-dimensional IR spectroscopy can be designed to eliminate the diagonal peaks and expose only the crosspeaks needed for structure determination. Proc Natl Acad Sci USA 98(20):11265–11270. https://doi.org/10.1073/pnas.20141 2998

34. Cervetto V, Helbing J, Bredenbeck J, Hamm P (2004) Double-resonance versus pulsed Fourier transform two-dimensional infrared spectroscopy: an experimental and theoretical comparison. J Chem Phys 121(12):5935–5942. https://doi.org/10.1063/1.1778163

35. Consani C, Aubock G, van Mourik F, Chergui M (2013) Ultrafast tryptophan-to-heme electron transfer in myoglobins revealed by UV 2D spectroscopy. Science 339(6127):1586–1589. https://doi.org/10.1126/science.1230758
36. Tanimura Y, Mukamel S (1993) 2-Dimensional femtosecond vibrational spectroscopy of liquids. J Chem Phys 99(12):9496–9511. https://doi.org/10.1063/1.465484
37. Wefers MM, Kawashima H, Nelson KA (1995) Automated multidimensional coherent optical spectroscopy with multiple phase-related femtosecond pulses. J Chem Phys 102(22):9133–9136. https://doi.org/10.1063/1.468862
38. Mukamel S (2000) Multidimensional femtosecond correlation spectroscopies of electronic and vibrational excitations. Annu Rev Phys Chem 51:691–729. https://doi.org/10.1146/annurev.physchem.51.1.691
39. Kurnit NA, Hartmann SR, Abella ID (1964) Observation of photon echo. Phys Rev Lett 13(19):567. https://doi.org/10.1103/PhysRevLett.13.567
40. Hahn EL (1950) Spin echoes. Phys Rev 80(4):580–594. https://doi.org/10.1103/PhysRev.80.580
41. Hamm P, Zanni MT (2011) Concepts and methods of 2D infrared spectroscopy. Cambridge University Press, Cambridge
42. Pastirk I, Lozovoy VV, Dantus M (2001) Femtosecond photon echo and virtual echo measurements of the vibronic and vibrational coherence relaxation times of iodine vapor. Chem Phys Lett 333(1–2):76–82. https://doi.org/10.1016/s0009-2614(00)01334-8
43. Pshenichnikov MS, deBoeij WP, Wiersma DA (1996) Coherent control over Liouville-space pathways interference in transient four-wave mixing spectroscopy. Phys Rev Lett 76(25):4701–4704. https://doi.org/10.1103/PhysRevLett.76.4701
44. Woerner M, Kuehn W, Bowlan P, Reimann K, Elsaesser T (2013) Ultrafast two-dimensional terahertz spectroscopy of elementary excitations in solids. New J Phys. https://doi.org/10.1088/1367-2630/15/2/025039
45. Tseng CH, Matsika S, Weinacht TC (2009) Two-dimensional ultrafast fourier transform spectroscopy in the deep ultraviolet. Opt Express 17(21):18788–18793. https://doi.org/10.1364/oe.17.018788
46. Krebs N, Pugliesi I, Hauer J, Riedle E (2013) Two-dimensional Fourier transform spectroscopy in the ultraviolet with sub-20 fs pump pulses and 250–720 nm supercontinuum probe. New J Phys. https://doi.org/10.1088/1367-2630/15/8/085016
47. Desilvestri S, Fujimoto JG, Ippen EP, Gamble EB, Williams LR, Nelson KA (1985) Femtosecond time-resolved measurements of optic phonon dephasing by impulsive stimulated Raman-scattering in alpha-perylene crystal from 20 to 300K. Chem Phys Lett 116(2–3):146–152. https://doi.org/10.1016/0009-2614(85)80143-3
48. Dhar L, Rogers JA, Nelson KA (1994) Time-resolved vibrational spectroscopy in the impulsive limit. Chem Rev 94(1):157–193. https://doi.org/10.1021/cr00025a006
49. Rullière C (2005) Femtosecond laser pulses: principles and experiments. Advanced texts in physics, 2nd edn. Springer, New York
50. Gaumnitz T, Jain A, Pertot Y, Huppert M, Jordan I, Ardana-Lamas F, Worner HJ (2017) Streaking of 43-attosecond soft-X-ray pulses generated by a passively CEP-stable mid-infrared driver. Opt Express 25(22):27506–27518. https://doi.org/10.1364/oe.25.027506
51. Mukamel S (1995) Principles of nonlinear optical spectroscopy, vol 6. Oxford series in optical and imaging sciences. Oxford University Press, New York
52. Savolainen J, Ahmed S, Hamm P (2013) Two-dimensional Raman-terahertz spectroscopy of water. Proc Natl Acad Sci USA 110(51):20402–20407. https://doi.org/10.1073/pnas.1317459110
53. Takeuchi S, Ruhman S, Tsuneda T, Chiba M, Taketsugu T, Tahara T (2008) Spectroscopic tracking of structural evolution in ultrafast stilbene photoisomerization. Science 322(5904):1073–1077
54. Kraack JP, Wand A, Buckup T, Motzkus M, Ruhman S (2013) Mapping multidimensional excited state dynamics using pump-impulsive-vibrational-spectroscopy and pump-degenerate-four-wave-mixing. Phys Chem Chem Phys 15(34):14487–14501
55. Hauer J, Buckup T, Motzkus M (2007) Pump-degenerate four wave mixing as a technique for analyzing structural and electronic evolution: multidimensional time-resolved dynamics near a conical intersection. J Phys Chem A 111(42):10517–10529
56. Kraack JP, Buckup T, Motzkus M (2013) Coherent high-frequency vibrational dynamics in the excited electronic state of all-trans retinal derivatives. J Phys Chem Lett 4(3):383–387

57. Buckup T, Kraack JP, Marek MS, Motzkus M (2013) Vibronic coupling in excited electronic states investigated with resonant 2D Raman spectroscopy. In: XVIIIth international conference on ultrafast phenomena. Vol 41 p 05018. https://doi.org/10.1051/epjconf/20134105018

58. Bredenbeck J, Helbing J, Behrendt R, Renner C, Moroder L, Wachtveitl J, Hamm P (2003) Transient 2D-IR spectroscopy: snapshots of the nonequilibrium ensemble during the picosecond conformational transition of a small peptide. J Phys Chem B 107(33):8654–8660. https://doi.org/10.1021/jp034552q

59. Kolano C, Helbing J, Kozinski M, Sander W, Hamm P (2006) Watching hydrogen-bond dynamics in a beta-turn by transient two-dimensional infrared spectroscopy. Nature 444(7118):469–472. https://doi.org/10.1038/nature05352

60. Chung HS, Ganim Z, Jones KC, Tokmakoff A (2007) Transient 2D IR spectroscopy of ubiquitin unfolding dynamics. Proc Natl Acad Sci 104(36):14237–14242. https://doi.org/10.1073/pnas.0700959104

61. Goodno GD, Dadusc G, Miller RJD (1998) Ultrafast heterodyne-detected transient-grating spectroscopy using diffractive optics. J Opt Soc Am B Opt Phys 15(6):1791–1794. https://doi.org/10.1364/josab.15.001791

62. Lepetit L, Cheriaux G, Joffre M (1995) Linear techniques of phase measurement by femtosecond spectral interferometry for applications in spectroscopy. J Opt Soc Am B Opt Phys 12(12):2467–2474. https://doi.org/10.1364/josab.12.002467

63. Tokmakoff A, Lang MJ, Larsen DS, Fleming GR (1997) Intrinsic optical heterodyne detection of a two-dimensional fifth order Raman response. Chem Phys Lett 272(1–2):48–54. https://doi.org/10.1016/s0009-2614(97)00479-x

64. Krebs N, Pugliesi I, Hauer J, Riedle E (2013) Two-dimensional Fourier transform spectroscopy in the ultraviolet with sub-20 fs pump pulses and 250–720 nm supercontinuum probe. New J Phys 15:17. https://doi.org/10.1088/1367-2630/15/8/085016

65. Dadusc G, Ogilvie JP, Schulenberg P, Marvet U, Miller RJD (2001) Diffractive optics-based heterodyne-detected four-wave mixing signals of protein motion: from "protein quakes" to ligand escape for myoglobin. Proc Natl Acad Sci USA 98(11):6110–6115. https://doi.org/10.1073/pnas.101130298

66. Kubarych KJ, Milne CJ, Lin S, Astinov V, Miller RJD (2002) Diffractive optics-based six-wave mixing: heterodyne detection of the full chi(5) tensor of liquid CS2. J Chem Phys 116(5):2016–2042. https://doi.org/10.1063/1.1429961

67. Cowan ML, Ogilvie JP, Miller RJD (2004) Two-dimensional spectroscopy using diffractive optics based phased-locked photon echoes. Chem Phys Lett 386(1–3):184–189. https://doi.org/10.1016/j.cplett.2004.01.027

68. Brixner T, Mancal T, Stiopkin IV, Fleming GR (2004) Phase-stabilized two-dimensional electronic spectroscopy. J Chem Phys 121(9):4221–4236. https://doi.org/10.1063/1.1776112

69. Brixner T, Stiopkin IV, Fleming GR (2004) Tunable two-dimensional femtosecond spectroscopy. Opt Lett 29(8):884–886. https://doi.org/10.1364/ol.29.000884

70. Shim SH, Strasfeld DB, Zanni MT (2006) Generation and characterization of phase and amplitude shaped femtosecond mid-IR pulses. Opt Express 14(26):13120–13130. https://doi.org/10.1364/oe.14.013120

71. Shim SH, Strasfeld DB, Ling YL, Zanni MT (2007) Automated 2D IR spectroscopy using a mid-IR pulse shaper and application of this technology to the human islet amyloid polypeptide. Proc Natl Acad Sci USA 104(36):14197–14202. https://doi.org/10.1073/pnas.0700804104

72. Shim SH, Zanni MT (2009) How to turn your pump-probe instrument into a multidimensional spectrometer: 2D IR and Vis spectroscopies via pulse shaping. Phys Chem Chem Phys 11(5):748–761. https://doi.org/10.1039/b813817f

73. Gundogdu K, Stone KW, Turner DB, Nelson KA (2007) Multidimensional coherent spectroscopy made easy. Chem Phys 341(1–3):89–94. https://doi.org/10.1016/j.chemphys.2007.06.027

74. Hornung T, Vaughan JC, Feurer T, Nelson KA (2004) Degenerate four-wave mixing spectroscopy based on two-dimensional femtosecond pulse shaping. Opt Lett 29(17):2052–2054. https://doi.org/10.1364/ol.29.002052

75. Vaughan JC, Hornung T, Stone KW, Nelson KA (2007) Coherently controlled ultrafast four-wave mixing spectroscopy. J Phys Chem A 111(23):4873–4883. https://doi.org/10.1021/jp0662911

76. Shim S-H, Strasfeld DB, Ling YL, Zanni MT (2007) Automated 2D IR spectroscopy using a mid-IR pulse shaper and application of this technology to the human islet amyloid polypeptide. Proc Natl Acad Sci USA 104(36):14197–14202. https://doi.org/10.1073/pnas.0700804104

77. Yabushita A, Lee YH, Kobayashi T (2010) Development of a multiplex fast-scan system for ultrafast time-resolved spectroscopy. Rev Sci Instrum. https://doi.org/10.1063/1.3455809
78. Helbing J, Hamm P (2011) Compact implementation of Fourier transform two-dimensional IR spectroscopy without phase ambiguity. J Opt Soc Am B Opt Phys 28(1):171–178. https://doi.org/10.1364/josab.28.000171
79. Draeger S, Roeding S, Brixner T (2017) Rapid-scan coherent 2D fluorescence spectroscopy. Opt Express 25(4):3259–3267. https://doi.org/10.1364/oe.25.003259
80. Helbing J, Hamm P (2011) Compact implementation of Fourier transform two-dimensional IR spectroscopy without phase ambiguity. J Opt Soc Am B 28(1):171–178. https://doi.org/10.1364/JOSAB.28.000171
81. Kauppinen J, Partanen J (2001) Fourier transforms in spectroscopy, 1st edn. Wiley-VCH, Berlin
82. Jonas DM (2003) Two-dimensional femtosecond spectroscopy. Annu Rev Phys Chem 54:425–463. https://doi.org/10.1146/annurev.physchem.54.011002.103907
83. Candes EJ, Wakin MB (2008) An introduction to compressive sampling. IEEE Signal Process Mag 25(2):21–30. https://doi.org/10.1109/msp.2007.914731
84. Roeding S, Klimovich N, Brixner T (2017) Optimizing sparse sampling for 2D electronic spectroscopy. J Chem Phys. https://doi.org/10.1063/1.4976309
85. Dunbar JA, Osborne DG, Anna JM, Kubarych KJ (2013) Accelerated 2D-IR using compressed sensing. J Phys Chem Lett 4(15):2489–2492. https://doi.org/10.1021/jz401281r
86. Almeida J, Prior J, Plenio MB (2012) Computation of two-dimensional spectra assisted by compressed sampling. J Phys Chem Lett 3(18):2692–2696. https://doi.org/10.1021/jz3009369
87. Tracy KM, Barich MV, Carver CL, Luther BM, Krummel AT (2016) High-throughput two-dimensional infrared (2D IR) spectroscopy achieved by interfacing microfluidic technology with a high repetition rate 2D IR spectrometer. J Phys Chem Lett 7(23):4865–4870. https://doi.org/10.1021/acs.jpclett.6b01941
88. Kanal F, Keiber S, Eck R, Brixner T (2014) 100-kHz shot-to-shot broadband data acquisition for high-repetition-rate pump-probe spectroscopy. Opt Express 22(14):16965–16975. https://doi.org/10.1364/oe.22.016965
89. Luther BM, Tracy KM, Gerrity M, Brown S, Krummel AT (2016) 2D IR spectroscopy at 100 kHz utilizing a Mid-IR OPCPA laser source. Opt Express 24(4):4117–4127. https://doi.org/10.1364/oe.24.004117
90. Bredenbeck J, Helbing J, Nienhaus K, Nienhaus GU, Hamm P (2007) Protein ligand migration mapped by nonequilibrium 2D-IR exchange spectroscopy. Proc Natl Acad Sci 104(36):14243–14248. https://doi.org/10.1073/pnas.0607758104
91. Ge NH, Hochstrasser RM (2002) Femtosecond two-dimensional infrared spectroscopy: IR-COSY and THIRSTY. PhysChemComm 5(3):17–26. https://doi.org/10.1039/B109935C
92. Rector KD, Fayer MD (1998) Vibrational echoes: a new approach to condensed-matter vibrational spectroscopy. Int Rev Phys Chem 17(3):261–306. https://doi.org/10.1080/0144235982 30063
93. Hamm P, Lim M, DeGrado WF, Hochstrasser RM (1999) The two-dimensional IR nonlinear spectroscopy of a cyclic penta-peptide in relation to its three-dimensional structure. Proc Natl Acad Sci 96(5):2036–2041. https://doi.org/10.1073/pnas.96.5.2036
94. Khalil M, Demirdöven N, Tokmakoff A (2003) Coherent 2D IR spectroscopy: molecular structure and dynamics in solution. J Phys Chem A 107(27):5258–5279. https://doi.org/10.1021/jp0219247
95. Ostrander JS, Knepper R, Tappan AS, Kay JJ, Zanni MT, Farrow DA (2017) Energy transfer between coherently delocalized states in thin films of the explosive pentaerythritol tetranitrate (PETN) revealed by two-dimensional infrared spectroscopy. J Phys Chem B 121(6):1352–1361. https://doi.org/10.1021/acs.jpcb.6b09879
96. Kraack JP (2017) Ultrafast structural molecular dynamics investigated with 2D infrared spectroscopy methods. Top Curr Chem 375(6):86. https://doi.org/10.1007/s41061-017-0172-1
97. Ferretti M, Hendrikx R, Romero E, Southall J, Cogdell RJ, Novoderezhkin VI, Scholes GD, van Grondelle R (2016) Dark states in the light-harvesting complex 2 revealed by two-dimensional electronic spectroscopy. Sci Rep. https://doi.org/10.1038/srep20834
98. Ferretti M, Novoderezhkin VI, Romero E, Augulis R, Pandit A, Zigmantasc D, van Grondelle R (2014) The nature of coherences in the B820 bacteriochlorophyll dimer revealed by two-dimensional electronic spectroscopy. Phys Chem Chem Phys 16(21):9930–9939. https://doi.org/10.1039/c3cp54634a

99. Wong CY, Alvey RM, Turner DB, Wilk KE, Bryant DA, Curmi PMG, Silbey RJ, Scholes GD (2012) Electronic coherence lineshapes reveal hidden excitonic correlations in photosynthetic light harvesting. Nat Chem 4(5):396–404. https://doi.org/10.1038/nchem.1302

100. Lim J, Palecek D, Caycedo-Soler F, Lincoln CN, Prior J, von Berlepsch H, Huelga SF, Plenio MB, Zigmantas D, Hauer J (2015) Vibronic origin of long-lived coherence in an artificial molecular light harvester. Nat Commun. https://doi.org/10.1038/ncomms8755

101. Scholes GD (2003) Long-range resonance energy transfer in molecular systems. Annu Rev Phys Chem 54(1):57–87. https://doi.org/10.1146/annurev.physchem.54.011002.103746

102. Mirkovic T, Ostroumov EE, Anna JM, van Grondelle R, Govindjee Scholes GD (2017) Light absorption and energy transfer in the antenna complexes of photosynthetic organisms. Chem Rev 117(2):249–293. https://doi.org/10.1021/acs.chemrev.6b00002

103. Kraack JP, Frei A, Alberto R, Hamm P (2017) Ultrafast vibrational energy transfer in catalytic monolayers at solid-liquid interfaces. J Phys Chem Lett 8(11):2489–2495. https://doi.org/10.1021/acs.jpclett.7b01034

104. Kraack JP, Sévery L, Tilley SD, Hamm P (2018) Plasmonic substrates do not promote vibrational energy transfer at solid–liquid interfaces. J Phys Chem Lett 9(1):49–56. https://doi.org/10.1021/acs.jpclett.7b02855

105. Cahoon JF, Sawyer KR, Schlegel JP, Harris CB (2008) Determining transition-state geometries in liquids using 2D-IR. Science 319(5871):1820–1823. https://doi.org/10.1126/science.1154041

106. Fayer MD (2009) Dynamics of liquids, molecules, and proteins measured with ultrafast 2D IR vibrational echo chemical exchange spectroscopy. Annu Rev Phys Chem 60(1):21–38. https://doi.org/10.1146/annurev-physchem-073108-112712

107. Ghosh A, Ostrander JS, Zanni MT (2017) Watching proteins wiggle: mapping structures with two-dimensional infrared spectroscopy. Chem Rev 117(16):10726–10759. https://doi.org/10.1021/acs.chemrev.6b00582

108. Christensson N, Kauffmann HF, Pullerits T, Mančal T (2012) Origin of long-lived coherences in light-harvesting complexes. J Phys Chem B 116(25):7449–7454. https://doi.org/10.1021/jp304649c

109. Monahan DM, Whaley-Mayda L, Ishizaki A, Fleming GR (2015) Influence of weak vibrational-electronic couplings on 2D electronic spectra and inter-site coherence in weakly coupled photosynthetic complexes. J Chem Phys. https://doi.org/10.1063/1.4928068

110. Fujihashi Y, Fleming GR, Ishizaki A (2015) Impact of environmentally induced fluctuations on quantum mechanically mixed electronic and vibrational pigment states in photosynthetic energy transfer and 2D electronic spectra. J Chem Phys. https://doi.org/10.1063/1.4914302

111. Duan HG, Nalbach P, Prokhorenko VI, Mukamel S, Thorwart M (2015) On the origin of oscillations in two-dimensional spectra of excitonically-coupled molecular systems. New J Phys. https://doi.org/10.1088/1367-2630/17/7/072002

112. Koeppe B, Tolstoy PM, Guo J, Nibbering ETJ, Elsaesser T (2011) Two-dimensional UV–Vis/NMR correlation spectroscopy: a heterospectral signal assignment of hydrogen-bonded complexes. J Phys Chem Lett 2(9):1106–1110. https://doi.org/10.1021/jz200285c

113. Tamimi A, Heussman DJ, Kringle LM, von Hippel PH, Marcus AH (2018) Measuring structure and disorder of (Cy3)2 dimer labeled DNA fork-junctions using two-dimensional fluorescence spectroscopy (2DFS). Biophys J 114(3):171A–171A

114. Goetz S, Li DH, Kolb V, Pflaum J, Brixner T (2018) Coherent two-dimensional fluorescence micro-spectroscopy. Opt Express 26(4):3915–3925. https://doi.org/10.1364/oe.26.003915

115. Pachon LA, Marcus AH, Aspuru-Guzik A (2015) Quantum process tomography by 2D fluorescence spectroscopy. J Chem Phys. https://doi.org/10.1063/1.4919954

116. Bakulin AA, Silva C, Vella E (2016) Ultrafast spectroscopy with photocurrent detection: watching excitonic optoelectronic systems at work. J Phys Chem Lett 7(2):250–258. https://doi.org/10.1021/acs.jpclett.5b01955

117. Karki KJ, Widom JR, Seibt J, Moody I, Lonergan MC, Pullerits T, Marcus AH (2014) Coherent two-dimensional photocurrent spectroscopy in a PbS quantum dot photocell. Nat Commun 5:5869. https://doi.org/10.1038/ncomms6869https://www.nature.com/articles/ncomms6869#supplementary-information

118. Nardin G, Autry TM, Silverman KL, Cundiff ST (2013) Multidimensional coherent photocurrent spectroscopy of a semiconductor nanostructure. Opt Express 21(23):28617–28627. https://doi.org/10.1364/OE.21.028617

119. Aeschlimann M, Brixner T, Fischer A, Kramer C, Melchior P, Pfeiffer W, Schneider C, Struber C, Tuchscherer P, Voronine DV (2011) Coherent two-dimensional nanoscopy. Science 333(6050):1723–1726. https://doi.org/10.1126/science.1209206
120. Liebel M, Toninelli C, van Hulst NF (2018) Room-temperature ultrafast nonlinear spectroscopy of a single molecule. Nat Photonics 12(1):45–49. https://doi.org/10.1038/s41566-017-0056-5
121. Mukamel S, Healion D, Zhang Y, Biggs JD (2013) Multidimensional attosecond resonant X-ray spectroscopy of molecules: lessons from the optical regime. Annu Rev Phys Chem 64(1):101–127. https://doi.org/10.1146/annurev-physchem-040412-110021
122. Biggs JD, Zhang Y, Healion D, Mukamel S (2013) Multidimensional X-ray spectroscopy of valence and core excitations in cysteine. J Chem Phys 138(14):144303. https://doi.org/10.1063/1.4799266
123. Kochise B, Yu Z, Markus K, Weijie H, Shaul M (2016) Multidimensional resonant nonlinear spectroscopy with coherent broadband X-ray pulses. Phys Scr T169:014002
124. Diels JC, Rudolph W (2006) Ultrashort laser pulse phenomena: fundamentals, techniques, and applications on a femtosecond time scale. Optics and photonics, 2nd edn. Elsevier/Academic Press, Amsterdam; Boston
125. Schmuttenmaer CA (2004) Exploring dynamics in the far-infrared with terahertz spectroscopy. Chem Rev 104(4):1759–1779
126. Kampfrath T, Tanaka K, Nelson KA (2013) Resonant and nonresonant control over matter and light by intense terahertz transients. Nat Photonics 7(9):680–690
127. Cho MH (2008) Coherent two-dimensional optical spectroscopy. Chem Rev 108(4):1331–1418. https://doi.org/10.1021/cr078377b
128. Cho MH (2002) Ultrafast vibrational spectroscopy in condensed phases. PhysChemComm 5:40–58. https://doi.org/10.1039/b110898k
129. Dietze DR, Mathies RA (2016) Femtosecond stimulated Raman spectroscopy. ChemPhysChem 17(9):1224–1251
130. Kubarych KJ, Milne CJ, Miller RJD (2003) Fifth-order two-dimensional Raman spectroscopy: a new direct probe of the liquid state. Int Rev Phys Chem 22(3):497–532
131. Rodriguez Y, Frei F, Cannizzo A, Feurer T (2015) Pulse-shaping assisted multidimensional coherent electronic spectroscopy. J Chem Phys 142(21):212451
132. Voronine DV, Abramavicius D, Mukamel S (2011) Coherent control protocol for separating energy-transfer pathways in photosynthetic complexes by chiral multidimensional signals. J Phys Chem A 115(18):4624–4629
133. Fuller FD, Ogilvie JP (2015) Experimental implementations of two-dimensional Fourier transform electronic spectroscopy. Annu Rev Phys Chem 66(66):667–690
134. Bakulin AA, Morgan SE, Kehoe TB, Wilson MWB, Chin AW, Zigmantas D, Egorova D, Rao A (2016) Nat Chem 8:16–23. https://doi.org/10.1038/nchem.2371
135. Sarovar M, Ishizaki A, Fleming GR, Whaley KB (2010) Nat Phys 6:462–467. https://doi.org/10.1038/nphys1652
136. Yan C, Nishida J, Yuan R et al (2016) J Am Chem Soc 138(30):9694–9703. https://doi.org/10.1021/jacs.6b05589
137. Maj M, Lomont JP, Rich KL, Alperstein AM, Zanni MT (2018) Chem Sci 9:463. https://doi.org/10.1039/c7sc03789a
138. Lu J, Zhang Y, Hwang HY, Ofori-Okai BK, Fleischer S, Nelson KA (2016) PNAS 113(42):11800–11805
139. Mehlenbacher RD, McDonough TJ, Grechko M et al (2015) Nat Commun. https://doi.org/10.1038/ncomms7732
140. Guo Z, Molesky BM, Cheshire TP, Moran AM (2015) J Chem Phys 143:124202
141. Roeding S, Klimovich N, Brixner T (2017) J Chem Phys 146:084201. https://doi.org/10.1063/1.4976309

Topics in Current Chemistry (2018) 376:28

Affiliations

Jan Philip Kraack[1] · Tiago Buckup[2]

✉ Tiago Buckup
tiago.buckup@pci.uni-heidelberg.de

Jan Philip Kraack
philip.kraack@chem.uzh.ch

1 Department of Chemistry, University of Zürich, Winterthurerstrasse 190, 8057 Zurich, Switzerland

2 Physikalisch-Chemisches Institut, Universität Heidelberg, Im Neuenheimer Feld 229, 69120 Heidelberg, Germany

Top Curr Chem (Z) (2018) 376:10
https://doi.org/10.1007/s41061-017-0180-1

Electronic Couplings in (Bio-) Chemical Processes

Margherita Maiuri[1] · Johanna Brazard[2]

Received: 7 September 2017 / Accepted: 1 December 2017 / Published online: 20 March 2018
© Springer International Publishing AG, part of Springer Nature 2018

Abstract During the last two decades, 2D optical techniques have been extended to the visible range, targeting electronic transitions. Since the report of the very first 2D electronic measurement (Hybl et al. in J Chem Phys 115:6606–6622, [2001]), two-dimensional electronic spectroscopy (2DES) has allowed fundamentally new insights into the structure and dynamics of condensed-phase systems (Ginsberg et al. in Acc Chem Res 42:1352–1363, 2009; Jonas in Annu Rev Phys Chem 54:425–463, 2003), producing experiments that measure correlations among electronic states of an absorbing species within complex systems. 2DES is used to investigate photophysical phenomena involving electronic or vibrational couplings in multi-chromophoric systems [energy transfer in photosynthesis is one great example of how 2DES can disentangle various energy transfer pathways (Brixner et al. in Nature 625–628, 2005; Engel et al. in Nature 446:782–786, 2007; Collini et al. in Nature 463:644–647, 2010)], but also ultrafast photochemical processes in which the tracked molecules change permanently or are heterogeneous (Ruetzel et al. in Proc Natl Acad Sci 111:4764–4769, 2014; Consani et al. in Science 339:1586–1589, 2013). We divide this chapter according to some of the major areas that have been established thanks to 2DES in the following fields: heterogeneity of systems, excitation energy transfer

Chapter 2 was originally published as Maiuri, M. & Brazard, J. Top Curr Chem (Z) (2018) 376: 10. https://doi.org/10.1007/s41061-017-0180-1.

✉ Margherita Maiuri
margherita.maiuri@polimi.it

Johanna Brazard
johanna.brazard@ipcms.unistra.fr

[1] Dipartimento di Fisica, Politecnico di Milano, Piazza Leonardo da Vinci 32, 20133 Milano, Italy

[2] Institut de Physique et Chimie des Matériaux de Strasbourg, Université de Strasbourg - CNRS UMR 7504, 67034 Strasbourg, France

mechanisms, photo-induced coherent oscillations associated with electronic and vibrational couplings, and complex chemical reactions (Fig. 1).

Keywords Two-dimensional electronic spectroscopy · Heterogeneity · Energy transfer · Vibronic coupling · Photoreactivity

1 Introduction

Two-dimensional electronic spectroscopy (2DES) can be defined as the "ultimate" time-resolved nonlinear optical experiment, since it measures both the real (absorptive) and imaginary (dispersive) parts of the complex third-order nonlinear polarization response [9]. 2DES has shown to be more powerful than one-dimensional four-wave-mixing measurements like transient absorption (often called pump–probe spectroscopy). In a formal description of transient absorption measurements based on the framework of nonlinear optics, the pump beam electric field interacts twice with the sample to photo-excite a fraction of the chromophores into electronic excited states. These two interactions simply represent the product of the transition dipole connecting the ground state to an excited state with its complex conjugate to give the probability of pump absorption. A probe pulse arrives after a variable delay T to stimulate a signal field that is detected as a change in transmission of the probe.

2DES experiments provide simultaneously high temporal and spectral resolution [10]. The excitation frequency is resolved in 2DES by delaying the two ultrafast excitation pulses with an interferometric resolution. Any other third-order nonlinear spectroscopy (i.e., transient absorption) is contained in the 2DES map (Fig. 2) where the emission detection is followed along the waiting time. The main breakthrough of 2DES resides in the disentanglement of processes in systems with multiple interacting components [3, 4, 11–14].

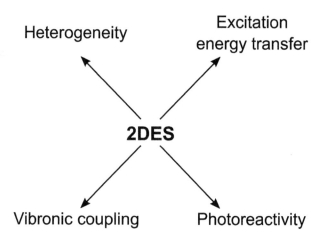

Fig. 1 Main fields impacted by two-dimensional electronic spectroscopy (2DES) in condensed phase. The major discoveries of each field will be described in different paragraphs

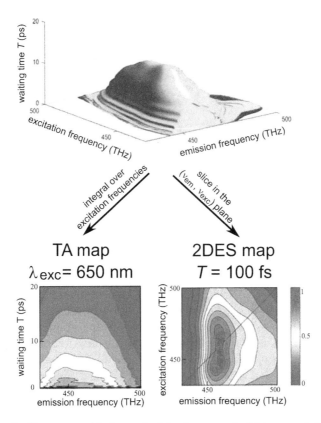

Fig. 2 Third-order 3D spectrum of the isomer P_r of the chromoprotein Cph1 represented in a 3D isosurface, the transient absorption spectra for an excitation centered at 650 nm (*left panel*), and the absorptive 2DES map for a waiting time of 100 fs (*right panel*) are extracted. The data are from the Ref. [15]

As excitation and emission frequencies are resolved, 2DES data are represented in 2DES maps for a waiting time (T). In a 2DES map, the peaks on the diagonal, i.e., at equal excitation and emission frequencies, are the signature of the populations of the individual transitions; whereas the peaks out of the diagonal, i.e., at different excitation and emission frequencies, probe couplings and relaxation dynamics between electronic states. With respect to other nonlinear spectroscopic techniques, 2DES has the following advantages:

(a) It is possible to separate, and thus distinguish, contributions to the nonlinear signal that are spectrally overlapped in one-dimensional experiments. Analysis of cross-peaks reveals whether the different transitions seen in the sample absorption spectrum arise from the same or different molecular species and can quantify electronic couplings and correlations between different excited states [3, 4, 11].

(b) 2DES distinguishes inhomogeneous and homogeneous broadening by the analysis of the lineshapes of the spectroscopic signals, enabling the individual levels to be singled out in strongly congested spectra [3, 15–19].

(c) 2DES can follow the parallel pathways by which the coupled electronic dynamics evolve after photoexcitation in real time. This makes the 2DES technique a particularly powerful tool for tracking energy and electron transfer processes [4, 13, 20].

Our aim here is to provide examples highlighting the real advantages of using 2DES. As physical chemists, we know that to understand a photo-initiated reaction, one should initially assign spectral signatures to reactants, products—eventually intermediates—and then propose a kinetic model to explain the evolution of the species, based on the observed spectral changes. However, in complex condensed-phase systems, spectral signatures are often overlapped and difficult to untangle. For example, steady-state measurements (e.g., linear absorption, fluorescence, anisotropy, and circular dichroism) supply information on the electronic structure. However, homogeneous line broadening can significantly obscure physical insights such as the distinction between homogeneous and inhomogeneous broadening and sometimes cryogenic temperature experiments are not able to reach the desirable spectral resolution. Further insights can be gained by time-resolved techniques, such as transient absorption and pump–probe spectroscopies, which can also track photo-induced dynamics, as energy and electron transfers. However, these techniques have intrinsic resolution limitations on either high temporal or high-frequency resolved-capability. In this context, 2DES has emerged as an optical technique that can accomplish many of the goals of conventional spectroscopies, also overcoming all the limitations mentioned above.

2 Implementation

2DES implementation faces two main technical challenges: (1) it requires a careful control of the interferometric stability between pulse pairs [21], which need to be phase-locked within a small fraction of their carrier wavelengths (i.e., a precision of few nanometers is required for ~ $\lambda/100$ at 550 nm) [3, 22, 23]; (2) to fully exploit its benefits, 2DES calls for ultra-broadband pulses, with the duration of just a few optical cycles, which are challenging to generate and control. Thus, the phase stability in a pair of broadband pulses has been the limiting factor that explains why 2DES was developed after 2D-IR spectroscopy [24]. However, thanks to the technological advancement of ultrafast optics, 2DES techniques are rapidly evolving and have today various applications, providing access to systems that contain electronic transitions spanning from the ultraviolet to the near-infrared and beyond. Technically, one can build a 2DES apparatus following three different types of geometries: (1) the non-collinear so-called "box-car" geometry (Fig. 3a), and (2) the collinear,

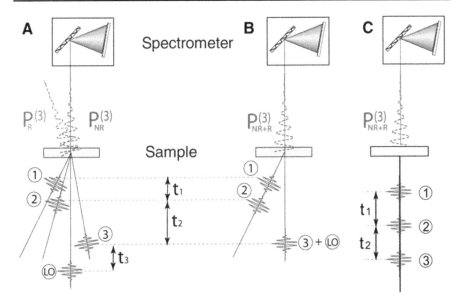

Fig. 3 **a** Non-collinear heterodyne and **b** collinear heterodyne and **c** fully collinear homodyne 2DES implementation. Rephasing ($P_R^{(3)}$) and non-rephasing ($P_{NR}^{(3)}$) signals are emitted in a new direction or in the same direction for the non-collinear and the collinear geometry, respectively. The signal is heterodyned by a fourth pulse, called local oscillator (LO). Adapted from [41], with the permission of AIP Publishing

so-called "pump-probe" geometry (Fig. 3b) and (3) the fully collinear geometry (Fig. 3c).

The box-car geometry is based on diffractive optics to generate four identical pulses and a set of crystals with variable thickness (wedges), placed in each individual path beam to control their relative delays [25, 26]. Depending on the arrival order of the two pump beams, the rephasing and non-rephasing signals are detected. The rephasing signal is recorded when the first pump arrives first, whereas the non-rephasing signal is measured when the second pump beam arrives first. The absorptive 2DES map is the sum of the rephasing and non-rephasing signals. Further box-car implementations substitute the diffractive optics with beam-splitters pairs (as for 2D-IR) to overcome the limit of the spatial chirp bandwidth introduced by the diffractive optics [27–29]. Even if homodyne detection is possible, heterodyne detection is preferred in most of 2DES setups [21, 30]. In a heterodyne detection, a weak signal is interfering with a strong "local oscillator" field to amplify the signal and enable the extraction of the complex signal field. The frequency of the mixing product is the sum or the difference of the frequencies of the signal and the local oscillator. The main advantages of the non-collinear geometry are the possibility to measure separately the non-rephasing and the rephasing signals and the possibility to reach high signal-to-noise ratios.

The collinear geometry exploits adaptive optics, as pulse shapers, to turn a conventional pump-probe setup into a 2DES apparatus by creating the first two phase-locked pulses from a single pump pulse that propagate in the same direction, in contrast to the box-car geometry [31–33]. Several groups have shown the potential and the capability to perform "two-color" kind of experiments by exploiting the collinear geometry [34]. In these experiments, the excitation axis is given by a first identical pulse pair resonant with the electronic transition of the system under study, while the detection axis can be spectrally tuned and depends solely on the bandwidth of the third pulse [35–37]. The generation of the first pulse pair is usually obtained by pulse-shaping techniques [38, 39], however, recently a new compact Translating-Wedge-Based Identical Pulses eNcoding System TWINS [40, 41] device has been shown to have capability of creating intrinsically phase-locked pulses exploiting the birefringence of non-linear crystals. The main advantages of the collinear geometry are the phase stability and the possibility to carry two-colors experiments.

The fully collinear geometry uses only one beam, which is pulse shaped into phase coherent pulse trains [31, 42]. It is based on a homodyne detection. The main advantage of this technique is to provide a simplified experimental design.

A summary of the various implementations of 2DES, together with benefits and drawbacks, bandwidth limitations, and typical bandwidths used has been nicely reviewed by Ogilvie et al. in a recent review [43]. A detailed discussion of the several technical aspects on how to perform 2DES experiments goes beyond the scope of this book, however review papers and books are available describing various 2DES implementations, as can be seen in references [11, 38, 44].

3 Heterogeneity

One of the main challenges in condensed-phase experiments is the heterogeneity of the samples, which present very broad absorption spectra at room temperature. 2DES allows unraveling the presence of multi-chromophoric systems and sometimes disentangles their dynamics. Furthermore, the temporal evolution of 2D line shape allows distinguishing a dynamic to a static contribution of a molecular dipole transition. The inhomogeneous broadening of a system is interpretable from the shape of diagonal peaks on a 2DES map (Fig. 4).

Firstly, at the waiting time $T = 0$, any particular transition excited will not yet have undergone dynamical processes that could change its resonant frequency; thus, its resonant frequency will be detected at its excitation frequency. In the homogeneous limit, the absorptive signal along the diagonal is expected to have a 2D-Lorentzian shape, because the rephasing and non-rephasing signals are identical [45]. As the inhomogeneous broadening increases, the 2D line shape broadens along the diagonal axis while the antidiagonal remains unchanged, i.e., the rephasing signal is larger than the non-rephasing signal at early time. Thus, it is possible to quantify the linewidth broadening along the diagonal to determine how heterogeneous the system is [3, 46, 47]. The lineshape along the diagonal will have an elliptical shape whose profile along the diagonal frequency axis is proportional to the traditional 1D absorption spectrum [13]. The width perpendicular to the diagonal represents the homogeneous linewidth. The ellipticity of 2D line shape quantifies how much the system is inhomogeneous.

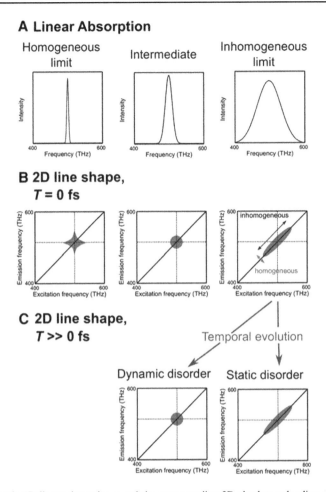

Fig. 4 Schematic 1D linear absorption **a** and the corresponding 2D absolute value line shapes **b** from the homogeneous to the inhomogeneous limit. **c** Schematic illustration of the temporal evolution (along the waiting time) of the 2D absolute value line shape in the inhomogeneous limit case

Secondly, at longer waiting time, spectral diffusion might occur [3]. In the so-called dynamic disorder case, the correlation between excitation and detection frequency will be lost and 2D line shape will become symmetric. Whereas in the static disorder case, 2D line shape will remain elongated along the diagonal [46, 47]. The time evolution of an asymmetric 2D line shape is discriminating between static and dynamic disorder.

The first example describes how 2DES probes the dynamic disorder of a system by tracking the time evolution of an asymmetric diagonal peak. Chlorophyll *a* is the smallest chromophore involved in photosynthesis [48]. Chlorophyll *a* is well studied for bio-mimetic applications [49]. The main goal is to understand how the complex interactions underlying the efficient energy conversion of absorbed photons into chemical energy to adapt this to biomimetic energy transduction devices. As shown on Fig. 5a,

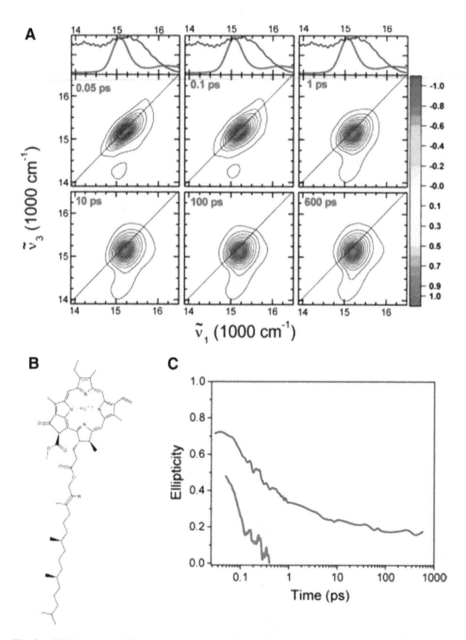

Fig. 5 **a** 2DES spectra of chlorophyll *a* in acetone. The *red numbers* indicate the waiting time in picoseconds. On the top panels, the absorption spectra and the laser spectrum are recalled in *red* and *blue*, respectively. **b** Molecular structure of chlorophyll *a*. **c** Ellipticity of the main peak as function of waiting time of chlorophyll *a* in cyclohexane (*red*) and acetone (*blue*) Adapted with permission from [18]. Copyright 2015 American Chemical Society

the broadband excitation spectrum spans over the entire absorption spectrum of chlorophyll *a*.

The monomeric chlorophyll *a* was probed in different solvents where an inhomogeneous broadening is always observed (Fig. 5) [18]. The elongated 2D line shape evolves towards a symmetric one in different time-scale depending on the solvent. This clearly unravels a spectral diffusion attributed to solvation. The slowest spectral diffusion was observed for H-bonding and/or viscous solvent. It was attributed to a spectral diffusion. Understanding these phenomena is a key parameter to optimize the biomimetic devices.

In Fig. 5a, there are two positive bands on the 2DES map of chlorophyll *a* in acetone at 50 fs: one elongated along the diagonal and another one out of the diagonal. The first one is attributed to the photobleaching and stimulated emission of the molecule and the second one to its stimulated emission from a vibronic energy level. From $T = 50$ fs to $T = 600$ ps, the photobleaching band gets more symmetric. This is clear evidence of a spectral diffusion. To quantify this inhomogeneity, the ellipticity of the photobleaching band was measured and plotted in Fig. 5c. The ellipticity is the ratio:

$$e = \frac{(D^2 - A^2)}{(D^2 + A^2)}$$

with D and A the 2DES spectrum diagonal and antidiagonal full width at half maximum, respectively [46]. An ellipticity close to 1 corresponds to the inhomogeneous limit, whereas $e = 0$ to a more homogeneous system. In all of the solvents tested, an initial inhomogeneous broadening was observed. In an apolar solvent like cyclohexane, the system evolves in a homogenous system in less than 1 ps, whereas in polar and even more in protic solvent, it takes nanoseconds to reach this homogeneity. This inhomogeneous broadening was interpreted as a spectral diffusion with different regimes:

1. Typical solvation response behavior in the fs/ps time-scale
2. Solvent dynamics modified by interactions with the solute
3. A strong function of solvent, being greater in H-bonding and viscous media.

Thus, it was possible to probe solvation and assign its dynamics directly by 2DES.

The second example illustrates the conformational heterogeneity in the ground state of chromoproteins: phytochromes (Fig. 6a). Phytochromes are red (P_r) and far-red (P_{fr}) light photoreceptors of plants and some cyanobacteria, fungi, and algae that regulate germination and flowering [50]. Phytochrome Cph1 from the cyanobacterium *Synechocystis* sp. PCC6803 has become the ubiquitous model of plant phytochrome [51]. More specifically, the photodynamics of the P_r isomer of phytochrome Cph1 has been controversial for years with two kinetics models describing multiphasic excited-state decay kinetics: (1) heterogeneous model, i.e., coexistence of ground state conformational subpopulations [52, 53], or (2) homogeneous model,

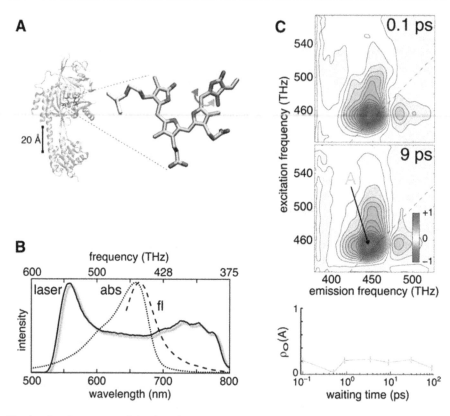

Fig. 6 a Protein structure of the phytochrome Cph1 and its bilin chromophore in the inset. **b** Steady-state absorption (*dot*) and fluorescence (*dash*) spectra of Cph1 in a PBS buffer. The spectrum of the laser is represented by a *black line*. **c** 2DES maps at a waiting time of 100 fs and 9 ps of Cph1. The *bottom panel* shows the ellipticity of peak A during the primary photoisomerization. Adapted with permission from [15]. Copyright 2017 American Chemical Society

i.e., a single ground state conformation, which undergoes an excited-state photoisomerization process—either branching on the excited state or relaxing through multiple sequential intermediates [54, 55]. Numerous time-resolved 1D studies support one of each model but none were able to clearly distinguish which one is correct.

2DES is an ideal probe of heterogeneous line broadening and was able to clearly answer to this controversy. The 2DES maps (Fig. 6c) present one main positive peak on the diagonal (photobleaching) and one negative peak out of the diagonal (excited state absorption). As described before, the ellipticity of the diagonal peak is an indicator of the inhomogeneous broadening. Thus, the ellipticity of the positive peak was followed along the waiting time T. It remains constant and small (~ 0.2) from 100 fs to 100 ps (Fig. 6c) [15]. This is clear evidence that the ground state is homogeneous and the heterogeneous model can be discarded.

It is worth noticing that the rephasing of the data is crucial to address the heterogeneity question. Indeed, the line shape of the rephasing and non-rephasing data alone has a meaningless lineshape due to an artifact known as phase twist [1, 43, 56, 57]. It is important to probe the ellipticity on the 2DES maps of the absorptive signal, i.e., after summing the rephasing and non-rephasing signals.

Due to the high degree of information contained within 2DES maps, it was possible to resolve a controversy that lasted for years in the literature. The photoisomerization of P_r isomer of the phytochrome Cph1 involves one homogeneous ground state. The photoisomerization takes place in the excited state. 2DES shows how it is easy to unravel this mechanism, which was impossible to interpret from 1D data.

The third example demonstrates how 2DES measurements can reveal insights into spectra that are completely dominated by inhomogeneous broadening. In this case, measuring non-rephasing and rephasing signals was crucial. It is a study on cubic CdSe nanocrystals where the nanocrystal size variation makes this system highly heterogeneous (Fig. 7) [12]. CdSe nanocrystals have applications in photovoltaics

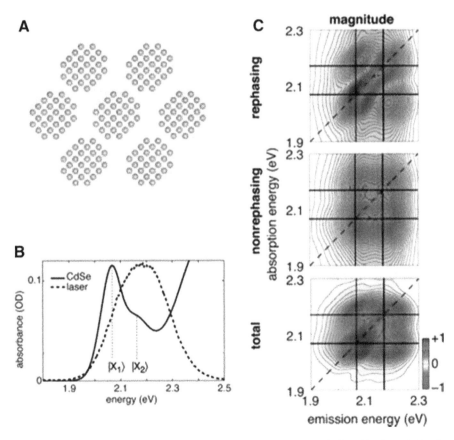

Fig. 7 **a** Representation of colloidal cadmium selenide (CdSe) nanocrystals in cubic phase. **b** Steady-state absorption spectrum of CdSe nanocrystals in toluene (*line*) and the excitation laser spectrum (*dash*). **c** Rephasing, non-rephasing and absorptive 2DES map in magnitudes of CdSe nanocrystals at a waiting time of 120 fs. Adapted with permission from [12]. Copyright 2012 American Chemical Society

devices, electronics, and others depend on the efficiency of the dynamics of each states and how they interact to each other [58]. 2DES can bring some clues on these phenomena.

The absorption spectrum of CdSe nanocrystals is broad with two low-energy exciton bands. During those experiments, both bands were excited and probed (Fig. 7b). The 2DES spectra show two diagonal and two anti-diagonal peaks corresponding to the two exciton bands and the two coupled exciton states, respectively. In comparison to the simulation, those 2DES spectra are really blurry because of the inhomogeneous broadening. This inhomogeneous broadening takes its origin in the size distribution of the nanocrystals in solution.

Figure 7 shows that analyzing the 2DES spectra by their rephasing, non-rephasing and absorptive signals give some insights into the inhomogeneous broadening. Indeed, photon-echo (rephasing signal) can improve the spectroscopic resolution by removing inhomogeneous line broadening [17]. After this analysis, the two diagonal and anti-diagonal peaks appear clearly on the spectra. This allows to determine that the binding energies of the $(|X_1X_2\rangle)$ and the $(|X_1X_1\rangle)$ biexcitons are similar. A value of 25 meV for this energy reproduces the peaks very well in simulations. The non-rephasing spectra do not contain photon-echo signal and therefore contain inhomogeneous contributions to both the diagonal and antidiagonal linewidths. We will see in the section photoreactivity that the disentanglement of those state dynamics is rendered possible by 2DES.

4 Excitation Energy Transfer (EET)

Excitation energy transfer (EET) in complex electronic systems has been successfully studied by 2DES. Among the advantages of 2DES over standard transient absorption measurement is the capability of distinguishing between weakly or strongly coupled systems (Fig. 8). In the case of a weakly coupled donor-acceptor system (Fig. 8a), EET can be tracked in a 2DES experiment as the appearance of a cross-peak at the donor (B)/ acceptor (A) frequencies in the 2DES map as a function of the waiting time T. In the strong coupling regime (Fig. 8b), the donor and acceptor share a common ground state and the excitation becomes delocalized over two excitonic states, α and β. Thus, at $T = 0$ the 2DES map reveals the existence of separate cross-peaks at the $\alpha\beta/\beta\alpha$ positions, which evolve along T. In transient absorption experiments, due to the lack of resolution over the excitation axis, these cases will be indistinguishable. In particular, in photosynthetic complexes, the interchromophore electrostatic interaction gives rise to Coulombic couplings, which in turn re-define energetically shifted delocalized and coupled states (namely exciton [59]) that ultimately determine the photosynthetic optical properties. Accordingly, the electronic absorption spectra are—in most cases—broad and congested, which is beneficial for maximizing light absorption, but makes the analysis of these systems extremely difficult.

2DES is the perfect tool to identify the chromophore couplings and track the energy transfer pathways in protein-pigment complexes. In addition to spectrally resolve excitation and emission frequencies with femtosecond resolution, this

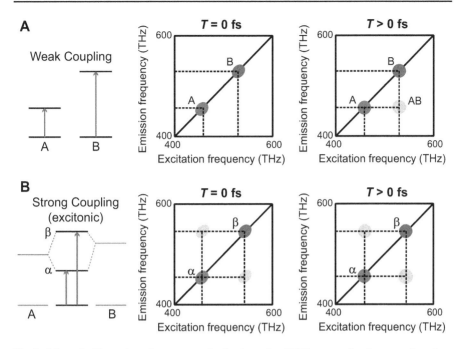

Fig. 8 Schematic illustration of energy transfer in absorptive 2DES spectra for the case of weak or strong coupling regime. **a** In the case of a weak coupling system, the correlation spectrum at $T = 0$ fs reveals two absorption species along the diagonal, while at $T > 0$ fs, a cross-peak appears (AB) and indicates energy transfer from B to A. **b** In the case of strong coupling, at $T = 0$ fs two instantaneous cross-peaks appear, at the $\alpha\beta/\beta\alpha$ positions, indicating the existence of a shared common ground state

technique has enabled discoveries about the structure and dynamics of various photosynthetic light-harvesting complexes. A complete picture on how the energy transfer cascade evolves from the light harvesting antenna complex to the reaction center has been recently shown in Fig. 9 for the photosynthetic apparatus of a green sulfur bacterium [60]. This study was made possible due to 2DES developments, which enable measurements in highly scattering environments as cells [60, 61], finally allowing determination of the full transfer network between electronic transitions. The steady-state spectrum of the apparatus (Fig. 9a) at 77 K reveals clear distinguishable features [48]: the high-energetic strongest band, which corresponds to the assembly of light harvesting antenna complex (chlorosome), three absorption peaks around 800 nm corresponding to the Fenna–Matthews–Olson (FMO) proteins that mediate the energy transfer from the chlorosome to the reaction center (RC), which shows a weak absorption band at the lowest energy (833 nm). The EET processes have been extensively studied in the isolated photosynthetic FMO subunits [4, 5, 62] (Fig. 9b), and we show here the main interchromophoric EET pathways measured in the isolated FMO complex (Fig. 9b, c) [62]. The 2DES absorptive maps reveal cross-peaks below and above the diagonal. While the cross-peak above the diagonal do not exhibit any strong evolution, a clear signature of coupling among

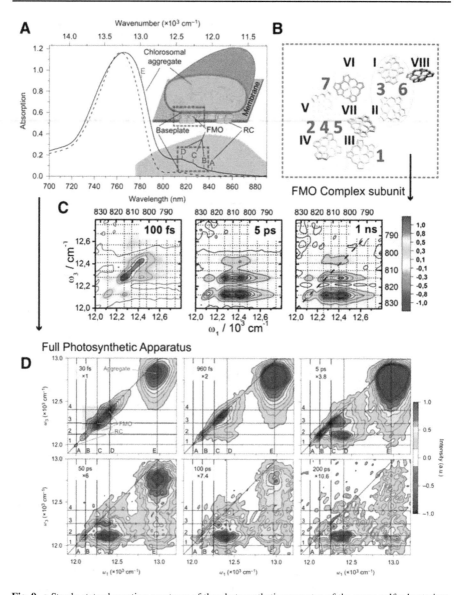

Fig. 9 **a** Steady-state absorption spectrum of the photosynthetic apparatus of the green sulfur bacterium and laser spectrum used in the 2DES experiment reported in **d**. **b** Arrangement of BChla pigments within the FMO sub-units. **c** Absorptive 2DES maps measured in the FMO complex at 77 K for increasing waiting times T with *dashed lines* indicating excitonic transition energies. **d** Absorptive 2DES maps on the entire apparatus described in **b** for increasing waiting times T. The *purple dashed lines* indicate the spectrum measured on the isolated FMO complex described in **a**. Adapted with permission from [62]. Copyright 2016 American Chemical Society. And adapted by permission from Macmillan Publishers Ltd: Nature Chemistry [60], copyright 2016

the chromophores, the ones below the diagonal, show a clear increase of the signal intensity, associated with the downhill EET.

In Fig. 9d we show a series of 2DES maps recorded at different T times for the whole photosynthetic cell [60]. The early map at $T = 30$ fs shows a series of diagonal peaks attributed to the different transition in the steady-state absorption spectrum. At later T times, one can clearly observe the formation and decay of cross-peaks below the diagonal, revealing how energy flows within and between the individual complexes in the apparatus. By examining the connectivity between different complexes and applying a global fitting method [60], it was possible to track the step-by-step energy flow through the entire unit (Fig. 9d) and observed for the first time that the FMO complex serves as energy conduit between the chlorosome and the RC.

A second example of EET process resolved by 2DES is reported in Fig. 10 [63]. In this work, the isolated light-harvesting antenna complex (namely LH1) of a purple bacterium is studied with a two-color 2DES apparatus, where both transition in the near-infrared and in the visible spectra region were covered with two different femtosecond broadband pulses. The absorption spectrum of the LH1 complex is constituted of two chromophores (Fig. 10a), a carotenoid—namely, spirilloxanthin (Spx)—whose 0–0 first optically allowed $S_0 \rightarrow S_2$ transition peaks at 540 nm, and a bacterio-chlorophyll (BChl) named B890 due to its Q_y band peaking at 881 nm (the higher energy Q_x band of the B890 peaks at 585 nm). By exciting in the visible range and detecting over both visible and near-IR ranges, it was possible to follow all the photoinduced processes, namely, i) the Spx internal conversion, ii) the BChl $Q_x \rightarrow Q_y$ internal conversion, and iii) the Spx $\rightarrow$ B890 EET process by tracking the formation of several cross-peaks in the 2DES maps. In the degenerate 2DES experiment (Fig. 10b), the internal conversion for the bright S_2 state of the Spx to the dark S_1 state is observed as the appearance of a negative cross-peak (color-coded in blue) on within the first few hundreds of femtoseconds, assigned to the formation of an excited state absorption from the S_1 state, at the 545/620 nm excitation/detection cross-peak. Moreover, a second negative feature is assigned to the formation of a parallel long-lived state, named S*, at the 545/570 nm excitation/detection cross-peak [64]. The degenerate 2DES map also shows that the diagonal peak of the Q_x of the B890 (585/585 nm excitation/detection) has become less pronounced after $T = 65$ fs, indicating that population in the B890 moiety, has started the internal conversion process to Q_y, reducing the stimulated emission contribution to the positive signal on the diagonal.

The two-color 2DES maps (excitation axis in the visible range and detection in the near-IR) in Fig. 10c show two positive cross-peaks at 875-nm detection wavelength that appear at the excitation wavelengths of 545 and 585 nm, corresponding to resonances of S_2 and Q_x, respectively. The center wavelength of the positive bands at 875 nm identifies these features as pure photobleaching contributions from Q_y, showing that at these early T times, the population is still entirely on Q_x and the optical probes which are specific for population in Q_y. At later T times, the further evolution of the 2DES maps consists in the growth of the positive cross-peaks at 875 nm and the appearance of two excited-state absorption peaks on the short wavelength side, around 845 nm. Both the features at 875 and 845 nm are specific optical probes for population on Q_y, showing that both the Spx $\rightarrow$ B890 EET and

Fig. 10 a Arrangement of the LH1 complex and supramolecular architecture of the BChl B890 (purple) ▶ and the Car Spx (*orange*) together with the absorption spectrum of LH1 complex extracted from Rsp. rubrum (*black line*) and pulse spectra used in the 2DES experiments in the visible (NOPA1, *green line*) and near-IR (NOPA2, *red line*) wavelength regions. **b** Degenerate 2DES maps of the LH1 complex from Rsp. rubrum, for different waiting times T, following excitation by a sub-10-fs visible pulse, resonant with the $S_0 \rightarrow S_2$ transition of the Car and the Qx transition of the BChl. **c** Two-color 2DES maps of the LH1 complex for different waiting times T, with same excitation axis as in B but detection axis in the near-IR. Adapted from [63], with the permission of AIP Publishing

the internal conversion from Q_x to Q_y are occurring on this time scale. The observation of energy transfers and internal conversion parallel pathways was only possible thanks to the two-color 2DES experiment.

In the third example (Fig. 11), a pulse generated via continuum filamentation spanning from 500 to 1300 nm is used to collect 2DES-WL (white light) spectra to resolve energy transfer within a network of semiconducting carbon nanotubes (CNTs) [36]. CNTs thin films are being explored for energy harvesting and optoelectronic devices because of their exceptional transport and optical properties. In this study, the nanotubes in the film are in close contact, allowing the energy to flow through the films. The film is composed primarily of four different diameter nanotubes, defined by their chiral indices: (7,5), (7,6), (8,6) and (8,7). Each tube exhibits an S_1 transition in the near-infrared (corresponding to the optical bandgap) and an S_2 transition in the visible. The 2DES-WL map described in Fig. 11b, exhibits several peaks, both on the diagonal as off diagonal. Each of the electronic transitions in the absorption spectrum creates a pair of diagonal peaks in the 2DES-WL spectrum. The negative peak on the diagonal corresponds to photobleaching (color-coded in blue) of the direct bandgap transition, while the blue-shifted positive peak along the detection axis is from an excited state absorption (color-coded in red). Moreover, the cross-peaks appearance correlates the different tubes.

We discuss as an example, only a sub-quadrant of the 2DES-WL map. We choose the S_1/S_1 quadrant. Excitation of S_1 lies below the ionization threshold and so no charges, but only excitons are generated. The cross-peaks appear on the lower half of the diagonal, indicating that energy transfer occurs downhill from larger to smaller bandgap tubes. The cross-peaks raise simultaneously, indicating that exciton transfer is equally probable between any downhill combination of nanotubes. Moreover, the cross-peaks are round, whereas the diagonal peaks are elongated, which imply uncorrelated energy transfer. This means that a photoexcited exciton created on a random nanotube has equal probability of transferring to another nanotube anywhere within the acceptor's distribution of energies. Thus, the inhomogeneity of the bandgap transitions does not impact energy transfer. Thus, it is equally likely for an S_1 exciton to relax to the next smallest bandgap nanotube as it is to relax directly to the smallest bandgap nanotube.

Based on the similarity of the transfer rates and a lack of correlation, the transfer was found to be independent of the spectral overlap, ruling out Förster resonance energy transfer between bright states as the rate limiting mechanism. These

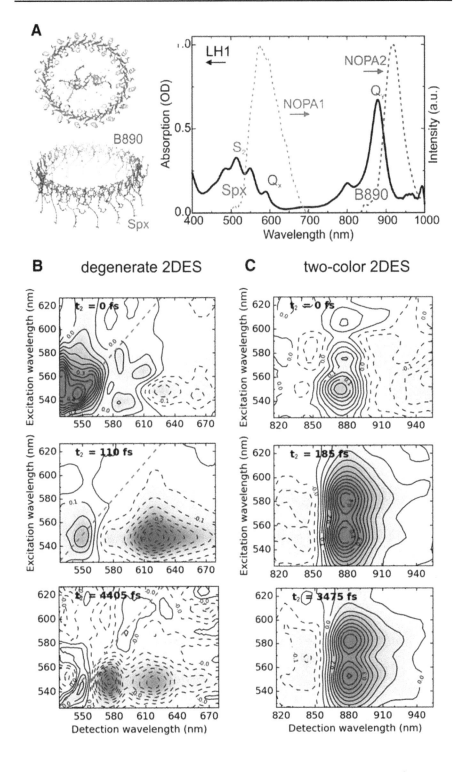

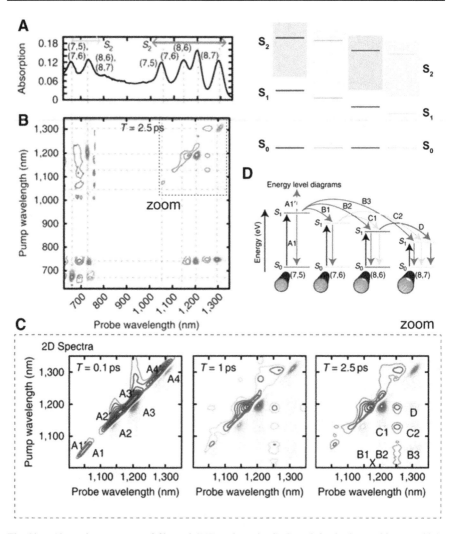

Fig. 11 **a** Absorption spectrum of films of CNTs, where the S_0–S_1 and the S_0–S_2 transitions are highlighted for the different tubes, according to their chirality. **b** 2DES map covering four quadrants generated by the combination of two pulses, spanning the visible and the near-IR range. **c** Zoom of the 2DES map on the top left quadrant in **b** and progression of three 2DES maps at various waiting times T. **d** Energy level scheme of the CNTs, highlighting the pathways observed in **c**. Adapted by permission from Macmillan Publishers Ltd: Nature Communications [36], copyright 2015

observations were only possible thanks to the accurate analysis of cross-peaks and it provided a basis for understanding the photo-physics of energy flow in CNT-based devices.

5 Vibronic Coupling

The study of electronic coupling in complex systems benefits enormously from 2DES, due to the observation of coherent oscillations detected on the cross-peaks of the 2DES maps [5, 6, 65] generated via broadband pulses. These oscillations can be observed as an amplitude modulation of the dynamics along the waiting time T measured at specific point on the excitation/detection 2DES map. The analysis of such oscillations requires the acquisition of various 2DES maps at different waiting times T and a subsequent Fourier transform along this axis of the residual oscillations after removing the slow-varying component associated to the specific cross-peaks trace [66] (Fig. 12). Understanding coherent oscillations in the 2DES maps gives profound insight of the system under study at a quantum mechanical level. Here we do not aim at describing the quantum mechanical formalism, but give some intuitive insight into how coherent oscillations can be observed and interpreted.

Coherent oscillations were observed early on 2DES measurements performed on the photosynthetic Fenna–Matthews–Olson (FMO) complex (described previously in Fig. 9). Interestingly, the oscillatory beating lasted for a time scale longer than the energy transfer rate (> 1 ps) [5]. This initial observation sparked interest in coherent phenomena in photosynthetic complexes and how such oscillations could have a role in the energy transfer processes. Accordingly, there has been a growing interest in investigating the origin of these long-lived oscillations and their relation to energy transfer in the system [67]. 2DES was used to observe robust oscillations in the FMO complex [68], antenna complexes of marine algae [6], the major light-harvesting antenna LHCII [69], and the reaction centers (RCs) of higher plants [70, 71], at low temperature and even at room temperature (Fig. 13).

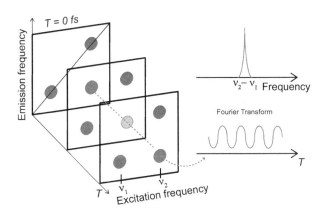

Fig. 12 Progression of the 2DES map of a model dimer system of Fig. 8 along the waiting time T. The amplitudes of the cross-peaks oscillate at the difference frequency of the two states. A cut through different waiting time T maps at the cross-peak position shows the residual signal amplitude modulation as a function of T cross-peak. An oscillation of the cross peak amplitude is observed at the difference frequency between the two states, and the Fourier transform of the signal amplitude gives a peak at the measured oscillation frequency

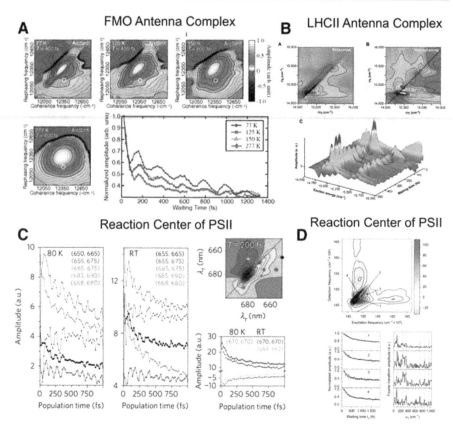

Fig. 13 Examples of 2DES studies where coherent oscillations have been observed in photosynthetic complexes, specifically in: **a** the FMO complex. Reprinted from Copyright 2010 National Academy of Sciences [68]; **b** LHCII complex. Reprinted with permission from Copyright 2009 American Chemical Society [69]; **c–d** reaction center RC of PSII. Reprinted by permission from Macmillan Publishers Ltd: Nature Chemistry [70], copyright 2014. Reprinted by permission from Macmillan Publishers Ltd: Nature Physics [71], copyright 2014

These oscillations have been attributed to electronic, vibrational, and vibronic origins in various systems. Gradually, the consensus has been reached that the long lasting oscillations and their frequencies must be explained by mixing of electronic and vibrational coherences [72–74]. In particular, vibrational modes with frequencies resonant with electronic energy gaps were suggested to be important for both spectroscopic signals and energy-transfer dynamics [75].

The following example (Fig. 14) shows a recent experimental and theoretical verification of coherent electronic–vibrational (vibronic) coupling as the origin of long-lasting coherence, measured in a model system for artificial light harvesting, a molecular J-aggregate [76]. In this tubular system (Fig. 14a, b), a polarization-controlled 2DES approach offered a great tool to distinguish the specific optical responses, leading to fewer region of the 2DES map, which show oscillatory

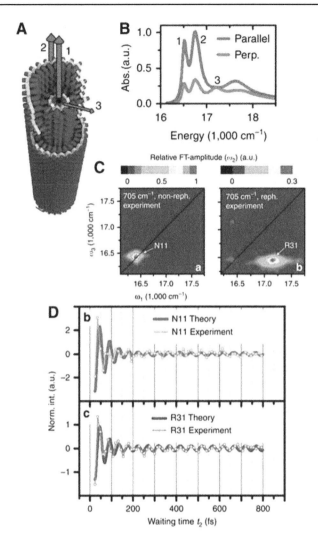

Fig. 14 **a** Molecular structure of the tubular J aggregated, where the main electronic transition dipole moments are drawn along their directions. **b** Absorption spectrum of **a** measured in parallel and perpendicular configuration. **c** The Fourier-transform (FT) amplitude maps of non-rephasing (non-reph.) and rephasing (reph.) at $\nu_1 = 700\,cm^{-1}$, which reveal the presence of a non-rephasing diagonal peak N11 and a rephasing cross-peak R31. **d** The time traces of N11 (*red*) and R31 (*blue*) in normalized intensity against waiting time *T*: experimental results shown as *circles* and theoretical simulation shown as a full line. Adapted from Nature communications [76]

components. In this system, a quasi-resonance exists between the vibrational frequencies ν_1 and ν_2 and the energy splitting of the electronic transition, which, in principle, makes it complicated to distinguish the contribution from electronic or vibrational origin. The resonance promotes an electronic–vibrational coupling that gives origin to vibronic (excited states) and vibrational (ground state) coherences, which can both lead to long-lived beating signals in 2DES maps.

By retrieving the 2D Fourier-transform amplitude maps of selected oscillation at frequency ν_1 close to 700 cm^{-1}, the authors observed a very defined coherence pattern for the non-rephasing and rephasing spectra. Specifically, the oscillations appear on the diagonal peak for the non-rephasing map and as a cross-peak on the rephasing map (Fig. 14c). To describe these features, a vibronic model is employed that describes the coupling of the two electronic bands (specifically 1 and 3 in Fig. 14b) with a quasi-resonant vibrational mode with frequency ν_1. The model predicts an initial fast decoherence, sustained by a long-lived vibrational character of the underdamped vibration (Fig. 14d). The polarization-implemented sequence of this 2DES result provides a strong foundation for understating vibronic coupling and its implication in energy transfer processes.

Coherent oscillations are clearly observed and assigned in inorganic complexes as well, which are characterized by narrower spectral lineshapes and reduction of ensemble disorder. A clear example of electronic coherence is described in Fig. 15 for the case of colloidal semiconductor nanoplatelets (NPLs) [77], where the electronic oscillations are not hidden by vibrational coherences or ensemble dephasing (Fig. 15). The absorption spectra of the CdSe and CdSe/CdZnS NPLs are shown in Fig. 15a, b together with the excitation laser spectra used for the 2DES experiments.

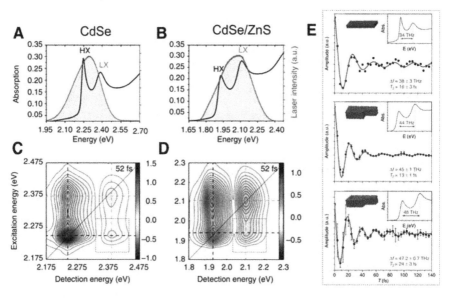

Fig. 15 Absorption spectra of CdSe (**a**) and CdSe/CdZnS (**b**) NPLs showing the HX and LX exciton transitions together with the laser spectra used in the 2DES. **c**, **d** 2DES maps (total and magnitude signal) for a waiting time $T = 52$ fs. **e** Coherent oscillations along the waiting time T of the lower cross-peak of the rephasing 2DES map (population relaxation removed) of CdSe NPL (*top*), CdSe/CdZnS NPL (*middle*) and CdSe/CdZnS NPL (*bottom*) are shown as *black dots*. Adapted by permission from Macmillan Publishers Ltd: Nature Communications [77], copyright 2015

Figure 15c, d, show the plot of the room-temperature 2DES map of the two samples for a waiting time of $T = 52$ fs.

Diagonal and cross-peaks are clearly resolved at the electronic transitions involving different hole bands and labeled heavy-hole and light-hole excitons (HX and LX, respectively). The presence of the cross-peaks is assigned to the strong coupling between the two transitions expected because they share the same electronic state.

Monitoring the evolution of the amplitude of the peaks appearing in the 2DES maps as a function of time allowed the study of coherent superpositions of exciton states. Oscillations are evident and visible on both the diagonal and the lower cross-peaks in the CdSe NPL and in the CdSe/ZnS NPL heterostructure. The authors reported the results for the oscillations observed at the lower cross-peak position in three different samples (CdSe, heterostructure CdSe/ZnS and CdSe/ZnS with thinner core) (Fig. 15e). The oscillations last for about 150 fs. The frequency of the oscillations in these two peaks matches the HX–LX frequency difference measured by the absorption spectrum and they are assigned to electronic coherences. The HX–LX electronic coherence dephases for the three different samples with different time constants. From these results, it is evident that the pure homogeneous line broadening of the colloidal NPLs enables clear observation of the electronic coherence.

6 Photoreactivity

2DES provides a key advantage over methods such as pump-probe spectroscopy for unraveling ultrafast kinetic processes because it decouples the frequency of excitation with the time resolution of the experiment. This renders possible since the excitation is resolved by an interferometric measurement between the two pump beams. Thus the discrimination between a sequential and a parallel mechanism is achievable, even if the states are too entangled and/or their dynamics are too fast [7].

For a sequential mechanism (Fig. 16a), a loss of population (photobleaching) is expected for the species A and a cross-peak will appear at the excitation frequency of A and emission of B. The gain of population (photoproduct) will result in a signal with the opposite sign [78, 79]. In the case of an equilibrium between A and C (Fig. 16b), two diagonal peaks will be detected as a photobleach for A and C and two off-diagonal peaks will be of the opposite sign at the excitation of A and emission of C and excitation of C and emission of A. For instance, if A gives B and C with A and C being in an equilibrium (Fig. 16c), two diagonal peaks are expected in A and C, and three off-diagonal peaks are expected in excitation of A and emission of B and C, and excitation of C and emission of A. Those cases are some extreme cases to describe a 2DES map and will be useful to interpret the data.

The first example is the electron and hole relaxation dynamics in cadmium telluride (CdTe) nanorods (NRs) in toluene (Fig. 17a inset) [80]. Indeed, conventional transient absorption spectroscopy cannot disentangle those phenomena because of the overlap of the transitions and their very fast relaxation time scales. We saw in the Heterogeneity paragraph that those systems are highly heterogeneous. Semiconductor nanocrystals are very interesting materials because the quantum confinement effect leads to: i) discrete energy levels, ii) tunable energy bandgap depending on

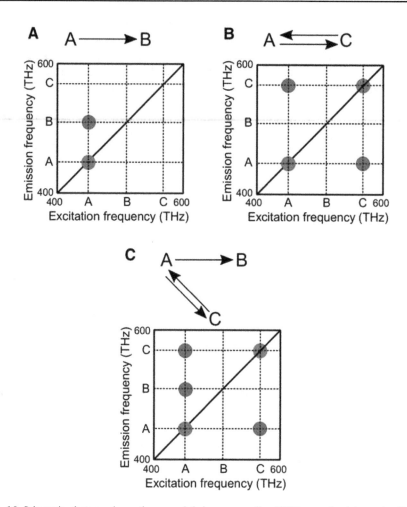

Fig. 16 Schematic photoreaction pathways and their corresponding 2DES maps for A becoming B (**a**), A and C are in equilibrium (**b**) and A becoming B and A is in equilibrium with C (**c**)

their size and iii) enhancement of the Coulomb interaction [81]. Their properties can be easily tuned by advanced synthetic methods. Therefore, they have been applied to a variety of fields: optics, photovoltaics, sensing, and electronics [82]. For instance, in photovoltaic and photocatalytic applications, the electron and hole dynamics contribute separately to the photon-to-electron conversion efficiency. It is crucial to understand and disentangle the dynamics of those species in order to improve the final materials.

Femtosecond transient absorption spectroscopy is a powerful technique to follow the electron/hole dynamics in real time, but it is very difficult to disentangle them. Indeed, the states are very close in energy and a broadband excitation (femtoseconds pulses) leads to an overlap of their transient signals, while narrowband pulses allow

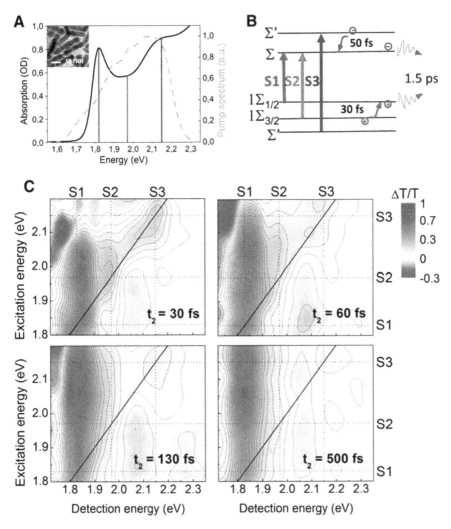

Fig. 17 a Steady-state absorption spectrum of cadmium telluride (CdTe) nanorods (NRs) in toluene (*black line*) and the pump laser spectrum (*green dashed line*). The three lowest excitonic states are indicated by colored lines (S_1 in *pink*, S_2 in *green* and S_3 in *purple*). The *inset* shows a TEM image of the NRs (average dimensions of 21 nm length and 5 nm width). **b** Summary of the identified relaxation pathways and respective time scales. **c** 2DES maps of CdTe NRs in toluene at different waiting times (30, 60, 130, and 500 fs). Adapted with permission from [80]. Copyright 2017 American Chemical Society

state selectivity but the temporal resolution is lost. 2DES is capable of disentangling the excitonic states and to distinguish between electron and hole relaxation dynamics.

Figure 17c shows 2DES maps of the differential transmission ($\Delta T/T$) of CdTe NRs at different waiting times [80]. At $T = 30$ fs, diagonal and cross-peaks are observed: i) three distinctive diagonal peaks are assigned to the photobleaching of S_1, S_2, and S_3 excitons, ii) positive S_2/S_1 (excitation/emission) and S_3/S_1 cross-peaks along the vertical at 1.8-eV emission energy are the signature of an ultrafast relaxation of S_3 and S_2 into S_1 in 30 fs, and iii) a positive signal at the S_1/S_2 cross-peak due to the shared electron level of the two excitons. The other 2DES maps allow tracking the photoreaction dynamics. At $T = 60$ and 130 fs, the relaxation from S_3 and S_2 to S_1 keeps happening (growth of the amplitude of the S_3/S_1 and S_2/S_1 cross-peaks). The amplitude of the S_2/S_2 peak decreases and the one of the S_3/S_3 stays stable. The dynamics of hot hole and electrons is faster than 500 fs, because the 2DES maps at 130 and 500 fs are similar.

By monitoring the decay dynamics of the diagonal and cross-peaks, a proposed mechanism of the ultrafast hole and electron dynamics and their relaxation pathways for the first three excitons of CdTe NRs is given in Fig. 17b. A hot hole thermalization from $1\Sigma_{3/2}$ to $1\Sigma_{1/2}$ in the range of 30 fs, while higher energetic electrons after S_3 pumping relax from Σ' to Σ within 50 fs time constant. Those results were unraveled only by 2DES and can design complex hybrid nanostructures that aim to separately control the electron and hole dynamics.

The second example is a new approach described by Ruetzel et al. [7] that allows tracking of the putative mechanisms involved in complexes photochemical reactions. The main advantage of 2DES is to be able to probe the reactants axis (excitation axis) at the same time that the products one (emission axis). The ring-open-6-nitro-BIPS (Fig. 18a) in acetonitrile exists in a mixture of two stable merocyanine isomers, which photoconvert through a photoisomerization process. The photoisomerization was reported but the identification of the mechanism remained unknown because of the spectral overlap of those isomers [83, 84]. Figure 18a shows the absorption spectrum of the different isomers of 6-nitro-BIPS: the dominant *trans–trans–cis* isomer (TTC) and only 10% of merocyanine molecules exist in the *trans–trans–trans* isomer (TTT). The TTC isomer has an absorption peak at 557 nm, whereas the absorbance peak of TTT isomer is centered at 595 nm.

The photoisomerization was recorded for a long waiting time. In order to be able to disentangle the mechanism, the 2DES data are Fourier-transformed along the waiting time, T, to obtain a third-order 3D spectrum, represented on Fig. 18b. The main contribution is observed in a plane at $\nu_T = 0$ cm^{-1}. By contrast, the observed vibrational motion leads to separated cross-peaks in the 3D spectrum at 176 cm^{-1}, which corresponds to a period of 190 fs centered at the crossing of TTC excitation and TTT photobleaching emission. A second isolated cross-peak appears around $\nu_T = 363$ cm^{-1} (period of 90 fs) and is slightly shifted away from the crossing point toward higher excitation frequencies, indicating that more excess energy is required to initiate it. The location in the 3D spectrum reveals that oscillations of cross-peaks connecting the two isomers give rise to these signatures. This analysis was possible only by measuring 2DES data.

Figure 18c shows slices of the 3D spectrum in the excitation/emission plane for the two main vibrational modes together with the phase. The main contributions are situated at TTC excitation and TTT emission wavenumbers. A vibrational wavepacket motion in the excited state absorption band of TTT is enhanced in the rectangle IV. Both modes are observable at the red edge of the stimulated emission

Fig. 18 a Steady-state absorption spectrum of ring-open 6-nitro-BIPS in acetonitrile. The predicted absorption spectra of the two isomers *trans–trans–cis* configuration (TTC) and *trans–trans–trans* configuration (TTT) are represented in *red dots* and *blue dashes*, respectively. **b** 3D isosurface representation of the 3D spectrum of the 6-nitro-BIPS sample. **c** Slices of the 3D spectrum in the $(\nu_{pump}, \nu_{probe})$ plane for $\nu_T = 164$ (*left*) and $\nu_T = 360$ cm^{-1} (*right*) and the phase for $\nu_T = 164$ cm^{-1} at a vertical cut at the TTC excitation wavenumber (*center*). Reprinted from: Copyright 2014 National Academy of Sciences [7]

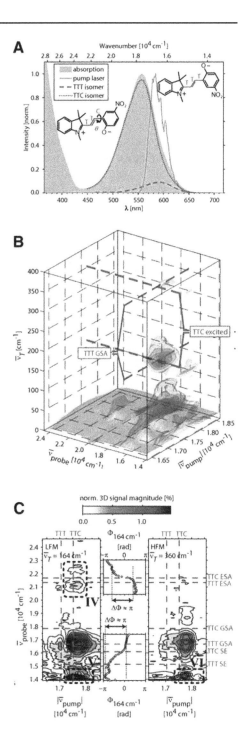

(marks V and VI). Those oscillations take place in the excited state. Moreover, a clear phase shift is observed at the minimum of the potential energy surface according to the coherent wavepacket evolution. Thus, the formation of the TTT involves wavepacket motions on the excited potential energy surface S_1. The photoproduct is formed on the timescale of a few vibrational periods in the S_1 state.

This analysis clearly probes an ultrafast photoisomerization dynamics where the main reaction coordinates are a torsional angle and a second not-further-specified vibrational coordinate. Those two modes were clearly observed by wave-packet dynamics on the excited state. After TTC excitation, most molecules relax back to the ground state. However, a small portion undergoes a photoisomerization *cis-trans* on the excited state. 2DES can isolate molecular reactive modes involved in an ultrafast photochemical process.

7 Conclusions and Perspectives

As we have shown in this chapter, 2DES allows to study i) heterogeneity in the ground state, ii) electronic coupling between chromophores during excitation energy transfer, iii) coherent oscillations in photoinduced dynamics, iv) intermediate or dark states in a photoreactivity cycle.

Furthermore, lots of experimental efforts have been recently done in pushing the limit of 2DES, expanding the ways one can excite and detect the non linear emitted signals (Fig. 19). For example, the third-order signal can be read via fluorescence

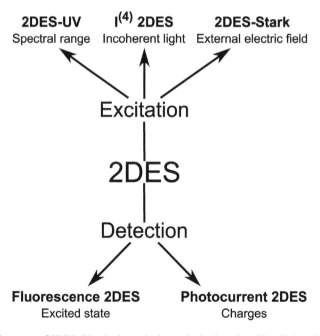

Fig. 19 Developments of 2DES either in the excitation or in the detection sides. Each technique is based on a more complex setup and bring more insight into the studied dynamics

[85–87] or photocurrent [65, 88] detection schemes. 2DES technique has been also implemented in the ultraviolet spectral range [89] or adding external perturbations such as an electric field (2DES Stark [90]). Moreover, some attempts have been made to study the coherent effects observed in 2DES experiment using incoherent light as excitation source [91–94].

7.1 2DES-UV

2DES was successfully implemented in the deep-UV region in the collinear [95] and non-collinear [89] geometry. Those developments were not possible until recently, because the control of the dispersion of UV pulses and the stability of their phase (stability of 2–3 nm) are more challenging in UV.

Various biological samples absorbed in UV, especially DNA or proteins. For instance, it is well known that the photodeactivation of DNA nucleobases is faster than the one of DNA polynucleotides [96, 97]. However, the nature of this fast photodeactivation remains controversial: the most accepted explanation is an ultrafast deactivation through a conical intersection between the excited state and ground state potential energy surfaces [98]. 2DES-UV experiments clearly show that the deactivation pathway is universal for all the nucleobases and is a two-step mechanism: the initial bright state $\pi\pi^*$ is depopulated to a dark state $n\pi^*$ and then to the ground state through two conical intersections [89].

Biological applications will benefit from this new development. Indeed, all the advantages of 2DES can be retrieved in UV region: heterogeneity, excitation energy transfer, coherent oscillation, and photoreactivity.

7.2 Incoherent 2DES

The large majority of 2DES experiments involved the use of coherent femtosecond pulses. Recently, some attempts have been made to perform 2DES using quasi-incoherent light—$I^{(4)}$ 2DES, where $I^{(4)}$ refers to four incoherent excitation fields. Incoherent excitation light combined with time-domain interferometry can produce a strong four-wave mixing signal in the phase-matched direction and this signal can be resolved in phase and amplitude using a local oscillator field. The light used in this series of experiments is referred to as noisy light, and can be continuous light, but often is constituted by pulses on the order of hundreds of nanoseconds, still "continuous" with respect to the femto-to-picosecond dynamics. Moreover, the coherent oscillations can be measured and then Fourier transformed to produce 2DES maps. The 2DES maps measured with this technique have similar information as conventional 2DES maps [91–94].

7.3 2DES Stark

Recently, a new method combining 2DES and Stark spectroscopy has been proposed and implemented successfully [90]. This method has the main advantages of distinguishing kinetic processes as energy and charge transfer, which is difficult in complex systems due to lack of clear spectral signatures of charge transfer states. Stark spectroscopy enables the identification of charge transfer states, their coupling to other charge transfer and exciton states, and their involvement in charge separation processes. Conventional Stark spectroscopy will benefit from the implementation of Stark-2DES for the following reasons:

1. the correlation excitation/detection map obtained by 2DES should be able to push the information about the change in dipole moment and polarizability of transitions available from conventional Stark spectroscopy. 2DES should thus reveal coupling of charge transfer states to other electronic transitions of the system.
2. the additional waiting time dimension of 2DES may enable identification of charge separation processes and the states that drive them.

7.4 Fluorescence 2DES

Phase-modulation two-dimensional fluorescence spectroscopy was developed by detecting the fluorescence out of a 2DES experiment [99]. Four phase-modulated collinear pulses have controlled delays and the signal is retrieve by phase cycling procedure where the phase-modulation fluorescence 2DES is based on a phase selective detection scheme of the nonlinear signals.

This technique have been used for the characterization of the excitonic coupling of dimer probes, and to determine dimer conformation. For instance, it was applied to an electronically coupled porphyrin dimer in a biological membrane [85, 100] or to probe the conformation-dependent electronic coupling of a dinucleotide of 2-aminopurine [101].

Fluorescence 2DES is a powerful technique to disentangle the dynamics of excited states and the electronic couplings involved in dimers.

7.5 2DES Photo-Current

An interesting approach to conventional optical heterodyne-detection techniques employs the detection of the 2DES through photocurrent [65, 88, 102]. A sequence of four collinear pulses creates the fourth-order population, which is detected by utilizing acousto-optic phase modulation of the first two excitation pulses in combination with phase-synchronous detection using a lock-in amplifier. This approach offers several advantages compared to conventional 2DES methods:

1. simultaneous detection of the rephasing, non-rephasing, and two-quantum signals,
2. possibility to obtain diffraction-limited spatial resolution due to fully collinear geometry.

2DES photo-current is particularly useful for the study of photovoltaic materials [103] and other photo-devices, since it provides direct access to ultrafast dynamics of the device under typical operating conditions.

References

1. Hybl JD, Albrecht Ferro A, Jonas DM (2001) Two-dimensional Fourier transform electronic spectroscopy. J Chem Phys 115:6606–6622. https://doi.org/10.1063/1.1398579
2. Ginsberg NS, Cheng Y-C, Fleming GR (2009) Two-dimensional electronic spectroscopy of molecular aggregates. Acc Chem Res 42:1352–1363. https://doi.org/10.1021/ar9001075
3. Jonas David M (2003) Two-dimensional femtosecond spectroscopy. Annu Rev Phys Chem 54:425–463. https://doi.org/10.1146/annurev.physchem.54.011002.103907
4. Brixner T, Stenger J, Vaswani HM et al (2005) Two-dimensional spectroscopy of electronic couplings in photosynthesis. Nature 625–628
5. Engel GS, Calhoun TR, Read EL et al (2007) Evidence for wavelike energy transfer through quantum coherence in photosynthetic systems. Nature 446:782–786. https://doi.org/10.1038/nature05678
6. Collini E, Wong CY, Wilk KE et al (2010) Coherently wired light-harvesting in photosynthetic marine algae at ambient temperature. Nature 463:644–647. https://doi.org/10.1038/nature08811
7. Ruetzel S, Diekmann M, Nuernberger P et al (2014) Multidimensional spectroscopy of photoreactivity. Proc Natl Acad Sci 111:4764–4769. https://doi.org/10.1073/pnas.1323792111
8. Consani C, Auböck G, van Mourik F, Chergui M (2013) Ultrafast tryptophan-to-heme electron transfer in myoglobins revealed by UV 2D spectroscopy. Science 339:1586–1589. https://doi.org/10.1126/science.1230758
9. Mukamel S (1995) Principles of nonlinear optical spectroscopy. Oxford University Press, New York
10. Mukamel S (2000) Multidimensional femtosecond correlation spectroscopies of electronic and vibrational excitations. Annu Rev Phys Chem 51:691–729
11. Cho M (2008) Coherent two-dimensional optical spectroscopy. Chem Rev 108:1331–1418. https://doi.org/10.1021/cr078377b
12. Turner DB, Hassan Y, Scholes GD (2012) Exciton superposition states in CdSe nanocrystals measured using broadband two-dimensional electronic spectroscopy. Nano Lett 12:880–886. https://doi.org/10.1021/nl2039502
13. Lewis KLM, Ogilvie JP (2012) Probing photosynthetic energy and charge transfer with two-dimensional electronic spectroscopy. J Phys Chem Lett 3:503–510. https://doi.org/10.1021/jz201592v
14. Anna JM, Song Y, Dinshaw R, Scholes GD (2013) Two-dimensional electronic spectroscopy for mapping molecular photophysics. Pure Appl Chem 85:1307–1319. https://doi.org/10.1351/PAC-CON-12-10-21
15. Bizimana LA, Epstein J, Brazard J, Turner DB (2017) Conformational homogeneity in the P_r isomer of phytochrome Cph1. J Phys Chem B 121:2622–2630. https://doi.org/10.1021/acs.jpcb.7b02180
16. Hybl JD, Albrecht AW, Gallagher Faeder SM, Jonas DM (1998) Two-dimensional electronic spectroscopy. Chem Phys Lett 307–313
17. Roberts ST, Loparo JJ, Tokmakoff A (2006) Characterization of spectral diffusion from two-dimensional line shapes. J Chem Phys 125:084502. https://doi.org/10.1063/1.2232271
18. Moca R, Meech SR, Heisler IA (2015) Two-dimensional electronic spectroscopy of chlorophyll a: solvent dependent spectral evolution. J Phys Chem B 119:8623–8630. https://doi.org/10.1021/acs.jpcb.5b04339
19. Gellen TA, Lem J, Turner DB (2017) Probing homogeneous line broadening in CdSe nanocrystals using multidimensional electronic spectroscopy. Nano Lett 17:2809–2815. https://doi.org/10.1021/acs.nanolett.6b05068
20. Chenu A, Scholes GD (2015) Coherence in energy transfer and photosynthesis. Annu Rev Phys Chem 66:69–96. https://doi.org/10.1146/annurev-physchem-040214-121713
21. Lepetit L, Chériaux G, Joffre M (1995) Linear techniques of phase measurement by femtosecond spectral interferometry for applications in spectroscopy. J Opt Soc Am B 12:2467–2474. https://doi.org/10.1364/josab.12.002467
22. Goodno GD, Dadusc G, Miller RD (1998) Ultrafast heterodyne-detected transient-grating spectroscopy using diffractive optics. JOSA B 15:1791–1794
23. Miller RJD, Paarmann A, Prokhorenko VI (2009) Diffractive optics based four-wave, six-wave, ..., ν-wave nonlinear spectroscopy. Acc Chem Res 42:1442–1451. https://doi.org/10.1021/ar900040f

24. Hamm P, Zanni MT (2011) Concepts and methods of 2D infrared spectroscopy. Cambridge University Press, Cambridge
25. Brixner T, Stiopkin IV, Fleming GR (2004) Tunable two-dimensional femtosecond spectroscopy. Opt Lett 29:884–886
26. Augulis R, Zigmantas D (2011) Two-dimensional electronic spectroscopy with double modulation lock-in detection: enhancement of sensitivity and noise resistance. Opt Express 19:13126–13133
27. Selig U, Langhojer F, Dimler F et al (2008) Inherently phase-stable coherent two-dimensional spectroscopy using only conventional optics. Opt Lett 33:2851–2853
28. Heisler IA, Moca R, Camargo FVA, Meech SR (2014) Two-dimensional electronic spectroscopy based on conventional optics and fast dual chopper data acquisition. Rev Sci Instrum 85:063103. https://doi.org/10.1063/1.4879822
29. Bizimana LA, Brazard J, Carbery WP et al (2015) Resolving molecular vibronic structure using high-sensitivity two-dimensional electronic spectroscopy. J Chem Phys 143:164203
30. Levenson MD, Eesley GL (1979) Polarization selective optical heterodyne detection for dramatically improved sensitivity in laser spectroscopy. Appl Phys Mater Sci Process 19:1–17
31. Tian P, Keusters D, Suzaki Y, Warren WS (2003) Femtosecond phase-coherent two-dimensional spectroscopy. Science 300:1553–1555
32. Keusters D, Tan H-S, Warren (1999) Role of pulse phase and direction in two-dimensional optical spectroscopy. J Phys Chem A 103:10369–10380. https://doi.org/10.1021/jp992325b
33. Gallagher Faeder SM, Jonas DM (1999) Two-dimensional electronic correlation and relaxation spectra: theory and model calculations. J Phys Chem A 103:10489–10505. https://doi.org/10.1021/jp9925738
34. Tekavec PF, Myers JA, Lewis KL, Ogilvie JP (2009) Two-dimensional electronic spectroscopy with a continuum probe. Opt Lett 34:1390–1392
35. Tekavec PA, Lewis KLM, Fuller FD et al (2012) Toward broad bandwidth 2-D electronic spectroscopy: correction of chirp from a continuum probe. IEEE J Sel Top Quantum Electron 18:210–217. https://doi.org/10.1109/JSTQE.2011.2109941
36. Mehlenbacher RD, McDonough TJ, Grechko M et al (2015) Energy transfer pathways in semiconducting carbon nanotubes revealed using two-dimensional white-light spectroscopy. 6:6732. https://doi.org/10.1038/ncomms7732 https://www.nature.com/articles/ncomms7732#supplementary-information
37. Tekavec PA, Myers JA, Lewis KLM et al (2010) Effects of chirp on two-dimensional Fourier transform electronic spectra. Opt Express 18:11015–11024
38. Shim S-H, Zanni MT (2009) How to turn your pump-probe instrument into a multidimensional spectrometer: 2D IR and Vis spectroscopies via pulse shaping. Phys Chem Chem Phys 11:748–761. https://doi.org/10.1039/b813817f
39. Seiler H, Palato S, Schmidt BE, Kambhampati P (2017) Simple fiber-based solution for coherent multidimensional spectroscopy in the visible regime. Opt Lett 42:643. https://doi.org/10.1364/OL.42.000643
40. Brida D, Manzoni C, Cerullo G (2012) Phase-locked pulses for two-dimensional spectroscopy by a birefringent delay line. Opt Lett 37:3027–3029
41. Réhault J, Maiuri M, Oriana A, Cerullo G (2014) Two-dimensional electronic spectroscopy with birefringent wedges. Rev Sci Instrum 85:123107. https://doi.org/10.1063/1.4902938
42. Seiler H, Palato S, Kambhampati P (2017) Coherent multi-dimensional spectroscopy at optical frequencies in a single beam with optical readout. J Chem Phys 147:094203. https://doi.org/10.1063/1.4990500
43. Fuller FD, Ogilvie JP (2015) Experimental implementations of two-dimensional fourier transform electronic spectroscopy. Annu Rev Phys Chem 66:667–690. https://doi.org/10.1146/annurev-physchem-040513-103623
44. Cho M, Brixner T, Stiopkin I et al (2006) Two dimensional electronic spectroscopy of molecular complexes. J Chin Chem Soc 53:15–24. https://doi.org/10.1002/jccs.200600002
45. Ernst RR, Bodenhausen G, Wokaun A (1987) Principles of nuclear magnetic resonance in one and two dimensions. Oxford University Press, Oxford
46. Lazonder K, Pshenichnikov MS, Wiersma DA (2006) Easy interpretation of optical two-dimensional correlation spectra. Opt Lett 31:3354–3356
47. Lazonder K, Pshenichnikov MS, Wiersma DA (2007) Two-dimensional optical correlation spectroscopy applied to liquid/glass dynamics. In: Ultrafast phenomena XV. Springer, pp 356–358

48. Blankenship RE (2002) Molecular mechanisms of photosynthesis. Wiley-Blackwell. https://doi. org/10.1002/9780470758472
49. Lukas AS, Zhao Y, Miller SE, Wasielewski MR (2002) Biomimetic electron transfer using low energy excited states: a green perylene-based analogue of chlorophyll *a*. J Phys Chem B 106:1299–1306. https://doi.org/10.1021/jp014073w
50. Rockwell NC, Lagarias JC (2010) A brief history of phytochromes. ChemPhysChem 11:1172–1180. https://doi.org/10.1002/cphc.200900894
51. Lamparter T (2004) Evolution of cyanobacterial and plant phytochromes. FEBS Lett 573:1–5. https://doi.org/10.1016/j.febslet.2004.07.050
52. Kim PW, Rockwell NC, Freer LH et al (2013) Unraveling the primary isomerization dynamics in cyanobacterial phytochrome Cph1 with multipulse manipulations. J Phys Chem Lett 4:2605–2609. https://doi.org/10.1021/jz401443q
53. Kim PW, Rockwell NC, Martin SS et al (2014) Dynamic inhomogeneity in the photodynamics of cyanobacterial phytochrome Cph1. Biochemistry (Mosc) 53:2818–2826. https://doi.org/10.1021/bi500108s
54. Dasgupta J, Frontiera RR, Taylor KC et al (2009) Ultrafast excited-state isomerization in phytochrome revealed by femtosecond stimulated Raman spectroscopy. Proc Natl Acad Sci 106:1784–1789
55. Spillane KM, Dasgupta J, Lagarias JC, Mathies RA (2009) Homogeneity of phytochrome Cph1 vibronic absorption revealed by resonance Raman intensity analysis. J Am Chem Soc 131:13946–13948. https://doi.org/10.1021/ja905822m
56. Brixner T, Mančal T, Stiopkin IV, Fleming GR (2004) Phase-stabilized two-dimensional electronic spectroscopy. J Chem Phys 121:4221. https://doi.org/10.1063/1.1776112
57. Turner DB, Wilk KE, Curmi PMG, Scholes GD (2011) Comparison of electronic and vibrational coherence measured by two-dimensional electronic spectroscopy. J Phys Chem Lett 2:1904–1911. https://doi.org/10.1021/jz200811p
58. Talapin DV, Lee J-S, Kovalenko MV, Shevchenko EV (2010) Prospects of colloidal nanocrystals for electronic and optoelectronic applications. Chem Rev 110:389–458. https://doi.org/10.1021/cr900137k
59. Van Amerongen H, Valkunas L, Van Grondelle R (2001) Photosynthetic excitons, 2000
60. Dostál J, Pšenčík J, Zigmantas D (2016) In situ mapping of the energy flow through the entire photosynthetic apparatus. Nat Chem 8:705–710. https://doi.org/10.1038/nchem.2525 http://www.nature.com/nchem/journal/v8/n7/abs/nchem.2525.html#supplementary-information
61. Dahlberg PD, Fidler AF, Caram JR et al (2013) Energy transfer observed in live cells using two-dimensional electronic spectroscopy. J Phys Chem Lett 4:3636–3640. https://doi.org/10.1021/jz401944q
62. Thyrhaug E, Žídek K, Dostál J et al (2016) Exciton structure and energy transfer in the Fenna–Matthews–Olson complex. J Phys Chem Lett 7:1653–1660. https://doi.org/10.1021/acs.jpclett.6b00534
63. Maiuri M, Réhault J, Carey A-M et al (2015) Ultra-broadband 2D electronic spectroscopy of carotenoid-bacteriochlorophyll interactions in the LH1 complex of a purple bacterium. J Chem Phys 142:212433. https://doi.org/10.1063/1.4919056
64. Polívka T, Sundström V (2009) Dark excited states of carotenoids: consensus and controversy. Chem Phys Lett 477:1–11. https://doi.org/10.1016/j.cplett.2009.06.011
65. Moody G, Cundiff ST (2017) Advances in multi-dimensional coherent spectroscopy of semiconductor nanostructures. Adv Phys X 2:641–674. https://doi.org/10.1080/23746149.2017.1346482
66. Fassioli F, Dinshaw R, Arpin PC, Scholes GD (2014) Photosynthetic light harvesting: excitons and coherence. J R Soc Interface. https://doi.org/10.1098/rsif.2013.0901
67. Scholes GD, Fleming GR, Chen LX et al (2017) Using coherence to enhance function in chemical and biophysical systems. Nature 543:647–656. https://doi.org/10.1038/nature21425
68. Panitchayangkoon G, Hayes D, Fransted KA et al (2010) Long-lived quantum coherence in photosynthetic complexes at physiological temperature. Proc Natl Acad Sci 107:12766–12770. https://doi.org/10.1073/pnas.1005484107
69. Calhoun TR, Ginsberg NS, Schlau-Cohen GS et al (2009) Quantum coherence enabled determination of the energy landscape in light-harvesting complex II. J Phys Chem B 113:16291–16295. https://doi.org/10.1021/jp908300c
70. Fuller FD, Pan J, Gelzinis A et al (2014) Vibronic coherence in oxygenic photosynthesis. Nat Chem. https://doi.org/10.1038/nchem.2005

  Springer

71. Romero E, Augulis R, Novoderezhkin VI et al (2014) Quantum coherence in photosynthesis for efficient solar-energy conversion. Nat Phys 10:676–682. https://doi.org/10.1038/nphys3017
72. Christensson N, Kauffmann HF, Pullerits T, Mančal T (2012) Origin of long-lived coherences in light-harvesting complexes. J Phys Chem B 116:7449–7454. https://doi.org/10.1021/jp304649c
73. Tiwari V, Peters WK, Jonas DM (2013) Electronic resonance with anticorrelated pigment vibrations drives photosynthetic energy transfer outside the adiabatic framework. Proc Natl Acad Sci 110:1203–1208. https://doi.org/10.1073/pnas.1211157110
74. Abramavicius D, Valkunas L (2016) Role of coherent vibrations in energy transfer and conversion in photosynthetic pigment–protein complexes. Photosynth Res 127:33–47. https://doi.org/10.1007/s11120-015-0080-6
75. Dean JC, Mirkovic T, Toa ZSD et al (2016) Vibronic enhancement of algae light harvesting. Chem 1:858–872. https://doi.org/10.1016/j.chempr.2016.11.002
76. Lim J, Paleček D, Caycedo-Soler F et al (2015) Vibronic origin of long-lived coherence in an artificial molecular light harvester. Nat Commun 6:7755. https://doi.org/10.1038/ncomms8755
77. Cassette E, Pensack RD, Mahler B, Scholes GD (2015) Room-temperature exciton coherence and dephasing in two-dimensional nanostructures. Nat Commun 6:6086. https://doi.org/10.1038/ncomms7086
78. Kullmann M, Ruetzel S, Buback J et al (2011) Reaction dynamics of a molecular switch unveiled by coherent two-dimensional electronic spectroscopy. J Am Chem Soc 133:13074–13080. https://doi.org/10.1021/ja2032037
79. Ruetzel S, Kullmann M, Buback J et al (2013) Tracing the steps of photoinduced chemical reactions in organic molecules by coherent two-dimensional electronic spectroscopy using triggered exchange. Phys Rev Lett. https://doi.org/10.1103/PhysRevLett.110.148305
80. Stoll T, Branchi F, Réhault J et al (2017) Two-dimensional electronic spectroscopy unravels sub-100 fs electron and hole relaxation dynamics in Cd-chalcogenide nanostructures. J Phys Chem Lett 8:2285–2290. https://doi.org/10.1021/acs.jpclett.7b00682
81. Brus L (1986) Electronic wave functions in semiconductor clusters: experiment and theory. J Phys Chem 90:2555–2560
82. Banin U, Ben-Shahar Y, Vinokurov K (2014) Hybrid semiconductor-metal nanoparticles: from architecture to function. Chem Mater 26:97–110. https://doi.org/10.1021/cm402131n
83. Wohl CJ, Kuciauskas D (2005) Excited-state dynamics of spiropyran-derived merocyanine isomers. J Phys Chem B 109:22186–22191. https://doi.org/10.1021/jp053782x
84. Chibisov AK, Görner H (1997) Photoprocesses in spiropyran-derived merocyanines. J Phys Chem A 101:4305–4312
85. Lott GA, Perdomo-Ortiz A, Utterback JK et al (2011) Conformation of self-assembled porphyrin dimers in liposome vesicles by phase-modulation 2D fluorescence spectroscopy. Proc Natl Acad Sci 108:16521–16526
86. Wagner W, Li C, Semmlow J, Warren WS (2005) Rapid phase-cycled two-dimensional optical spectroscopy in fluorescence and transmission mode. Opt Express 13(10):3697–3706
87. Draeger S, Roeding S, Brixner T (2017) Rapid-scan coherent 2D fluorescence spectroscopy. Opt Express 25(4):3259–3267
88. Nardin G, Autry TM, Silverman KL, Cundiff ST (2013) Multidimensional coherent photocurrent spectroscopy of a semiconductor nanostructure. Opt Express 21:28617. https://doi.org/10.1364/OE.21.028617
89. Prokhorenko VI, Picchiotti A, Pola M et al (2016) New insights into the photophysics of DNA nucleobases. J Phys Chem Lett. https://doi.org/10.1021/acs.jpclett.6b02085
90. Loukianov A, Niedringhaus A, Berg B et al (2017) Two-dimensional electronic stark spectroscopy. J Phys Chem Lett 8:679–683. https://doi.org/10.1021/acs.jpclett.6b02695
91. Turner DB, Arpin PC, McClure SD et al (2013) Coherent multidimensional optical spectra measured using incoherent light. Nat Commun. https://doi.org/10.1038/ncomms3298
92. Turner DB, Howey DJ, Sutor EJ et al (2013) Two-dimensional electronic spectroscopy using incoherent light: theoretical analysis. J Phys Chem A 117:5926–5954. https://doi.org/10.1021/jp310477y
93. Ulness DJ, Turner DB (2015) Lineshape analysis of coherent multidimensional optical spectroscopy using incoherent light. J Chem Phys 142:212420. https://doi.org/10.1063/1.4917320
94. Ulness DJ, Turner DB (2017) Coherent two-quantum two-dimensional electronic spectroscopy using incoherent light. J Phys Chem. https://doi.org/10.1021/acs.jpca.7b0944

95. Borrego-Varillas R, Oriana A, Ganzer L et al (2016) Two-dimensional electronic spectroscopy in the ultraviolet by a birefringent delay line. Opt Express 24:28491. https://doi.org/10.1364/OE.24.028491

96. Middleton CT, de La Harpe K, Su C et al (2009) DNA excited-state dynamics: from single bases to the double helix. Annu Rev Phys Chem 60:217–239. https://doi.org/10.1146/annurev.physchem.59.032607.093719

97. Onidas D, Markovitsi D, Marguet S et al (2002) Fluorescence properties of DNA nucleosides and nucleotides: a refined steady-state and femtosecond investigation. J Phys Chem B 106:11367–11374. https://doi.org/10.1021/jp026063g

98. Improta R, Santoro F, Blancafort L (2016) Quantum mechanical studies on the photophysics and the photochemistry of nucleic acids and nucleobases. Chem Rev 116:3540–3593. https://doi.org/10.1021/acs.chemrev.5b00444

99. Tekavec PF, Lott GA, Marcus AH (2007) Fluorescence-detected two-dimensional electronic coherence spectroscopy by acousto-optic phase modulation. J Chem Phys 127:214307. https://doi.org/10.1063/1.2800560

100. Perdomo-Ortiz A, Widom JR, Lott GA et al (2012) Conformation and electronic population transfer in membrane-supported self-assembled porphyrin dimers by 2D fluorescence spectroscopy. J Phys Chem B 116:10757–10770. https://doi.org/10.1021/jp305916x

101. Widom JR, Johnson NP, von Hippel PH, Marcus AH (2013) Solution conformation of 2-aminopurine dinucleotide determined by ultraviolet two-dimensional fluorescence spectroscopy. New J Phys 15:025028. https://doi.org/10.1088/1367-2630/15/2/025028

102. Karki KJ, Widom JR, Seibt J et al (2014) Coherent two-dimensional photocurrent spectroscopy in a PbS quantum dot photocell. Nat Commun 5:5869. https://doi.org/10.1038/ncomms6869

103. Bakulin AA, Silva C, Vella E (2016) Ultrafast spectroscopy with photocurrent detection: watching excitonic optoelectronic systems at work. J Phys Chem Lett 7:250–258. https://doi.org/10.1021/acs.jpclett.5b01955

Top Curr Chem (Z) (2018) 376:24
https://doi.org/10.1007/s41061-018-0201-8

REVIEW

Towards Accurate Simulation of Two-Dimensional Electronic Spectroscopy

Javier Segarra-Martí[1] · Shaul Mukamel[2] · Marco Garavelli[3] · Artur Nenov[3] · Ivan Rivalta[1]

Received: 21 December 2017 / Accepted: 24 April 2018 / Published online: 1 June 2018
© Springer International Publishing AG, part of Springer Nature 2018

Abstract We introduce the basic concepts of two-dimensional electronic spectroscopy (2DES) and a general theoretical framework adopted to calculate, from first principles, the nonlinear response of multi-chromophoric systems in realistic environments. Specifically, we focus on UV-active chromophores representing the building blocks of biological systems, from proteins to nucleic acids, describing our progress in developing computational tools and protocols for accurate simulation of their 2DUV spectra. The roadmap for accurate 2DUV spectroscopy simulations is illustrated starting with benchmarking of the excited-state manifold of the chromophoric units in a vacuum, which can be used for building exciton Hamiltonians for large-scale applications or as a reference for first-principles simulations with reduced computational cost, enabling treatment of minimal (still realistic) multi-chromophoric model systems. By adopting a static approximation that neglects dynamic processes such as spectral diffusion and population transfer, we show how 2DUV is able to characterize the ground-state conformational space of dinucleosides and small peptides comprising dimeric chromophoric units (in their native environment) by tracking inter-chromophoric electronic couplings. Recovering the excited-state coherent vibrational dynamics and population transfers,

Chapter 3 was originally published as Segarra-Martí, J., Mukamel, S., Garavelli, M., Nenov, A. & Rivalta, I. Top Curr Chem (Z) (2018) 376: 24. https://doi.org/10.1007/s41061-018-0201-8.

✉ Marco Garavelli

✉ Artur Nenov

✉ Ivan Rivalta
  ivan.rivalta@ens-lyon.fr

[1] Université de Lyon, École Normale Supérieure de Lyon, CNRS, Université Claude Bernard Lyon 1, Laboratoire de Chimie UMR 5182, 69342 Lyon, France

[2] Department of Chemistry, University of California, Irvine, CA 92697-2025, USA

[3] Dipartimento di Chimica Industriale, Università degli Studi di Bologna, Viale del Risorgimento 4, 40136 Bologna, Italy

we observe a remarkable agreement between the predicted 2DUV spectra of the pyrene molecule and the experimental results. These results further led to theoretical studies of the excited-state dynamics in a solvated dinucleoside system, showing that spectroscopic fingerprints of long-lived excited-state minima along the complex photoinduced decay pathways of DNA/RNA model systems can be simulated at a reasonable computational cost. Our results exemplify the impact of accurate simulation of 2DES spectra in revealing complex physicochemical properties of fundamental biological systems and should trigger further theoretical developments as well as new experiments.

Keywords Nonlinear electronic spectroscopy · Theoretical simulations · Wavefunction methods · QM/MM computations · DNA/RNA nucleobases · Aromatic amino acids

1 Introduction

Two-dimensional (2D) optical spectroscopy based on multipulse laser sequences originated as an extension of the 2D nuclear magnetic resonance (2DNMR) technique [1] to the optical regime [2]. 2DNMR had an impressive impact in several fields, with first handover to the infrared (IR) regime (2DIR) now a well-established method often employed in the characterization of the structure and dynamics of complex molecular systems by directly mapping their vibrational couplings [3]. Mapping of electronic couplings by 2D electronic spectroscopy (2DES) [4] has become feasible thanks to advances [5, 6] in ultrafast optical techniques that allow the targeting of electronic transitions in the visible (Vis) range. 2DES in the Vis range (2DVis) has become increasingly popular over the last decade, showcasing its great potential by deciphering energy transfer processes in photosynthesis [7–10]. The desirable extension to the UV domain (2DUV), however, where many fundamental biomolecules display strong absorption bands, has been slow thus far, impeded by technical difficulties associated with the use of UV laser pulses [11]. Recent progress [12–18] in attaining interferometric phase stability and sufficient laser bandwidth has enabled access to the first examples of experimental 2DUV spectra [19–22]. Simulation studies have demonstrated that 2DUV of aromatic residues in the near UV (NUV) and the backbone in the far UV can effectively probe protein secondary structure [23, 24]. Among the most recent developments in multidimensional spectroscopy related to 2DES, it is important to highlight the advances in electronic-vibrational spectroscopy [25] and 2D Stark spectroscopy for characterizing dark charge-transfer states [26], along with the impressive potential of developing techniques in the X-ray regime [27].

The signals recorded in 2DES correlation plots refer to the third-order nonlinear response of the sample, and contain a wealth of information on the excited-state manifold of the chromophoric units and its photo-induced evolution in time, which are strongly related due to intra- and inter-chromophoric electronic couplings and coherence/decoherence effects in chemical and biophysical systems [28].

The electronic transitions involving excited-state absorptions, especially in 2DUV spectra, would encompass high-energy electronic levels whose nature is hitherto unexplored. The complexity of the information contained in 2DES maps and the involvement of high-energy electronic states calls for the implementation of computational tools based on a combination of ab initio electronic structure and nonlinear response formalism. Accurate simulation of 2DUV electronic spectroscopy may, in fact, lead to both interpretation and prediction of nonlinear spectra, enhancing the potential application of this technique in various fields. Here, we outline a route towards accurate simulation of 2DUV spectra of fundamental biological systems. After describing the general 2DES technique and the target UV-active bio-chromophores in this section, we present a basic background in Sect. 2, aimed at providing the theoretical foundations of our simulation protocols and showing the approximations adopted. In Sect. 3, the main results of our developments and applications are illustrated for both nucleic acid and protein model systems, showing the potential for 2DUV in characterizing the ground-state (GS) conformational space and excited-state dynamics in these target systems. Finally, an outlook summary and perspective are provided in Sect. 4.

2 The 2DES Technique

Two-dimensional electronic spectroscopy [29] is based on a sequence of three ultrashort laser pulses interacting with the sample, generating a third-order nonlinear polarization and emitting signal fields in phase-matched directions (Fig. 1). The major advantage of 2DES is the higher spectral resolution relative to 1D-PP time-resolved techniques, with the nonlinear signal containing information on system dynamics and electronic couplings spread over two frequency axes (pump and probe, Ω_1 and Ω_3, respectively; see Fig. 1b). The spectral resolution in two dimensions enables accurate characterization of inhomogeneous and homogeneous broadening processes, with the ability to separate these two contributions (not possible with 1D techniques), providing information about solvent reorganization timescales as well as enabling the detection of signals associated with coupling between electronic excitations and charge/energy transfer processes.

The heterodyne-detected three-pulse photon echo (3PPE) non-collinear scheme [5, 30] is the experimental setup that can fully resolve (in amplitude and phase) the third-order nonlinear response, by collecting a four-wave mixing signal in a background-free direction, which is heterodyned by a local oscillator (LO); see Fig. 1a. The signal field emitted can be detected in the so-called rephasing K_I ($k_{LO} = -k_1 + k_2 + k_3$) and non-rephasing K_{II} ($k_{LO} = +k_1 - k_2 + k_3$) phase-matching directions, as a function of three controlled excitation-pulse time delays (t_1, t_2 and t_3). Instead of 3PPE, the partially collinear pump–probe (PCPP) geometry [21, 31, 32] can be adopted by using a pair of collinear pump pulses (k_1, k_2) non-collinearly combined with a probe pulse (k_3), providing a nonlinear response signal which is heterodyned by the probe pulse itself (self-heterodyning). The PCPP experiment is disadvantaged by the strong background signal, and it intrinsically provides the combined $K_I + K_{II}$ "quasi-absorptive" response, with loss of information on specific rephasing and non-rephasing signals, although removal of slowly decaying dispersive contributions due to phase cancellation of the concurring K_I and K_{II} signals might help in resolving weak signals.

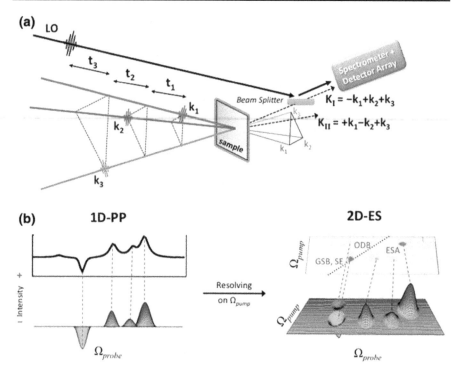

Fig. 1 Schematic representation of **a** heterodyne-detected three-pulse photon echo (3PPE) 2DES experiment (LO: local oscillator) showing both rephasing K_I ($k_{LO} = -k_1+k_2+k_3$) and non-rephasing K_{II} ($k_{LO} = +k_1 -k_2+k_3$) phase-matching directions, and **b** 1D pump–probe (1D-PP) and analogous 2D electronic spectra (2D-ES) at a fixed waiting time t_2, showing how the different shapes and signs of the nonlinear signals are properly resolved in the 2DES maps

For a given "population time" t_2 (also known as "waiting time", and analogous to the time delay between the pump and the probe pulse in 1D pump–probe [1D-PP] spectroscopy), a 2D spectrum (or map) is recorded as a function of "excitation frequency" Ω_1 (Ω_{pump}) and "detection frequency" Ω_3 (Ω_{probe}) by Fourier transformation with respect to the t_1 and t_3 time intervals. Thus, the nonlinear spectral signatures of the GS are obtained by setting the time t_2 to zero, while the state-specific response along excited-state dynamics can be monitored for $t_2 > 0$.

As illustrated in Fig. 1b, the induced nonlinear polarization measured in a 2DES map at any certain population time is richer than that of a 1D-PP spectrum, as by adding the spectral resolution along Ω_{pump}, it is possible to disentangle the contributions arising from different pump frequencies and the line broadening along both pump and probe frequencies. In general, a 2DES spectrum at $t_2 = 0$ contains diagonal peaks that correspond to ground-to-excited-state absorptions (ground-state bleaching, GSB) and stimulated emissions (SE), symmetric off-diagonal signals associated with coupling between ground- and excited-state electronic excitations (off-diagonal bleaching, ODB) and asymmetric off-diagonal peaks (with the opposite sign relative to diagonal peaks) due to excited-state absorptions (ESAs) to the high-lying states. For $t_2 > 0$, the s dynamics shape the 2DES spectra, with SE and ESA peaks exhibiting at short waiting times, fluctuations along Ω_{probe}, an indirect probe of the coherent vibrational dynamics along the photoactive state, and at longer waiting times, Stoke shifts associated with internal energy redistribution and dissipation in the environment, as well as intensity decay, a function of the finite excited-state lifetimes. For additional details on 2DES experimental techniques and map features, we recommend the review by Maiuri and Brazard in this series [33].

2.1 2DES in the Ultraviolet: A New Light in Photobiology

The emergence of novel UV pulse technologies [14] has enabled the use of pump and/or probe laser pulses in the UV window. These efforts may have a particular impact in monitoring biologically relevant processes [34]. Indeed, cyclic aromatic groups are often the UV-active chromophores that play a major role in photoinduced events in biological systems [35]. UV light, in fact, is absorbed by an incredibly large portion of biomolecules, ranging from proteins to nucleic acids. As shown in Fig. 2, adsorption by both protein and nucleic acid backbones occurs exclusively at high energies, i.e. above ca. 5 eV (i.e. at wavelengths below ca. 250 nm). Aromatic rings in proteins and DNA/RNA, including tryptophan (Trp), phenylalanine (Phe), tyrosine (Tyr) and histidine (His) amino acid side chains and adenine (Ade), guanine (Gua), cytosine (Cyt), thymine (Thy) and uracil (Ura) nucleobases, feature their most intense absorptions in the same spectral window as their backbones. However, all these aromatic units (except histidine) have distinctive absorption bands associated with $\pi \rightarrow \pi^*$ transitions at lower energies (i.e. around 4.4–5.0 eV, ca. 250–280 nm), thus providing a UV range where their excitations are clearly separated from those of the backbone. Notably, when considering even lower transition energies in the NUV range around 300–400 nm (i.e. 3.1–4.1 eV), quite important UV-active biomolecules are found, including essential cofactors such as plastoquinones, flavins and reduced nicotinamide adenine dinucleotide (NADH), to name a few. This evidence suggests that the low-energy UV window represents a promising option for monitoring many photoinduced biological phenomena involving aromatic units in proteins or DNA/RNA and eventually coenzymes.

Thus, since 2DES experiments may provide unique information on electronic coupling and related electron/energy transfer processes (see Sect. 1.1), applications of this nonlinear spectroscopic technique in the UV range could shed new light on a vast number of biological processes, from various enzymatic redox catalysis to photosynthesis, and could be used either to obtain structural information (in the GS) or to track photoinduced excited-state events. Notably, given its high spectral and temporal resolution, 2DUV spectroscopy can be used to disentangle complex decay processes such as in DNA/RNA polynucleotides, where multiple chromophores exhibit various competitive deactivation pathways, with timescales from sub-100 fs to nanoseconds. In this context, developing computational protocols for accurate simulation of 2DUV spectroscopy is

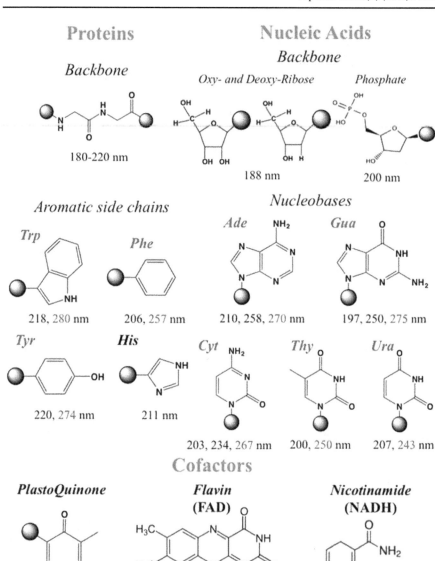

Fig. 2 UV-absorbing bio-chromophores in proteins and nucleic acids, together with associated absorption wavelengths. States involved in absorption out of the spectral region of backbones are highlighted in blue for the UV-chromophores reported in this review. Their respective backbone units, absorbing in the high-energy UV window, and important protein cofactors are also depicted for the sake of completeness

highly desirable, as together with current advances in experimental techniques, it would ultimately lead to a comprehensive understanding of these processes.

In the following sections, after presenting the basic theoretical background for 2DES and the approximation frameworks adopted in our studies, we review our recent efforts in developing computational tools for simulating 2DUV spectra of two major biological macromolecules, proteins and nucleic acids, focusing on the chromophoric units that provide absorption distinct from their backbones.

3 Theoretical Background

3.1 Density Matrix Formalism and 2DES Response Functions

A unified theory for nonlinear optical spectroscopy [2] has been developed using density matrix formalism to represent the state of matter and its evolution in the Liouville space in order to determine the nonlinear response generated by the light–matter interactions. As mentioned above, the source of the signal field recorded in 2DES experiments is the induced nonlinear polarization, which is defined as the expectation value of the dipole operator $\hat{\mu}$

$$P^{(n)}(t) = \left\langle \hat{\mu}\hat{\rho}^{(n)}(t) \right\rangle = \mathrm{Tr}\left[\hat{\mu}\hat{\rho}^{(n)}(t) \right] \tag{1}$$

where $\hat{\rho}(t)$ is the density matrix that describes the quantum state of an ensemble of optically active sites, i.e. chromophores (or more generally, molecules)

$$\hat{\rho} = \sum_{ab} \rho_{ab}|a\rangle\langle b| \tag{2}$$

where the sum runs over the electronic states of the chromophores, with density matrix diagonal elements representing populations and off-diagonal elements, coherences. The dipole operator for this ensemble is analogously defined as

$$\hat{\mu} = \sum_{ab} \mu_{ab}|a\rangle\langle b| \tag{3}$$

, and μ_{ab} is the transition dipole between states a and b.

The time evolution of the density matrix (Eq. 2) satisfies the Liouville-von Neumann equation,

$$i\hbar\frac{d\hat{\rho}}{dt} = [\hat{H}, \hat{\rho}] \tag{4}$$

where $\hat{H}$ is the system Hamiltonian. In the absence of an external field ($\hat{H} = \hat{H}_0$), the free evolution of an unperturbed density matrix, $\hat{\rho}^{(0)}(t)$, is conveniently described using the retarded Green's function (forward propagator) $G(t)$; see detailed description in the Appendix.

The inclusion of an external optical electric field (provided it is weak) is achieved using a perturbative approach, where the time-dependent external field perturbation $\hat{H}'(t)$, defined as

$$\hat{H}'(t) = -\hat{\mu} \cdot E(t) \tag{5}$$

, is included in the system Hamiltonian, i.e. $\hat{H} = \hat{H}_0 + \hat{H}'(t)$. The density matrix is expanded as

$$\hat{\rho}(t) = \hat{\rho}^{(0)}(t) + \hat{\rho}^{(1)}(t) + \cdots + \hat{\rho}^{(n)}(t) + \cdots = \hat{\rho}^{(0)}(t) + \sum_{n=1}^{\infty} \hat{\rho}^{(n)}(t), \tag{6}$$

where the superscript denotes the nth-order expansion. In this perturbative scheme, the Liouville-von Neumann equation can be solved by sorting in powers of $\hat{\rho}$, followed by an iterative integration of the resulting equations (see Appendix for the explicit form of the third-order density matrix).

Writing out the perturbation (Eq. 5) and reformulating the density matrix as a function of the time intervals t_i between the pulses (with $t_1 = \tau_2 - \tau_1, t_2 = \tau_3 - \tau_2, t_3 = t - \tau_3$), the nth-order polarization $P^{(n)}(t) = \langle \hat{\mu} \hat{\rho}^{(n)}(t) \rangle = \text{Tr}\left[\hat{\mu} \hat{\rho}^{(n)}(t)\right]$ becomes third-order,

$$P^{(3)}(t) = \text{Tr}[\hat{\mu}\hat{\rho}^{(3)}(t)] = \int_0^{\infty} dt_3 \int_0^{\infty} dt_2 \int_0^{\infty} dt_1 R^{(3)}(t_3, t_2, t_1) \times E(t - t_3)E(t - t_3 - t_2)E(t - t_3 t_2 - t_1)$$

$$\tag{7}$$

where the system-specific (third-order) response $R^{(3)}(t_1, t_2, t_3)$

$$R^{(3)}(t_1, t_2, t_3) = \left(\frac{i}{\hbar}\right)^3 \text{Tr}[\hat{\mu}G(t_3)[\hat{\mu}G(t_2)[\hat{\mu}G(t_1)[\hat{\mu}, \rho(0)]]]] \tag{8}$$

is formally separated from the incident electric fields. Equation (7) is the general equation for the computation of the nonlinear signals in 2D optical spectroscopy. Recording in a given phase-matched direction translates into the selective detection of subgroups of contributions in the Liouville space (eight in total), referred to as Liouville pathways, following directly from Eq. 8. For example, detection in the rephasing (K_1) phase-matching direction (see Fig. 1a) selects three contributions to the response

$$R_{k_1}^{(3)}(t_1, t_2, t_3) = \sum_{i=\text{GSB,ESA,SE}} R_{k_1,i}^{(3)}(t_1, t_2, t_3) \tag{9}$$

that are associated with different physical processes occurring in the system during the interaction with the external fields (see GSB, SE and ESA in Sect. 1.1) that can be represented by Feynman diagrams.

Typically, simulations of ultrafast spectroscopy assume temporally well-separated ultrashort laser pulses and work in the so-called impulsive limit (i.e. the limit in which the pulse duration is shorter than a single vibrational oscillation period). In that case, the third-order polarization $P^{(3)}(t_1, t_2, t_3)$ (Eq. 7) becomes equivalent to the nonlinear response of the system $R^{(3)}(t_1, t_2, t_3)$ (Eq. 8). This approximation simplifies the calculations, as undesired effects due to the temporal overlap of the pulses do not need to be considered. Furthermore, the obscuring of any coherent dynamics, which occurs on the same timescale as the duration of the pulses, is avoided, and the highest possible temporal resolution is achieved. The impulsive limit has been adopted throughout this review.

Simulation protocols for nonlinear optical spectroscopy deal with solving Eq. (7) or, more generally, Eq. 8 when working with realistic pulse shapes. This requires computing the interactions of the system with three time-dependent electric fields as well as its field-free evolution between them. In principle, Eq. 7 could be solved without any approximation, given knowledge of the multidimensional potentials associated with each electronic state involved in the evolution of the density matrix, enabling propagation of populations and coherences. In this case, non-adiabatic couplings and transition dipole moments along the potential energy surfaces (PES) would be required to allow population transfer and interaction with the field, respectively. It would then be possible to treat electronic and nuclear degrees of freedom fully quantum-mechanically by quantum dynamics, thereby resolving coherent oscillatory dynamics, line broadening (due to finite excited-state lifetimes) and vibrational progressions due to wave packet decoherence and recoherence. Furthermore, arbitrary pulses could be straightforwardly incorporated into the simulations, providing an opportunity to study the effects of the pulse's central frequency, bandwidth and polarization [36, 37].

The drawback to quantum dynamics simulations is the high cost of pre-computing the PES, a task that quickly becomes infeasible with an increasing number of degrees of freedom. As a workaround, simulations are often run on parameterized two- or three-dimensional potentials that approximate the exact PES through analytical expressions, with parameters chosen to fit the energetics of a few representative points along the PES. This approach was recently applied to simulate 2DES in Rhodopsin [38]. Methods such as multi-configuration time-dependent Hartree have facilitated simulations on multidimensional surfaces [39].

Another approach for solving Eqs. (7) and (8) involves the stochastic modeling of bath fluctuations in a semi-classical fashion (stochastic Liouville equations, SLE) [40, 41]. At the heart of SLE is the explicit inclusion of collective bath modes in the system's Hamiltonian. The Hamiltonian is formally time-independent, but depends on some classical time-dependent stochastic variables $\sigma(t)$, which represent the bath. The response function is calculated by averaging over an ensemble of realizations, propagated through trajectory-based molecular dynamics simulations, where at every subsequent time-step it is evaluated as

$$R^{(3)}(t_1, t_2, t_3) = \left(\frac{i}{\hbar}\right)^3 \mathrm{Tr}[\hat{\mu} G_{\sigma(t)}(t_3 + t_2 + t_1, t_2 + t_1)[\hat{\mu} G_{\sigma(t)}(t_2 + t_1, t_1)[\hat{\mu} G_{\sigma(t)}(t_1, 0)[\hat{\mu}, \rho(0)]]]]$$

(10)

where $G_{\sigma(t)}(\tau_i, \tau_j)$ is the Green's function solution of the field-free evolution between two time steps

$$\frac{d\hat{\rho}_{\sigma(t)}}{dt} = \frac{i}{\hbar}\left[\hat{H}_{0;\sigma(t)}, \rho_{\sigma(t)}\right]$$

(11)

Since bath dynamics is non-Markovian, solving Eq. 10 requires knowledge of the past evolution of the system. Bulk-induced dephasing effects are automatically included via ensemble averaging. Trajectory-based approaches face the challenge of the computational cost associated with the propagation of a swarm of trajectories (particularly those in the excited state) and the absence of a well-defined density matrix responsible for the quantum feedback on the classical bath during a coherence state evolution. Various trajectory-based implementations that address these issues have been documented in recent years [42–45].

For systems with a classical bath following Gaussian statistics and linear system–bath coupling, Eqs. (8) can be solved using the second-order cumulant expansion, i.e. the cumulant expansion of Gaussian fluctuations (CGF) [46]. This method makes it possible to calculate the shapes of electronic transition bands coupled to a bath (for fluctuations with arbitrary timescales) using the formalism of line shape functions, $g_{ij}(t)$; see Appendix for details. Within the CGF framework, the population transfer can be accounted for phenomenologically according to the Lindblad equation (see Appendix), with secular approximation to the Green's function enabling partitioning of the nonlinear response (Eqs. 9, 10) into population and coherence contributions. For instance, considering the manifold of ground (g) and excited (e, f) states, and just the ESAs detected in the rephasing phase-matching direction (with the expressions for GSB and SE given in Ref. [36]), the population ($e'=e$ during delay time t_2) contributions are stated as

$$R^{(3)}_{k_1\text{ESA},i} = +i \sum_{e',e,f} \mu^2_{fe'} \mu^2_{ge} G_{e'e',ee}(t_2) \times e^{-i(\varepsilon_f - \varepsilon_{e'})t_3 + (\varepsilon_e - \varepsilon_g)t_1 + \varphi^{\text{ESA},i}_{fe'e}(t_1, t_1+t_2, t_1+t_2+t_3, 0)}$$

(12)

where $G_{e'e',ee}(t)$ is the Green's function controlling the population transport (see Appendix for additional details), while the coherence ($e' \neq e$ during delay time t_2) contributions are instead stated as

$$R^{(3)}_{k_1\text{ESA},ii} = +i \sum_{e',e,f}^{e' \neq e} \sum_f \mu_{fe'} \mu_{fe} \mu_{ge'} \mu_{ge} e^{-i(\varepsilon_f - \varepsilon_e)t_3} \times e^{-i(\varepsilon_{e'} - \varepsilon_e)t_2 + (\varepsilon_e - \varepsilon_g)t_1 + \varphi^{\text{ESA},ii}_{fe'e}(t_1, t_1+t_2, t_1+t_2+t_3, 0)}$$

(13)

where ε_a, with $a \in \{g, e, f\}$, are the energies of the eigenstates of the system's Hamiltonian H_0, their energy differences ($\varepsilon_a - \varepsilon_b$) being the transition energies (TEs), μ_{ab} are the associated transition dipole moments (TDMs), and $\varphi_{e'fe}(\tau_4, \tau_3, \tau_2, \tau_1)$ is the phase function that describes spectral diffusion, thus translating the vibrational structure of the evolving electronic states into a series of oscillating diagonal and off-diagonal peaks. The mathematical formulation of the phase functions therefore depends on the level of sophistication adopted for describing the system vibrational dynamics [2, 47]. In any case, the phase function is based on the line shape functions, and two examples of phase functions built from line shape functions will be given in Sects. 3.3 and 3.4 (Eqs. 16 and 19, respectively).

3.2 Coupling Accurate Electronic Structure Computations to 2DES

Two-dimensional electronic spectroscopy targets the manifold of (localized and delocalized) excited states of the chromophoric units in the sample. A thorough characterization of the electronic structure of the chromophoric systems is thus required to properly interpret 2D electronic spectra, which generally represents a great challenge even for the most advanced computational techniques, despite a plethora of quantum mechanics (QM) methods currently available. In particular, the QM treatment of isolated large chromophoric moieties (usually absorbing in the Vis) or rather small UV-chromophores that still involve high-energy electronic levels and thus encompass a large number of excited states in the target manifolds, might already push the limits of what is currently computationally feasible. Treating multiple interacting chromophores, i.e. molecular aggregates, at the ab initio level then quickly becomes prohibitive with increasing number or size of chromophoric units. At the same time, as shown in Eqs. 12 and 13, TEs and TDMs are the fundamental ingredients for simulating the third-order nonlinear response recorded in 2DES maps, and even if largely approximate, their estimations cannot be circumvented. A common strategy for coping with these limitations is to adopt Frenkel exciton models that make simulations of 2DES spectra of realistic model systems computationally feasible [36, 48]. Exciton modeling works well for vibrational excited-state absorptions in 2DIR spectroscopy (namely overtone absorptions), since anharmonicity for overtones of a chromophore (local overtones) or between coupled chromophores can be reliably computed perturbatively (without explicit knowledge of the high-lying energy level manifold). In general, such a perturbative treatment is less effective for electronic transitions, since local excited states behave quite differently with respect to local overtones, showing large anharmonic couplings that cannot be described perturbatively. Still, when dealing with coupled electronic excitations between interacting chromophores, the TEs and TDMs of each isolated chromophore can be initially calculated at the QM level (usually in the gas phase and for a given geometry) and then employed as parameters to build the exciton Hamiltonian. The electronic couplings in multi-chromophoric systems can then be estimated, while neglecting electron exchange between chromophores. Clearly, the limited description provided by Frenkel exciton modeling has computational advantages that become evident only when just a few excited states (i.e. energy levels) are considered in the model Hamiltonian (i.e. few QM computations are initially performed). This implies that many of the excited states related to (potential) ESA signals in 2DES maps are often neglected in most conventional spectroscopy simulation protocols. This assumption generally holds if ESA signals are expected to fall outside the spectral window of interest. However, broadband transient absorption [49] and, more recently, also 2D [16, 50, 51] spectra show that this assumption breaks down regularly, especially in UV-active bio-chromophores. In fact, as we will show in the following sections, ESAs are ubiquitous, system-dependent, state-specific spectroscopic fingerprints that, for instance, make it possible to selectively study the excited-state dynamics of different decay channels. In cases such as charge-transfer states, ESAs might represent the only spectroscopic signature of a given state, being of particular interest for studying its photophysics.

The ability to accurately predict ESA signals would then open the door for simulating broadband 2D electronic spectroscopy [37, 52–61] as well as double quantum coherences spectroscopy [53, 62, 63].

As will be illustrated throughout Sect. 3, obtaining the electronic structure of biologically relevant chromophores from first principles, at a level of accuracy needed for comparison against experimental data, is far from trivial. A protocol for computing reliable TEs and TDMs of isolated bio-chromophores (or dimeric aggregates) is presented in Sect. 3.1, which comes with additional benefits: (a) the results can be used to expand currently available exciton models beyond the lowest transitions; (b) the results can be used to benchmark low-cost approaches that can then be applied for larger-scale computations; (c) the results essentially make simulations independent from the experiment, providing them with predictive power and allowing us to envision problem-driven experimental setups. To this aim, we take a step back in the theoretical descriptions given in Sect. 2.2 and start from a drastic approximation, i.e. treating the chromophoric system uncoupled from the bath of vibrations and thus as a closed quantum system having only electronic degrees of freedom. Within this framework the electronic states become eigenstates of the system's Hamiltonian, and their dynamics are reduced to a $e^{-i\varepsilon_f t}$ phase factor (herein referred to as a "static picture").

This approximation implies setting the phase functions $\varphi = 0$ and neglecting population transport $G(t)$ in Eqs. 12 and 13, which then simplify to

$$R_{k_1\text{ESA},i}^{(3)} = +i \sum_{e,f} \mu_{fe}^2 \mu_{ge}^2 e^{-i(\varepsilon_f - \varepsilon_e - i\gamma_{fe})t_3 + i(\varepsilon_e - \varepsilon_g - i\gamma_{eg})t_1} \tag{14}$$

and

$$R_{k_1\text{ESA},ii}^{(3)} = +i \sum_{e',e}^{e' \neq e} \sum_{f} \mu_{fe'} \mu_{fe} \mu_{ge'} \mu_{ge} e^{-i(\varepsilon_f - \varepsilon_e - i\gamma_{fe})t_3} e^{-i(\varepsilon_{e'} - \varepsilon_e - i\gamma_{e'e})t_2 + i(\varepsilon_e - \varepsilon_g i\gamma_{eg})t_1} \tag{15}$$

All dephasing processes are thus condensed into the phenomenological dephasing constants γ_{ab}, which induce homogeneous broadening of all $a \rightarrow b$ electronic transitions. Equations 14 and 15 allow us to simulate the nonlinear responses for systems with hundreds of excited states, assuming all the corresponding TEs ($\varepsilon_b - \varepsilon_a$) and TDMs ($\mu_{ab}$), a protocol known as the sum-over-states (SOS). Figure 3 shows in greater detail, for the specific case of an ESA signal recorded in the rephasing K_I phase-matching direction (Fig. 3a), how the nonlinear response is computed through Eq. 14. The final relation obtained for the third-order nonlinear response is accomplished by looking at the various contributions adding up in time, following the evolution of the system density matrix elements as described by the corresponding (ESA) Feynman diagram (Fig. 3b) and by making explicit the terms contributing to the density matrix evolutions during each light–matter interaction (Fig. 3c) during a 3PPE experiment (see Fig. 1a).

Recently [52], we combined this quasi-static (as all dynamic effects have been neglected) protocol with a hybrid scheme combining a QM electronic structure with a molecular mechanics (MM) treatment of environmental effects (i.e. the SOS//QM/MM

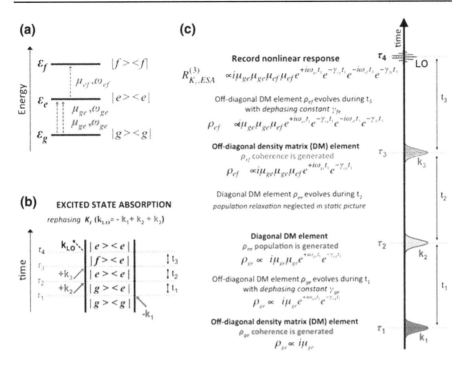

Fig. 3 **a** A three-level system, showing transition dipole moments (μ) and frequencies (ω) related to the ground (g) and excited (e, f) states and their energies ($\varepsilon_{g,e,f}$). **b** Feynman diagram of an ESA signal recorded in the rephasing K_I phase-matching direction. **c** Schematic representation of a 3PPE experiment (see also Fig. 1a), including explicit terms contributing to the density matrix evolutions during each light–matter interaction and the final relation for the third-order nonlinear response (see Eq. 14)

protocol) in order to generate 2DES spectra for several biological key players in their native environment. Some applications of the SOS//QM/MM protocol to bio-chromophores are showcased in the Results section, mainly evidencing how a number of bright excited states in the UV–Vis spectral range often represent the major contribution to 2DES maps and could embody specific spectroscopic fingerprints for tracking ground- or excited-state dynamics. Accurate characterization of the excited-state manifold by advanced electronic structure computations thus represents a pillar supporting the reliable simulation of 2D electronic spectra.

4 Results and Discussion

The basic theoretical background for understanding how to compute nonlinear electronic spectra has been illustrated in Sect. 2, showing how a complete 2DES simulation technique would need to combine high-precision electronic structure computations and non-adiabatic quantum dynamics simulations to resolve the elaborate nonlinear response equations. On one hand, the practical use of quantum dynamics makes it possible to accurately model spectral dynamics while relying only on approximate electronic potentials, capturing just the essence of the electronic structure complexity, which is reflected in the absence of ESA contributions. On the other hand, the electronic structure complexity can be straightforwardly incorporated within a static approach, recovering ESA contributions. However, as dynamics are neglected, this approach is not able to provide realistic spectral line shapes. Trajectory-based approaches offer a compromise by permitting us to adjust the level of precision applied to molecular dynamics and PES sampling. In the following section, we begin our discussion by utilizing the static approach to outline a protocol for the accurate computation of transition energies and dipole moments (Sect. 3.1), as well as for benchmarking low-cost methods. Sections 3.2 and 3.3 demonstrate how the latter can be employed to study GS conformational dynamics of oligopeptides and DNA nucleobase dimers within the framework of the SOS// QM/MM protocol. We then go beyond the static approximation and re-introduce dynamic features in Eqs. 14 and 15 within the CGF framework for the study of excited states. Specifically, coherent intramolecular vibrational dynamics is discussed in Sect. 3.3, while Sect. 3.4 covers population transfer.

4.1 Benchmarking the Excited-State Manifolds

Accurate prediction of the basic spectroscopic parameters that define the space of electronic transitions (i.e. TEs and TDMs) accessible by 2DES experiments is the first computational challenge. The excited-state manifolds of UV-active chromophores in the 2DUV energy range (from ca. 3.5 to 11 eV) comprise various types of electronic transitions, generally including single and double excitations with local or (inter- and intramolecular) charge-transfer character, and thus involving covalent as well as ionic (valence bond-like) excited states. Wave function approaches introduced in the 1980s, based on the combination of CASSCF multi-configurational wave functions [64] and PT2 perturbative energy corrections [65] (CASSCF// CASPT2), currently represent the most widely used methodology for handling such a variety of excited states on equal footing [66], generally providing good quantitative predictions of TEs and TDMs, with expected error of around 0.2 eV (ca. 1600 cm^{-1}). Application of the CASSCF//CASPT2 methodology to larger and larger molecules has become possible through many developments over the years, including the implementation of the restricted active space self-consistent field (RASSCF) methodology [67], efficient approximations for two-electron integral estimates [68] and large-scale parallelization [69], to cite some of those most widely used in the results reported here. The single-state PT2 treatment (hereafter "PT2") has likely been preferred to the more expensive (and in most cases infeasible) multistate PT2 approach. However, the accuracy of CASSCF//PT2 predictions strongly depends on the choice of active spaces and basis sets, two parameters that involve a critical increase in computational cost. Moreover, the number of excited states present in the energy ranges of 2DUV experiments increases dramatically with the size of the (multi-)chromophoric system under consideration, requiring the use of large state-averaging procedures to converge the multi-configurational wave functions, with a consequent further increase in computational cost. To cope with the computational feasibility issues of multi-configurational treatments, we have generally adopted the RASSCF methodology, as its flexibility makes it possible either to improve the electronic wave function of the largest feasible CASSCF computation, usually involving the full-valence π-orbital space, or to reduce the overall computational cost by limiting the CAS treatment to a relatively small active space (RAS2).

Indeed, with the RASSCF approach, extra virtual orbitals can be included in the active space where a restricted number of electrons are allowed (RAS3 space), increasing the number of electronic excitations and expanding the multi-configurational character of the electronic wave functions relative to what is already accounted for in a full-valence RAS2 space. This type of RAS scheme can be employed to benchmark $S_0 \rightarrow S_N$ vertical excitations of chromophore monomeric units. To this end, we have generally attempted to effect a direct comparison between vertical excitations in the Franck–Condon (FC) region, i.e. at the GS equilibrium geometries, and experimentally recorded cross-sections in a vacuum, when available. The choice of the gas phase reflects the need for comparison in the absence of environmental effects that would both complicate the computational modeling strategy and reduce the experimental spectral resolution (as line broadening is larger in solution), and is aimed at assessing the prediction of the overall distribution of the absorption bands in the whole UV range, keeping in mind that vibronic effects [70] are not accounted for in this study.

Benzene and phenol molecules are the chromophores related to the Phe and Tyr amino acid aromatic side chains (see Fig. 2). As shown in Fig. 4, CASSCF//PT2 computations of the $S_0 \rightarrow S_N$ vertical excitations (in the FC region) indicate that the full-valence active spaces of benzene and phenol, i.e. CAS(6,6) and CAS(8,7), respectively, are not sufficiently accurate when compared with the values of experimental cross-sections in a vacuum [58]. Instead, increasing the active space by allowing single and double excitations to additional extravalence RAS3 orbitals enables an appropriate description of the dynamic σ–π polarization and significantly improves accuracy, reaching convergence with eight extra orbitals and quantitative agreement with experiments. It is worth mentioning that these computations have been performed applying a computational recipe to eliminate Rydberg contamination in the valence states after adding a set of diffuse and uncontracted basis functions in the center of charge of the molecules (ANO-L-aug) [71, 72], which has been further used to simulate dimers of amino acid side chains and other chromophores.

While the $S_0 \rightarrow S_N$ excitations do not cover all possible excitations involved in a 2DUV experiment, because the $S_N \rightarrow S_M$ excitations would strongly contribute as ESA signals in the 2D spectra, the energy range considered in this benchmark study involves quite large TEs (with energies up to 11 eV from the GS). In fact, considering a one-color 2DUV experiment with pump and probe frequencies centered at the $S_0 \rightarrow S_1$ transition energy (4.5–4.7 eV), the S_M excited states considered here already comprise those that could be involved as ESA signals (i.e. the $S_1 \rightarrow S_M$ transitions, with $M < 9$) and are properly described at the RASSCF//PT2

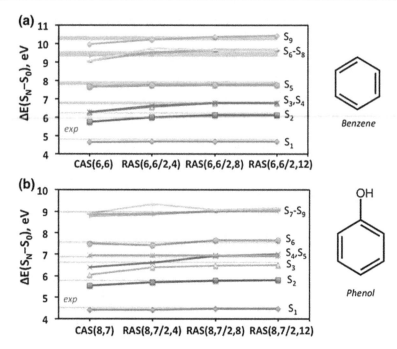

Fig. 4 CASSCF//PT2 and RASSCF//PT2 vertical $S_0 \rightarrow S_N$ excitation energies (in eV) of benzene and phenol monomers in a vacuum obtained using the large ANO-L(432,21)-aug basis set at different levels of theory. Minimal active space CAS(X,Y) results are compared to restricted active space (RAS2/RAS3) calculations with an increasing number of virtual orbitals (4, 8 and 12) and experimental values (gray bars, indicating ranges of experimental absorption maxima recorded). Reproduced from data reported in Ref. [58]

level. This outcome suggests that theoretical predictions of vertical excitations in the UV can be reliably obtained by extended RAS schemes. Notably, convergence was shown by TDMs as well as TEs, thus providing robust theoretical and reference RAS schemes for employment in further studies.

When extending the benchmark study of benzene and phenol monomers to the remaining aromatic protein chromophores, i.e. the indole moiety of the Trp amino acid, and to the canonical nucleobases of DNA and RNA, we found a lack of experimental data. In fact, the gas-phase experimental cross-sections of indole, Ade, Gua, Cyt, Thy and Ura are limited to energies up to 6.4 [73], 7.7 [57, 74], 4.4 [75], 7.1 [75–77], 7.4 [78] and 6.6 eV [75], respectively. Therefore, after a preliminary comparison of $S_0 \rightarrow S_N$ vertical excitations against the available cross-section values, the benchmark study of these chromophores was conducted monitoring the $S_N \rightarrow S_M$ excitations, giving rise to ESA signals that mainly contribute to the 2D electronic spectra. In particular, the S_N states are chosen among the lowest-lying and brightest (i.e. with largest $S_0 \rightarrow S_N$ TDM) excited states of each chromophore, as these states are generally the targets of experimental 2DUV spectra. This approach provides a direct visualization of the influence of the level

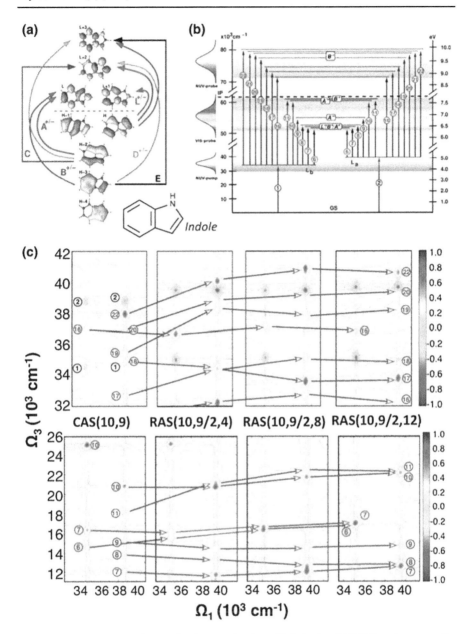

Fig. 5 Benchmark computations on indole: **a** molecular orbitals in the CAS(10,9) active space displaying a labeling scheme for **b** the various types of transitions associated with ESAs in the Vis and UV probing (Ω_3 frequency) windows, as arising from the primary $\pi\pi^*$ (L_b and L_a) states (in the pump frequency range $\Omega_1 = 34,000\text{-}40,000$ cm^{-1}). **c** 2D electronic spectra simulated at the CAS(10,9), RAS(10,9|2,4), RAS(10,9|2,6) and RAS(10,9|2,12) levels of theory, using the ANO-L(432,21)-aug basis set. Note that in this RAS nomenclature, the empty RAS1 subspaces are omitted for simplicity. Negative (blue) signals are related to bleaching, while positive (red) peaks arise from $S_{1,2} \rightarrow S_M$ ESAs. Reproduced from data reported in Ref. [53]

of theory adopted on the resulting 2D map, which is shaped by both TE and TDM computed values.

Figure 5 summarizes the benchmark study of indole [53] performed with a large basis set [ANO-L(432,21)-aug], indicating (as in benzene and phenol) that going beyond a CASSCF treatment with full-π valence active space (by introducing at least eight extravalence orbitals in the RAS3 subspace) is required in order to converge the energies and dipole moments of the electronic transitions involved and, thus, the overall aspect of the 2D spectra (see Fig. 5c). Both one-color (UV-pump and UV-probe, 2DUV–UV) and two-color (UV-pump and Vis-probe, 2DUV–Vis) 2D spectra are reported in Fig. 5c, showing a broad set of ESA and bleaching signals that vary in position and intensity as a function of the level of theory. Overall, in the 2DUV spectra of indole, two signal traces related to the first two ($\pi\pi^*$) excited states, labeled L_b and L_a in Platt notation [79], are monitored, as they are quite close in energy, and their transitions from the GS have similar TDMs. In particular, 2DUV–UV spectra (see Fig. 5) feature off-diagonal bleaching signals (ODB, see Fig. 1) related to couplings between L_b and L_a states. It is worth mentioning that these signals are characteristic 2DUV fingerprints of excited states localized on the same molecule. They can be seen as strongly coupled "chromophores", where the energy of their "bi-exciton" (i.e. the state given by the sum of the two electronic transitions) is significantly shifted from the sum of the single-exciton energies due to the large anharmonicity constant, which cannot be computed perturbatively with exciton Hamiltonians.

Here, we consider the 2D map converged with respect to the levels of theory when the intensities of all the peaks are maintained and their positions do not vary more than 1600 cm^{-1}, i.e. the expected error in TE for this methodology (0.2 eV). It is worth mentioning that such type of benchmark study requires a full assignment of the 2D peaks in order to assess the variations in the 2D maps according to the theoretical level employed. This work is particularly tedious, as it requires labeling of all relevant transitions (Fig. 5b) according to the molecular orbitals (MO) involved in the electronic excitations.

Figure 6 shows the benchmark computations for the Ura, Thy and Cyt pyrimidine nucleobases and the effect of the larger active space on the 2DUV–UV spectra [59]. All these canonical nucleobases feature the same π-orbital valence AS, i.e. CAS(10,8), while the RAS3 spaces involving 4, 8 and 12 virtual orbitals and the large ANO-L(432,21)-aug basis set have been employed, as for the benchmarks of amino acid side chains (Figs. 4, 6, 5). While some ESA signals show more AS-dependent fluctuations than others, the overall the 2DUV–UV spectra converge with the addition of eight extravalence orbitals. Cyt is particularly interesting, as the first excited state is very bright but is not the brightest low-lying $\pi\pi^*$ state (as in the other pyrimidine bases), giving rise to two signal traces along two relatively close-lying pump frequencies. This occurrence offers the opportunity to monitor the off-diagonal GSB signals related to the two low-lying $\pi\pi^*$ states (as in indole and purine bases) and the performances of the various levels of theory in producing reliable 2DUV–UV maps.

Adenine and guanine are the canonical purine nucleobases, which feature larger aromatic heterocyclic structure than pyrimidine bases, with the pyrimidine ring

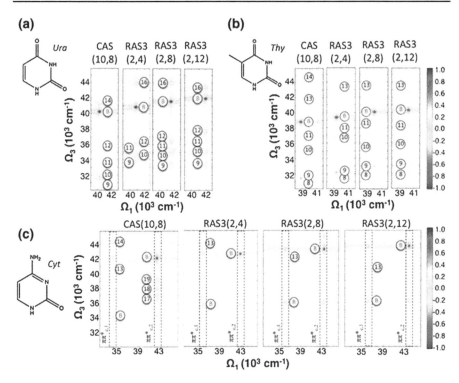

Fig. 6 Benchmark computations on pyrimidine nucleobases **a** uracil, **b** thymine and **c** cytosine, showing the 2DES maps in the UV-pump/UV-probe window (2DUV–UV) as computed at the CAS(10,8), RAS(10,8|2,4), RAS(10,8|2,8) and RAS(10,8|2,12) levels of theory and using the ANO-L(432,21)-aug basis set. Note that only the (variable) RAS3 spaces are indicated in the picture, for simplicity. Reproduced from data reported in Ref. [59]

fused to an imidazole ring. Adenine has just one more π-orbital in the valence active space than indole (MOs depicted in Fig. 5a), i.e. CAS(12,10), and similar MOs that give rise to analogous electronic transitions (Figs. 5, 7). As with indole, adenine features two close-lying $\pi\pi^*$ excited states, but in the purine base the GS $\rightarrow L_a$ transition has higher TDM than GS $\rightarrow L_b$. As shown in Fig. 7b, the benchmark of 2DUV–UV maps demonstrates remarkable changes when the active space is increased with respect to the full valence space, with positions of the Ω_1 traces more sensible to the level of theory than in indole, while convergence is still reached when eight extravalence orbitals are added in a RAS(12,10|2,8) scheme.

A comparison of the computed $S_0 \rightarrow S_N$ excitation energies with the available experimental cross-sections indicated unexpected discrepancies in the case of adenine [57] relative to other aromatic systems. Therefore, for this nucleobase, the benchmark study was extended to investigate the role of the geometry optimization methodology used to define the GS equilibrium structure. In particular, the lack of dynamic correlation at the CASSCF level introduces significant shortening of double bonds in the Ade heterocycle that are significantly elongated when optimizing the geometry at density functional theory (DFT) or second-order

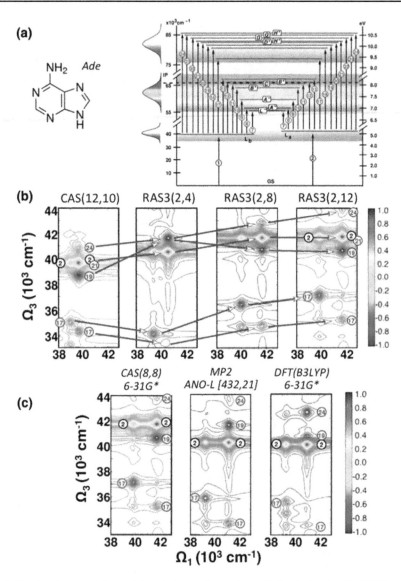

Fig. 7 Benchmark computations of the adenine excited-state manifold (**a**), showing the 2DES maps in the UV-pump/UV-probe window (2DUV–UV) as computed (**b**) at the CAS(12,10), RAS(12,10|2,4), RAS(12,10|2,8) and RAS(12,10|2,12) levels of theory and using the ANO-L(432,21)-aug basis set, or (**c**) at the RAS(12,10|2,12) level with the ANO-L(432,21)-aug basis set while using the GS geometry optimized at the CAS(8,8)/6-31G*, MP2/ANO-L(432,21)-aug and DFT(B3LYP)/6-31G* levels. Note that only the (variable) RAS3 spaces are indicated in the picture, for simplicity. Reproduced from data reported in Ref. [57]

Møller–Plesset perturbation theory (MP2) levels. As shown in Fig. 7c, indeed, the choice of geometry optimization method may have a significant effect on the simulations of 2D electronic spectra, and it should be considered as another factor that could affect comparisons with experimental data.

Thus far, we have reviewed the benchmark studies of the aromatic protein chromophores and the canonical nucleobases of DNA and RNA (with the exception of Gua, which is a work in progress), aimed at obtaining converged results among the highest (still computationally feasible) levels of multi-configurational treatment for each chromophore so that reference theoretical data are made available. These reference transition energies and dipole moments could indeed be used both as ab initio parameters in exciton Hamiltonians and as target values for benchmarking computationally cheaper approaches, in order to tackle 2DUV spectra simulation of larger (multi)chromophoric systems and/or several of their structural conformations. To this end, we have used computational recipes to account for the σ-π polarization effects when dealing with reduced active spaces [71, 72] (as necessary for increased system size). In the benchmark study reported above, we observed that reduced active space schemes (i.e. lacking virtual orbitals with higher orbital momenta) can dramatically underestimate transition energies, as they overestimate the dynamic correlation of ionic (in valence bond terms) states. In the search for cost-effective protocols to counteract or at least damp this effect, we came up with two semi-empirical procedures. On the one hand we observed that by deleting a number of virtual extravalence π^*-orbitals with higher angular momentum in the perturbation treatment, the number of dynamic correlations is reduced, producing a state-dependent blue-shift of the excitation energies, thus minimizing the mean deviation from the reference values in the whole excited-state manifold. On the other hand, we realized that real and imaginary shift parameters (originally introduced to cure intruder state problems) [80, 81] invoke a non-uniform decrease in the correlation contribution, much more pronounced for ionic than for covalent states. This makes their use well suited for our purposes, even if we need to resort to larger values than suggested in the literature (a detailed argumentation can be found in Ref. [82]). However, the choice of shift parameter is chromophore-dependent, and its application, while simpler with respect to orbital removal procedure, is limited to systems where transition energies of different chromophores can be corrected with a similar shift parameter, such as homodimeric systems. We must stress that the application of either of the two protocols makes sense only in the framework of semi-empirical parameterization against a reliable reference data set, and their use is otherwise discouraged.

As an example case study, here we report a benchmark for benzene and phenol and their dimer in the gas phase and for their corresponding amino acid side chains (Phe and Tyr, respectively) in a model tetrapeptide in water solution [58].

As shown in Fig. 8, the reference TEs computed for benzene and phenol (see Fig. 4), i.e. those at RAS(6,6|2,12) and RAS(8,7|2,12) levels with ANO-L(432,21)-aug basis set, respectively, can be used as target values for determining computationally cheaper approaches that could accurately simulate 2DES spectroscopy of related multi-chromophoric systems. In fact, reference computations with large RAS schemes are unaffordable even for the smallest multimeric system comprising benzene and phenol monomers, i.e. the benzene–phenol dimer. The first step towards reducing the computational costs is thus to develop a recipe that at least allows the use of minimal (still full-valence) active spaces (mAS), i.e. CAS(6,6) and CAS(8,7) for benzene and phenol, respectively, and relatively small basis sets, without compromising TEs and TDMs estimates. As shown in Figs. 4 and 8, the full-valence mAS in combination with the large basis set,

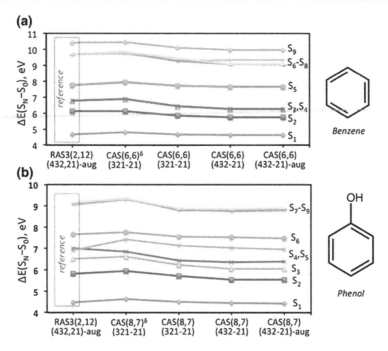

Fig. 8 CASSCF//PT2 and RASSCF//PT2 vertical $S_0 \rightarrow S_N$ excitation energies (in eV) of benzene and phenol monomers in a vacuum obtained at various levels of theory. Reference data are those computed at the RAS3(2,12)//PT2/ANO-L(432,21)-aug levels and are compared to cheaper computational approaches, including minimal (full-valence) active spaces (mAS), i.e. CAS(6,6) and CAS(8,7) for benzene and phenol, respectively, with various ANO-L basis sets and refined mAS (r-mAS)$^\delta$ with the smallest ANO-L(321-21) basis set. Reproduced from data reported in Ref. [58]

i.e. ANO-L(432,21)-aug, results in a large underestimation of several excited-state energies (i.e. those with ionic character), and this discrepancy is essentially due to $\sigma-\pi$ polarization effects associated with the smaller active space size, with basis set size having a minor effect. Deleting a number of virtual extravalence π^*-orbitals with higher angular momentum (e.g. six for benzene and seven for phenol) in the perturbation treatment reduces the dynamic correlations, significantly improving the performances of the mAS (Fig. 8). This active space refinement procedure allows us to obtain reasonably accurate excited-state energies of benzene and phenol monomers in the gas phase. Thus such type of refined-mAS (r-mAS)$^\delta$, i.e. CAS(6,6)$^\delta$ and CAS(8,7)$^\delta$ for benzene and phenol, respectively, could be employed in conjunction with the ANO-L(321,21) basis set, yielding a significant reduction in computational cost and allowing us to move forward to the study of small multimeric systems.

Extending the refined-mAS approach from the benzene and phenol monomers to their dimer has proven to work well for a non-interacting aggregate in the gas phase, a model system with distant (~ 11 Å) monomers that allows direct comparison with reference calculations and experimental data on monomers (Fig. 9) [58]. The full-valence mAS for the benzene–phenol dimer is the CAS(14,13) space, i.e. the sum of the two mAS of the monomers, which already comprises a large number of configuration state functions (CSFs), i.e. 736'164 CSFs. Simulation of the 2DUV–UV

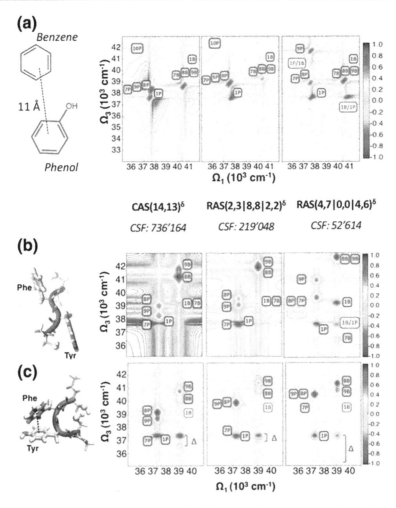

Fig. 9 Comparison of simulated one-color 2DUV–UV spectra of benzene–phenol dimers with different active spaces, including CAS(14,13)$^\delta$ and RAS(2,3|8,8|2,2)$^\delta$ and RAS(4,7|0,0|4,6)$^\delta$ schemes and using the ANO-L(321,21) basis set. The numbers of CSFs are also indicated for each level of theory. Spectra are shown for **a** a non-interacting dimer in the gas phase, and **b** non-interacting and **c** stacked dimers in the water-solvated CFYC tetrapeptide model. The signal labels (nB, nP) refer to benzene (B) or phenol (P) Sn excited states, while cross-peaks arising from "artificial" quartic coupling (Δ) are indicated as 1B/1P and 1P/1B. Reproduced from data reported in Ref. [58]

spectrum of the non-interacting dimer with the refined-mAS CAS(14,13)$^\delta$ provides a 2D map that is consistent with the sum of the 2D maps of the monomers computed at the CAS(6,6)$^\delta$ and CAS(8,7)$^\delta$ levels [58]. This preliminary check indicated that the CAS(14,13)$^\delta$ 2DUV–UV spectrum of the non-interacting dimer, in turn, could be used as a reference spectrum, to further reduce the computational cost. The CAS(14,13)$^\delta$ computation of the dimer, indeed, almost reaches the limit of a reasonable computational time with current computer capabilities, and more efficient protocols would be beneficial. In this context, we have adopted the RASSCF scheme to reduce the computational cost. In fact, a small (or even empty) RAS2 space could be adopted, while including the remaining valence orbitals in the RAS1/RAS3 active spaces, where a restricted number of holes are allowed in the RAS1 space and the corresponding electrons are promoted to the RAS3 orbitals. This type of RAS1|RAS2|RAS3 scheme has been tested to minimize the active space and the computational cost for simulation of the non-interacting benzene and phenol dimer. Thus, here, we apply the RASSCF protocol to reduce the active space (and consequently the computational cost) by limiting the number of excitations, opposite to what is done for the reference computations, where the number of excitations has been increased for improving the accuracy.

Figure 9 shows the comparison of simulated one-color 2DUV–UV spectra of benzene–phenol dimers obtained with different refined active spaces, including the refined-mAS, i.e. CAS(14,13)$^\delta$, and the computationally cheaper RAS(2,3|8,8|2,2)$^\delta$ and RAS(4,7|0,0|4,6)$^\delta$ schemes, while using the relatively small ANO-L(321,21) basis set. As compared to the dimer refined-mAS, the RAS spaces involve much less CSF, with the RAS(2,3|8,8|2,2) and the RAS(4,7|0,0|4,6) spaces corresponding to 219'048 and 52'641 CSF, respectively. As shown in Fig. 9, the former RAS scheme provides a reliable one-color spectrum when compared to that of the CAS(14,13)$^\delta$, while the latter cannot not properly describe some of the states located in the UV probing region (e.g. phenol signals related to the $S_0 \rightarrow S_{7-9}$ transitions, namely 7P, 8P and 9P in Fig. 9a), as well as mixed doubly excited states that give rise to off-diagonal signals (1B/1P or 1P/1B in Fig. 9a). In fact, the electronic structure of the dimer involve several of these mixed double excitations, given by the collective one-electron excitations on both monomers and appearing at the energy sum of the localized single excitations. The electronic coupling between two interacting monomers, denoted as "quartic" coupling (Δ) [2, 83], shifts the energies of mixed states relative to the sum of the corresponding single excitations, yielding off-diagonal cross-peaks in the 2DUV–UV maps. Thus, quartic coupling and corresponding off-diagonal 2D signals are expected to be absent (or negligible) in the non-interacting benzene–phenol dimer. This is correctly observed in the CAS(14,13)$^\delta$ and RAS(2,3|8,8|2,2)$^\delta$ spectra, but it is broken in the RAS(4,7|0,0|4,6)$^\delta$ simulations, indicating that too great a reduction of the active space degrades the description of the excited-state manifold and the corresponding 2D maps.

In order to validate the benchmark study on the gas-phase non-interaction and to establish an efficient protocol for 2DUV spectra simulations of realistic systems, the cysteine-phenylalanine-tyrosine-cysteine (CFYC) tetrapeptide (Fig. 9b, c), solvated in water solution, was considered as a model protein system [54]. As will be shown in the next section, the folding/unfolding dynamics of this tetrapeptide involve configurations with interacting and non-interacting benzene and phenol chromophores, occurring as aromatic side chains of the phenylalanine (Phe, F) and tyrosine (Tyr, Y) amino acid residues, respectively. In Fig. 9, we report two representative structures of these configurations and their corresponding 2DUV–UV spectra, simulated with our SOS//QM/MM approach [52]. Notably, the comparison of various levels of theory shown for the non-interacting dimer in the gas phase (Fig. 9a) is completely preserved for the non-interacting aromatic side chains in the realistic solvated tetrapeptide model, with the computed RAS(2,3|8,8|2,2)$^\delta$ spectrum showing the same accuracy as that of the CAS(14,13)$^\delta$, while a cheaper RAS scheme yields artificial quartic coupling. When considering a configuration with interacting aromatic side chain in a stacked conformation (Fig. 9c), an analogous trend is found. In this case, a sizeable electronic quartic coupling is expected, and it is found equally at $CAS(14,13)^\delta$ and $RAS(2,3|8,8|2,2)^\delta$ levels, while the energy shift (Δ) is significantly overestimated at the $RAS(4,7|0,0|4,6)^\delta$ level.

Here, a successful benchmark study aimed at identifying efficient multi-configurational/multi-reference approaches in terms of the computational cost/accuracy ratio has been demonstrated in detail for benzene and phenol chromophores involved in protein systems. Analogous studies have been performed for indole [82] and adenine [57] chromophores, while the remaining nucleobases are currently under investigation. Overall, we believe that such theoretical benchmark studies will provide a full set of parameters and computational recipes that will enable the construction of excitonic model Hamiltonians for application in large systems (such as full proteins or large DNA/RNA sequences) and simulating 2DUV spectra of dimeric and small oligomeric systems under realistic conditions with unprecedented accuracy.

In the next sections, we will illustrate how accurate simulation of 2DUV spectroscopy could be used as a powerful tool to investigate physicochemical properties of biological systems.

4.2 2DES for Tracking GS Conformational Dynamics

In Sect. 3.1, we illustrated how 2DUV spectroscopy holds the potential to resolve electronic couplings associated with the interaction between UV chromophores in a model protein system containing two aromatic side chains. In particular, the energy shift (Δ) of mixed doubly excited states due to quartic coupling clearly affects the 2DUV–UV spectrum of the solvated CFYC tetrapeptide, enabling discrimination between peptide conformations with interacting and non-interacting chromophores. The CFYC peptide was chosen for modeling protein folding/unfolding dynamics because its terminal cysteine residues can form a disulfide bond that holds the tetrapeptide in a cyclic *closed* conformation while, if this bond breaks, CFYC will naturally unfold in an *open* conformation (Fig. 10). QM/MM geometry optimization of the closed CFYC, solvated in water solution, yields a structure in which the two aromatic side chains are rather close to other, in a so-called T-stacked conformation [52]. On the other hand, the open CFYC conformation is associated with unstacked (i.e. non-interacting) chromophoric units (Fig. 10a).

As shown in Fig. 10b, the excited-state manifolds of both closed T-stacked and open unstacked QM/MM optimized structures have been fully characterized at the $CAS(14,13)^\delta$ level [37, 52] showing how they are strongly differentiated by the chromophoric electronic couplings, which have a twofold effect: (i) red-shift of the mixed states by quartic coupling, and (ii) strong red-shift of charge-transfer (CT) states (from Phe to Tyr and vice versa) with a concomitant increase in the TDMs associated with the corresponding $S_{1\text{-}2} \rightarrow CT$ transitions. The former would cause off-diagonal (negative) signals to appear in the 2DUV–UV maps (see Fig. 9c), while the latter would show up exclusively at lower energies, i.e. in 2DUV–Vis spectra. Figure 10c shows the SOS//QM/MM computed 2DUV–Vis spectra for closed T-stacked and open unstacked conformations. In these spectra, few (positive) ESA

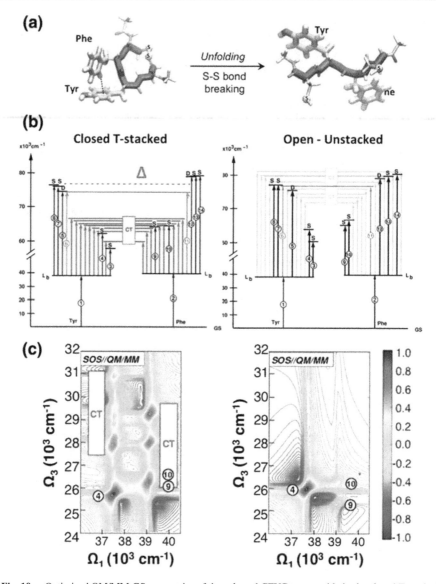

Fig. 10 **a** Optimized QM/MM GS geometries of the solvated CFYC tetrapeptide in the closed T-stacked and (upon disulfide bond breakage) open unstacked conformations. **b** Excited-state manifolds of the two CFYC conformations computed at the CAS(14,13)$^\delta$//PT2/ANO-L(432-2) level, with local (single and double) excitations in black, mixed double excitations in green, and bright or dark CT states in red or light red, respectively. **c** Simulated two-color 2DUV–Vis spectra of both CFYC conformations, with ESAs labels as defined in the scheme of panel b. Reproduced from data reported in Ref. [37]

signals arising from the $S_{1-2} \rightarrow S_M$ transitions (with S_1 and S_2 being the first $\pi\pi^*$ states of Tyr and Phe, respectively, namely L_b in Platt notation) are present for non-interacting side chains, while they are accompanied by a large number of ESA signals in the T-stacked conformation, due to bright $S_{1-2} \rightarrow CT$ transitions. It is worth noting that the appearance of these important ESA signals in the 2D maps would be completely missed in standard exciton modeling, with SOS//QM/MM computations of chromophore aggregates here showing their relevance.

In another example, we benchmarked low-cost methods for indole [82]. The first two $(\pi\pi^*)$ excited states (L_b and L_a) of indole (see Fig. 5) have essentially opposite character: L_b is covalent and apolar, with low oscillator strength, while L_a is ionic, polar and bright. Thus, indole represents a more intricate challenge compared to benzene and phenol, which possess a single (covalent) state absorbing in the NUV. Understandably, additional compromises had to be made in the calibration procedure. Nevertheless, a RAS(4,5|0,0|4,4) level of theory followed by a single-state RASPT2 energy correction utilizing an imaginary level shift parameter of 0.5 a.u. (Figure 11a) shows reasonable agreement with the reference data obtained with RAS(10,9|2,12) level of theory (Figs. 5c, 11a) for states below 8 eV, with significantly reduced computational effort. Employing the cost-efficient protocols developed for computing the electronic structures of indole and phenol, we applied the SOS//QM/MM scheme to the Trp-cage peptide, a common protein model for studying protein folding/unfolding, which contains Trp and Tyr side chains. As for the Phe- and Tyr-containing CFYC tetrapeptide (Fig. 10), two-color 2DUV–Vis spectroscopy can be used to detect ESA signatures of chromophore–chromophore interactions. We provide an indirect way to detect Tyr and Phe, whose absorption bands remain hidden under the (more) intense envelope of Trp in linear absorption experiments. A signal-free probing window (between 18,000 and 24,000 cm^{-1}) along the Ω_1 trace of Trp seems to facilitate this undertaking. Specifically, weak CT signals are resolved in this spectral window in a folded Trp-cage conformation with the two chromophores at a distance of about 5 Å. This outcome indicates that an experimentally detected enhancement of the ESA signal in Tyr- and Trp-containing peptides, in the (otherwise) signal-free Vis spectral window between 18,000 and 24,000 cm^{-1}, should be regarded as a clear signature of chromophore–chromophore proximity of 5 Å or below.

From the above examples, we can thus conclude that 2DUV spectroscopy can be used to distinguish between folded and unfolded structures of protein models with interacting and non-interacting UV-chromophores, respectively. The question then arises as to whether, given the high temporal (femtosecond) resolution of 2DUV experiments, this technique could also be used to monitor folding/unfolding dynamics of protein models and, more generally, whether it represents a useful tool for tracking GS conformational dynamics. We have exploited the SOS//QM/MM computational recipes described above to tackle these questions by simulating the 2DUV spectra along the unfolding dynamics of the CFYC tetrapeptide [54]. Figure 12 shows the structural changes relative to the inter-chromophore distance (d) and angle (between vectors normal to the aromatic planes, α) observed during a ratchet-and-pawl biased molecular dynamics (rMD) simulation [84], indicating how, starting from a closed T-stacked conformation (with $d = 4.5$–6 Å and $\alpha > 60°$), the CFYC peptide unfolds ($d > 6.5$ Å), passing through twisted offset stacked conformations (with $d = 5$–6.5 Å and $\alpha = 40$–$60°$). Notably, given the restricted flexibility of the peptide backbone, the two aromatic side chains are never observed in a parallel

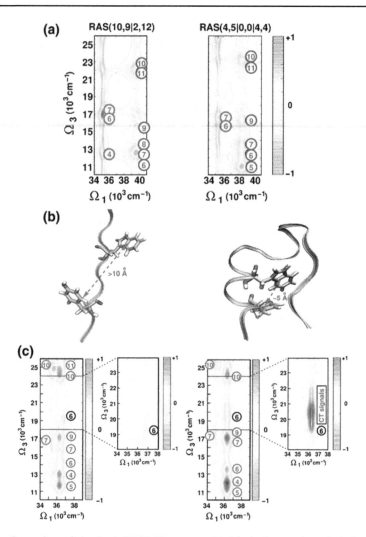

Fig. 11 a Comparison of simulated 2DUV–Vis spectra of indole in the gas phase, including the reference RAS(10,9|2,12) (left panel) and the low-cost RAS(4,9|0,0|4,7) level (right panel). Structures (**b**) and 2DUV–Vis spectra (**c**) of unfolded (left panel) and folded (right panel) Trp-cage from representative snapshots. ESAs associated with the indole side chain of Trp (L_b in red, L_a in blue) or with the phenol side chain of Tyr (black) are labeled. Probing of the spectral region around $\Omega_3 = 18{,}000$–$24{,}000$ cm^{-1} is zoomed in to highlight the weak signatures of charge-transfer states emerging in the stacked conformation. Labeling in (**a**) and (**c**) according to Fig. 5b. Reproduced from data reported in Ref. [82]

(namely π-) stacking conformation (i.e. with $d < 6$ Å and $\alpha < 30°$). The 2DUV–UV spectra computed for selected snapshots along the rMD simulation indicate a good correlation between the inter-chromophore distance and the quartic electronic coupling, with smaller d accompanied by larger shifts Δ (see Fig. 12b), which is less sensitive to the relative orientation of the chromophores, with T-stacked or twisted offset conformations possessing large quartic couplings. Significant correlations are also observed in the 2DUV–Vis spectra [54], where the energetic positions

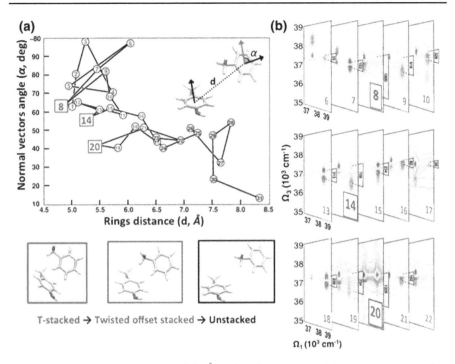

Fig. 12 **a** Inter-chromophore distances d (in Å) and angles α (measured between vectors normal to the aromatic planes, in degrees) for selected snapshots along the biased rMD simulation of the CYFC tetrapeptide unfolding dynamics. Color code according to the three-step unfolding/unstacking process (T-shaped in red, twisted offset stacked in blue, and unstacked in black), with representative structural arrangements of the UV chromophores depicted within boxes. **b** Simulated one-color 2DUV–UV spectra of selected structures, with snapshots having the shortest inter-chromophore distances highlighted. Reproduced from data reported in Ref. [54]

and brightness of the CT states are affected by inter-chromophore distances, their π-orbital overlaps and environmental effects explicitly accounted for in our SOS// QM/MM approach.

Considering the benchmark study of the protein aromatic chromophores described in the previous section, which provides converged results that are consistent with available experimental data, we can conclude that our predictions for two-dimensional electronic spectra strongly support the use of the 2DUV technique as an analytical tool for monitoring GS conformational dynamics of protein systems containing aromatic UV-chromophores.

We expect that the adoption of 2DUV will be equally well suited for tracking of GS conformational changes in nucleic acids. Dinucleoside monophosphates, i.e. dimers of DNA/RNA nucleobases, represent excellent model systems for exploring this hypothesis from our theoretical perspective. In fact, they are realistic biomolecules that are treatable by our QM approaches [57, 60] and assume various stacking conformations in water solution [85], currently representing model systems for studying the complex excited-state dynamics of nucleic acids

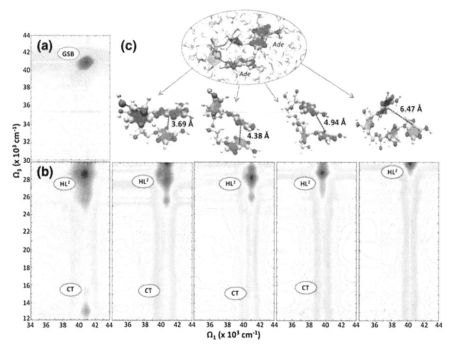

Fig. 13 Simulated 2DUV–UV (**a**) and 2DUV–Vis (**b**) spectra of the water-solvated ApA dinucleoside monophosphate, considering the conformational space sampled with unbiased MD in the GS. Panel (**c**) shows the molecular structure of the ApA system and a few selected representative conformations displaying the different intermolecular interactions possible. Reproduced from data reported in Ref. [61]

[86]. As an example case study, we very recently [61] performed 2DUV spectra simulations of the adenine–adenine monophosphate dinucleoside (ApA) in water solution. Figure 13 shows the computed 2DUV–UV and 2DUV–Vis for a set of structures extracted from an unbiased MD simulation of the solvated ApA in the GS, featuring various intermolecular arrangements that are known to deeply affect the photophysical properties of the adenine moieties [87]. Computations were performed using the SOS//QM/MM approach and adopting the RAS(4,12|0,0|4,6) active space, i.e. the cheapest (but still reliable) multi-configurational scheme calibrated in our benchmark study of adenine monomer and homodimer [57]. By adopting this efficient scheme we could achieve computing at a reasonable cost the excited-state manifold of multiple MD snapshots. Still, the electronic structures computations must be computed for selected (most representative) conformations among those thermally accessible in the ApA conformational space. The presence of explicit solvent molecules in the QM/MM treatment and the inclusion of multiple conformations (Fig. 13c) give rise to inhomogeneous broadening of the 2D signals. In particular, the GSB signals found in the 2DUV–UV spectrum of ApA (Fig. 13a) feature the characteristic broadening along the diagonal due to interactions with the environment and between the chromophores. In fact, the couplings among adenine chromophores

introduce energy splitting in the main signal trace along Ω_1, i.e. that associated with the fundamental $S_0 \rightarrow S_{3-4}$ transitions (with S_3 and S_4 being the lowest $\pi\pi^*$ bright state of the two adenines, namely ^1L_a), which results in an effective additional broadening of the GSB signals. Analogous broadening along Ω_1 is observed in the 2DUV–Vis spectrum (Fig. 13b) for the dominant ESA signals arising from excitations to the adenine doubly excited state (referred to as HL^2), lying at Ω_3 around 28,000–30,000 cm^{-1}.

The solvated ApA dinucleoside is a highly flexible system, featuring conformations ranging from T-shaped to quasi-planar π-stacked, to completely unstacked adenines along the MD trajectory. Cluster analysis has been used to group molecular conformations and select those that are most representative, a few of which are depicted in Fig. 13c. Generally, short inter-chromophore distance in the 3.5–4.5 Å range (measured as C5–C5 distance) is associated with rotated π-stacked structures that yield substantial broadening in 2D maps along Ω_1 (i.e. ^1L_a trace splitting), while also featuring CT signals in the low-energy Vis probing window (at Ω_3 around 16,000 cm^{-1}). At inter-chromophore distances of around 5 Å, the ApA structures can be found in T-shaped conformations, usually with the amino group of one adenine pointing towards the 5-ring of the opposite adenine moiety. Generally, these T-stacked conformations can be differentiated from the others because the presence of ESA signals from CT states is not accompanied by broadening along Ω_1 in this case. As expected, unstacked conformations behave as non-interacting dimeric systems, featuring dark CT states (thus not contributing to the 2DUV–Vis maps) and ESA signals with negligible Ω_1 broadening.

Overall, these results demonstrate how 2DUV spectroscopy is theoretically able to characterize the conformational space of a dinucleoside monophosphate by exploiting the two-dimensional spectral resolution, representing a powerful alternative or complement to standard pump–probe experiments for elucidating the role of structural arrangements on nucleic acid photophysics and photochemistry. However, the spectral line shapes reported in the examples above are not realistic, as they account for only partial contributions to the spectral line broadening. Various effects (coupling to nuclear degrees of freedom, coupling to environment, finite excited-state lifetimes) are the source of dephasing-induced broadening. In real 2D, electronic spectra broadening could obscure the spectral fingerprints predicted by ab initio simulations. Therefore, inclusion of dynamic effects is indispensable to move towards accurate 2DES simulation, and this is the focus of the next two sections.

4.3 Excited-State Coherent Vibrational Dynamics Resolved by 2DES

Real quantum systems are not closed. Upon interaction with the incident electric field, the electron density of the molecular system rearranges instantaneously in a new discrete quantum state. The changed electron density exerts a force on the nuclei, which are set in motion, and vibrational dynamics is initiated. As the nuclei are much heavier than electrons, their vibrations have vibrational periods from 10 fs (fast hydrogen stretch vibrations) to a few hundred femtoseconds (for wagging, twisting and other slow vibrational modes). Nonlinear spectroscopy with ultrashort sub-10-fs pulses (an order or magnitude shorter than most of the vibrational periods) are capable of detecting this vibrational dynamic, which manifests in coherent oscillations of the spectral signatures. These coherences provide insight into the photoactive vibrational modes of the system and help in understanding its reactivity.

We begin this section by re-introducing a linear coupling of the electronic degrees of freedom to a Gaussian bath, while still neglecting population transfer. Regarding the description of ESAs, this corresponds to adding the phase function $\varphi_{e'fe}(\tau_4, \tau_3, \tau_2, \tau_1)$ to Eqs. (14) and (15), where it is worth noting the absence of population transfer $e' = e$ in Eq. 14. The phase function acquires the following complex form:

$$
\begin{aligned}
\varphi_{e'fe}(\tau_4, \tau_3, \tau_2, \tau_1) = & -g_{e'e'}(\tau_{43}) - g_{ff}(\tau_{32}) - g_{ee}(\tau_{21}) \\
& - g_{e'f}(\tau_{42}) - g_{e'f}(\tau_{43}) - g_{e'f}(\tau_{32}) \\
& - g_{e'e}(\tau_{41}) - g_{e'e}(\tau_{42}) - g_{e'e}(\tau_{31}) - g_{e'e}(\tau_{32}) \\
& - g_{fe}(\tau_{31}) - g_{fe}(\tau_{32}) - g_{fe}(\tau_{21})
\end{aligned}
\tag{16}
$$

where $g_{ij}(\tau_{ij})$ are line shape functions (see Appendix) and $\tau_{ij} = \tau_i - \tau_j$. Equation 16 depends in a non-trivial way on the three delay times t_1, t_2 and t_3 between the pulses, and it captures coherences which survive during the entire duration of the multipulse experiment. This model can describe bath fluctuations of arbitrary timescales. In the following, we show an example were we focus on the spectral signatures of the strong coupling to a bath of discrete high-frequency intramolecular vibrational modes. The line shape function for this model is formulated on the basis of the multidimensional uncoupled displaced harmonic oscillator (DHO) [2]:

$$
g_{ij}^{DHO}(t) = \sum_k \frac{\omega_k \tilde{d}_{ik} \tilde{d}_{jk}}{2} \left[\coth\left(\frac{\omega_k}{2k_B T}\right)(1 - \cos(\omega_k t)) + i \sin(\omega_k t) \right]
\tag{17}
$$

where the one-dimensional potential for the ith state along each normal mode k is fully characterized by two parameters, frequency ω_k and relative displacement $\tilde{d}_{ik}$. In the simulations, a composite line shape function is used, constructed by combining the DHO line shape function with the line shape function of the semi-classical Brownian oscillator (see Appendix), which describes the coupling to a continuum of low-frequency modes (of the environment), inducing decoherence on the timescale of a few tens of femtoseconds and giving rise to the homogeneous broadening of the peaks. Note that the omission of a mechanism for population transfer decay implies infinite excited-state lifetimes. However, in reality, excited states have finite lifetimes. When ultrafast (< 100 fs) decay channels are present, the dephasing due to population transfer may induce additional signal broadening, which in the present framework is treated phenomenologically.

Pyrene (see Fig. 14a) is a polycyclic aromatic hydrocarbon that has attracted attention for its prominent photophysical properties such as its remarkably long

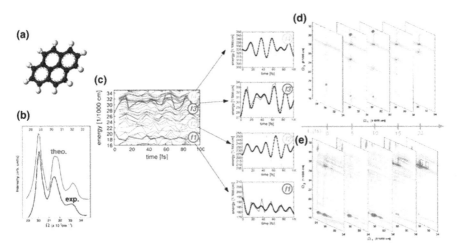

Fig. 14 **a** Structure of pyrene; **b** comparison of recorded (black) and simulated (red) linear absorption spectrum; **c** state density up to 34,000 cm^{-1}, highlighting the energy profiles of the bright states f–f_4, fingerprints of the L_a state, with superimposed fits according to Eq. 18 (black dotted lines); (**d**, **e**) simulated quasi-absorptive 2D electronic spectra of pyrene for waiting times t_2 in the range 0–20 fs obtained through pumping at the frequency of the L_a transition and supercontinuum probing in the UV–Vis region within the "quasi"-static picture (**d**) and with bath fluctuations included (**e**). Color code: GSB (blue), ESA (red). Reproduced from data reported in Ref. [93]

fluorescence lifetime and high fluorescence quantum yields [88]. It is used to probe solvent polarity [89] and constitutes the basis of commercial dyes, fluorescence probes for electro-chemiluminescence and many larger photoactive molecules [90–92]. Pyrene features well-separated absorption bands (see Fig. 14b) with clear FC progressions in the NUV (lowest bright state L_a at 320 nm) and in the deep UV (second bright state at 280 nm), making it an excellent model for assessing novel spectroscopic techniques. Pyrene has been the subject of some of the first documented broadband 2DES spectra in the NUV [16] and deep UV [18], therefore representing a perfect target for benchmarking our theoretical protocols against 2D experimental data [93].

Here, we review 2DES simulations which resolve the coherent spectral dynamics initiated immediately after photoexcitation of the first bright excited state (L_a) of pyrene, exhibiting a vibrational progression with a fundamental (most intense) transition and two overtones between 28,500 cm^{-1} (~350 nm) and 33,500 cm^{-1} (~300 nm) (see Fig. 14b). In particular, we aim at reproducing the spectra for waiting times of 0–20 fs in a broadband probe window ranging from 16,000 cm^{-1} (~600 nm) to 32,000 cm^{-1} (~300 nm). These simulations require knowledge of (i) the electronic structure of pyrene up to 65,000 cm^{-1} (ca. 8 eV), thus covering not only the L_a state (labeled e) but also the manifold of higher-lying states (labeled f) that fall within the envelope of the probe pulse (i.e. 16,000–32,000 cm^{-1} above the L_a state); (ii) the FC active modes of the L_a state, source of the vibrational dynamics appearing as coherences in the 2D spectra; and (iii) the response of the states from the f-manifold to the vibrational dynamics in the L_a state. ESA bands are an indirect probe of the coherent vibrational dynamics in the photoactive state, featuring high sensitivity by oscillations of signals intensities and/or spectral positions as a function of the delay time.

To obtain the required information listed above, 60 excited states were computed simultaneously at the RASSCF(4,8|0,0|4,8)//PT2 level of theory, where the RAS comprise all valence π-orbitals of the molecule, allowing up to quadruple excitations only in the RAS1/RAS3 spaces. Coherent vibrational dynamics were described with the DHO model. The frequency (ω_k) and displacement ($\tilde{d}_{lk}$) parameters used to construct the line shape functions g_{ij} (Eq. 17) were extracted from a 100-fs mixed quantum–classical molecular dynamics simulation treating electrons quantum-mechanically by solving the time-dependent Schrödinger equation, while applying classical Newtonian dynamics to the nuclei. The dynamics were initiated in the L_a state of pyrene at the FC point without initial kinetic energy. The choice of a reference Hamiltonian acting on the classical bath during the coherence propagation, as is the case for the delay times t_1 and t_3, is not unique [94], as the *bra* and the *ket* sides of the density matrix are subject to different electronic potentials. We chose to propagate the nuclear degrees of freedom as though interacting with the L_a state. The frequency and displacement parameters themselves were obtained by fitting the electronic gaps $E_e(t)$–$E_g(t)$ (giving rise to GSB and SE signals) and $E_f(t)$–$E_e(t)$ (giving rise to ESAs), computed at the aforementioned RASSCF(4,8|0,0|4,8)//PT2 level along the quantum–classical dynamics, to the analytical expression for the classical time-dependent fluctuation of the L_a–S_n energy gap (n belonging to a state from either the GS, g, or the f-manifold)

$$E_h(t) - E_e(t) = -\sum_k \omega_k^2(\tilde{d}_{ek} - \tilde{d}_{hk})\tilde{d}_{ek}\cos(\omega_k t) + \sum_k \frac{\omega_k^2(\tilde{d}_{ek} - \tilde{d}_{hk})^2}{2} + (\omega_{hg} + \omega_{eg})$$

(18)

with $h \in \{g,f\}$, which made it possible to extract the mass-weighted displacement coefficients d_{ek} and d_{fk}, as well as the electronic transition energies ω_{eg} and ω_{fg} (the parameters d_{gk} and ω_{gg} describing the GS per definition, are zero). In the adopted DHO framework, the spectral dynamics of the ESA are a function of the dynamics in the photoactive state (namely S_3 for pyrene), i.e. of the relative displacement of the higher-lying excited-state PES relative to the PES of the photoactive state, and can induce positive or negative frequency correlations [95]. By definition, it is restricted to the normal modes describing the molecular dynamics in the e-manifold (i.e. if d_{ek} is zero, then so is d_{fk}). This is an implication of the missing state-specific modes. We emphasize that the extraction of the energy fluctuations of states from the f-manifold must be performed in a diabatic representation (i.e. by following the excited-state wave function rather than the adiabatic root) in order to stay in the framework of the uncoupled harmonic oscillators, as due to the high state density in the Vis and NUV, PES of higher-lying states intersect, thus rendering the resulting adiabatic potentials highly anharmonic (see Fig. 14c). To achieve this goal, we selected at the FC point the states from the manifold of higher-lying states f within the probed spectral window and with significant oscillator strength out of the L_a state (i.e. the reference diabatic states, shown in red, green, blue and cyan in Fig. 14c), and tracked the temporal evolution of the wave function associated with each reference state along the dynamics by searching for the adiabatic state with the greatest overlap with the reference. The energy profiles extracted for the four brightest transitions are depicted in Fig. 14c, showing clear oscillatory dynamics properly captured by the fitting procedure (black dotted lines). Note that the energy state f_1 (red) oscillates out of phase with respect to f_2, f_3 and f_4.

Figure 14d, e compares simulated 2DES spectra between 0 fs and 20 fs simulated in the "quasi"-static picture (Fig. 14d) and with bath fluctuations included (Fig. 14e). The dynamic evolution of the spectra in the static picture is approximated by assuming fast bath fluctuations during t_1 and t_3, inducing homogenous Lorentzian broadening of the signals. Within this approximation, the 2DES at a waiting time $t_2 \neq 0$ can be computed with the SOS protocol (note that with a single state in the e-manifold, the coherence term Eq. 14 vanishes), setting the pump pulse to interact with the e-manifold of the FC point, while setting the probe pulse to interact with the f-manifold of the snapshot reached along the trajectory at t_2. It is apparent that the "quasi"-static picture reproduces qualitatively the main spectral features: (a) the GSB around 30,000 cm^{-1} (shown in blue); (b) the oscillatory dynamics of the SE between 30,000 cm^{-1} and 25,000 cm^{-1}, with a period of ca. 20 fs (shown in blue); (c) the aforementioned three ESA contributions (shown in red) around 19,000 cm^{-1} (f_1), 24,000 cm^{-1} (f_2) and 32,000 cm^{-1} (f_4); and (d) the time-dependent fluctuation of the intensities, a consequence of the coordinate dependence of the TDM. However, the static approach fails to correctly describe the spectral line shapes. This becomes evident when slow bath fluctuations are taken into account (Fig. 14e). The vibrational progression in the L_a spectrum (Fig. 14b) translates into three traces along Ω_1 associated with the fundamental transition (28,500 cm^{-1}) and two overtones. As a consequence, the GSB and SE adopt a characteristic checkerboard pattern [96, 97]. Another remarkable difference is that the oscillations of the ESA peaks become less distinct. Signals showing more pronounced energy gap fluctuations (f_4) appear broadened along Ω_3 and, as a consequence of this broadening, exhibit reduced intensity compared to the static spectrum. Overall, the ESA associated with f_1 ($\Omega_3 \sim 17,500$ cm^{-1}) constitutes the most characteristic signature of the L_a state in the Vis range. In fact, it has been observed in transient pump–probe spectra [45, 98]. The weaker ESA around 25,000 cm^{-1} associated with state f_2 has been observed in 1D-PP experiments [99]. To the best of our knowledge, no transient spectra pumping the L_a state and probing beyond 32,000 cm^{-1} for short (< 100 fs) waiting times has been reported in the literature; therefore, the absorption associated with f_4 is yet to be detected experimentally and it is predicted by our simulations.

4.4 Spectral Characterization of Long-Lived ES Intermediates

Excited states have finite lifetimes. Decay to the GS occurs either via light radiation or non-radiatively. In the second scenario, population transfer between electronic states is facilitated through non-adiabatic (e.g. singlet–singlet transfer) or spin–orbit (e.g. singlet–triplet transfer) couplings, which are coordinate-dependent and become large in areas where the PES of electronic states intersect. Population transfer has a twofold effect on the appearance of the 2D electronic spectra: (i) homogeneous line broadening, and (ii) decay and simultaneous buildup of spectral signatures, the former process being associated with the population decrease in the initial state, the latter with the population growth in the final state. The CGF framework offers a prescript for incorporating population transfer, i.e. the so-called doorway–window (DW) factorization [100, 101]. When dephasing is much faster compared to transport, the secular Redfield approximation can be applied, providing separate expressions for coherences (Eq. 13) and populations (Eq. 12). Correlated bath dynamics in the coherence is described with the phase function $\varphi_{e'fe}(\tau_4, \tau_3, \tau_2, \tau_1)$ introduced in the previous section (Eq. 16). In contrast, when bath fluctuations decay more slowly than the coherence dephasing but faster than population transport, the population term (in Eq. 13) can be treated by applying the Markovian approximation to the time intervals between the pulses, with the phase function expressed as

$$
\begin{aligned}
\varphi_{e'fe}(t_3, t_1) = &-g_{ee}(t_1) - g_{e'e'}(t_3) - g_{ff}^*(t_3) + g_{fe'}(t_3) + g_{e'f}^*(t_3) \\
&+ 2i(\lambda_{e'e'} - \lambda_{fe'})t_3
\end{aligned}
\tag{19}
$$

where, in the limit $t_2 \rightarrow \infty$ (i.e. timescales for which memory is lost and the spectral dynamics is dominated by population transfer), the e'–f coherence evolution during t_3 introduces a broadening along Ω_3, which is a function of the reorganization energy λ_{ij}. Clearly, Eq. 19 depends on both states (e and e') involved in the population transfer, while f denotes the manifold of excited states coupled to state e'. Fluctuations in the initial coherence state g–e during t_1 and in the final coherence state e'–f during t_3 are formally treated exactly, but the dynamics are not coupled. More sophisticated treatments are possible when bath fluctuations and transport occur on the same timescale in order to retain memory effects during t_2 and to describe ultrafast population transfer events [47].

A limiting, and quite interesting, case for Eq. 19 is associated with the trapping of population in an excited state during the decay process on a timescale facilitating the dissipation of excess vibrational energy in the environment. In such case, the spectrum is dominated by the electronic structure of the equilibrium geometry of the excited state in which the population is trapped. Within this framework, the dynamics during t_3 can be calculated with the e' Hamiltonian as reference, thus effectively eliminating all terms depending on e', with the phase function taking the simplified form

$$
\varphi_{fe}(t_3, t_1) = -g_{ee}(t_1) - g_{ff}^*(t_3)
\tag{20}
$$

which depends only on the line shape function $g_{ee}(t_1)$ of the initial state and $g_{ff}^*(t_3)$ of the states coupled to the probe pulse. The 2D electronic spectra of long-lived intermediates in photochemical and photophysical processes can thus be obtained, in this case, at a reasonably low computational cost, as they do not require quantum dynamics simulations. The reduced computational effort would allow the focus to be directed entirely on the electronic structure, thereby providing invaluable insight into the spectral fingerprints of excited-state minima.

From experimental observations, it is known that in pyrene, after L_a excitation, an ultrafast population transfer towards a lower-lying state (namely L_b) takes place with a time constant of 85 fs [16]. This state is spectroscopically dark; therefore, the population remains trapped on a nanosecond timescale before it eventually decays to the GS through fluorescence with a high quantum yield [102, 103]. Recent excited-state molecular dynamics simulations [93] have revealed the mechanism of non-adiabatic $L_a \rightarrow L_b$ internal conversion. As shown in Fig. 15, after irradiation, carbon–carbon stretching modes with an oscillation period of 20 fs are activated. The momentum accumulated in these modes drives the system beyond the L_a minimum to a turning point on the PES, which lies in the vicinity of the L_a/L_b crossing region. Low-frequency out-of-plane vibrations induce a finite non-adiabatic coupling between the two states, the wave packet bifurcates, and a part continues its dynamics on the L_b surface while the rest remains on the L_a surface, oscillating back and returning to the crossing region after ca. 20 fs. Dissipation of excess vibrational energy (cooling) in the L_b state was found to occur with a time constant of 4 ps [99].

Considering the above-described mechanism and employing the RASSCF(4,8|0,0|4,8)//PT2 level of theory mentioned above for computing TEs and TDMs at the optimized geometry on the L_b PES, i.e. named $^1(L_b)_{min}$, we generated a 2D spectrum representative of waiting times in the ps regime, i.e. longer than the L_a lifetime but shorter than the L_b lifetime, which justify the use of the CGF protocol with the phase function presented in Eq. 20. Figure 15 shows the simulated 2DUV spectrum of pyrene, exhibiting a structured checkerboard pattern attributed to the GSB (L_b is dark, so there is no SE signal) between 28,000 cm^{-1} and 32,000 cm^{-1}, as well as two distinct ESA peaks at 21,000 cm^{-1} (labeled B) and 26,000 cm^{-1} (labeled

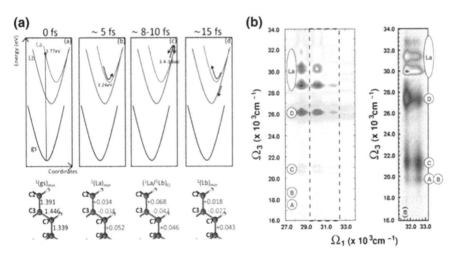

Fig. 15 a Schematic representation of the photophysics of pyrene based on quantum–classical dynamics. The dominant bond deformations are given, with the GS equilibrium geometry used as a reference. **b** Theoretical (left panel) and experimental (right panel) quasi-absorptive 2D electronic spectra of pyrene for a delay time t_2 in the picosecond range ($t_2 = 1$ ps in the experiment) obtained through pumping at the frequency of the L_a transition and supercontinuum probing in the UV–Vis region. Color code: GSB (red), ESA (blue). Note that the color code is inverted with respect to other figures so as to enable easy comparison with the experimentally reported spectrum. Reproduced from data reported in Ref. [93]

C, D). Further, less intense peaks (A, B) appear at around 18,000 cm^{-1}. To underscore that different higher-lying excited states are probed relative to those previously reported for the 2DUV spectra at the FC, i.e. $t_2 = 0$ (see Fig. 14d, e), we use letter-code (A–D) to label the ESA peaks. As a consequence of the Markovian approximation, the three traces along Ω_1 adopt an identical peak structure with a progressively decreasing intensity of the overtones. The physical interpretation behind this is that, on a picosecond timescale, excess energy acquired by exciting the overtones would be dissipated, and probing would become insensitive to the pump wavelength. Riedle and co-workers recently reported an experimental 2DUV spectrum of pyrene in methanol at time delay $t_2 = 1$ ps, recorded in a collinear pump–probe set-up, utilizing a narrowband pump–pulse pair, centered at around 32,000 cm^{-1}, i.e. at the second and third vibrational bands of the L_a absorption, and a supercontinuum probe pulse covering the UV–Vis spectral window (16,000–38,000 cm^{-1}) [93]. This result represents one of the few experimental examples employing UV–Vis broad pulses for 2D spectroscopy. The spectrum is shown in Fig. 15b and compared to our simulations (note the inverted color code with respect to the usual color coding adopted in other figures, in order to facilitate appropriate visual comparison with the experimental map). Remarkable agreement is achieved between theory and experiment with regard to the GSB and the positions and relative intensities of the ESA signals. Few differences could be found: (i) through the use of a narrowband pump–pulse pair, the fundamental (0–0) transition was suppressed along Ω_1 in the experiment; (ii) the ESA peaks were noticeably elongated along Ω_3, which could be rationalized by the short lifetimes of the higher-excited states. The pyrene example, where a simulated 2DUV spectrum adopting the excited-state equilibrium geometry was successfully compared with the experimental spectrum recorded at a waiting time of 1 ps, demonstrates that even when equilibration has not been completely carried out (time constant is 4 ps), the spectrum is dominated by the electronic structure of the excited-state equilibrium geometry, and theoretical simulations can predict/interpret 2D spectroscopic fingerprints of excited-state minima. This outcome paves the way for extending simulation of 2DUV spectra through our SOS//QM/MM to other biologically relevant systems, pursuing the idea that 2DUV fingerprints of excited-state minima can be computed at a reasonable computational cost and could be directly compared with experimental data recorded at appropriate t_2 delay times.

In this context, we have focused our efforts on 2DUV simulations for tracking the complex photophysics of nucleic acids, based on the extended benchmark studies illustrated in the previous sections. Recent efforts to obtain the first experimental 2DUV spectra of DNA nucleobase monomers [21, 22] indicate the concrete possibility of using this technique to understand the photophysics of more realistic DNA/RNA models, e.g. multimeric species in short single or double strands. As mentioned in Sect. 3.2, dinucleoside monophosphates, i.e. dimers of DNA/RNA nucleobases, represent excellent model systems to start exploring nucleic acid photophysics from both a theoretical and experimental standpoint. In fact, the presence of a neighboring base significantly affects the excited-state dynamics of monomeric systems due to excimer/exciplex formation and a range of charge/proton/hydrogen transfer/recombination events [104–107]. These have been shown to be extremely difficult to separate and assess with standard

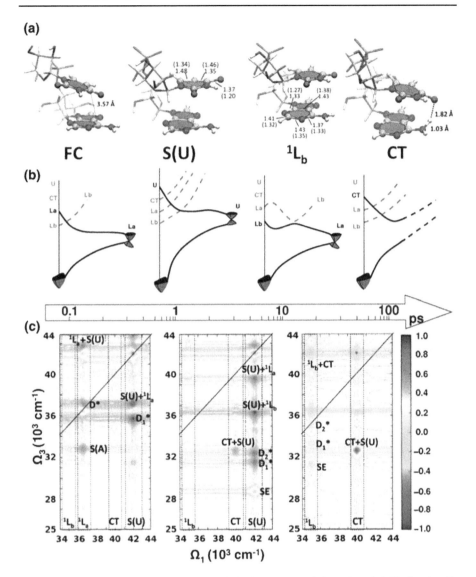

Fig. 16 a Ground- (FC) and excited-state optimized structures of the solvated ApU dinucleoside monophosphate, including S(U), ^1L_b and CT states. Representative inter-bases or bond distances (in Å) are reported, with GS values in parentheses for excited-state minima. **b** Schematic representation of the PES involved in excited-state decay channels. **c** Predicted time evolution of the 2DUV–UV spectra for stacked ApU conformations. The time arrow marks the approximate scale at which the spectra will evolve according to the potential energy calculations. Reproduced from data reported in Ref. [60]

1D pump–probe techniques, and their particular contributions to the overall de-excitation process are key to understanding photo-induced lesions and mutations [108, 109]. Figure 16 shows the simulated 2DES spectra of water-solvated adenine–uracil dinucleoside (ApU), a system exhibiting different excited-state relaxation mechanisms that would be intricately obscured in 1D-PP experiments. Early experiments had shown how adenine and uracil separately would feature ultrafast signals [110–112], as well as long-lived signals arising from partial population of their dark $^1n\pi^*$ states [113], but that an additional signal would arise in dimeric species, probably due to inter-nucleobase interactions [87]. This signal could not be clearly disentangled, but it was associated with the ability of the dimeric nucleosides/tides to attain stacking conformations that would lead to the formation of excimer/exciplex or charge-transfer events. In order to ascertain this possibility, we have proposed the use of 2DUV spectroscopy for characterizing, with high temporal and spectral resolution, the spectroscopic fingerprints of various excited-state minima along the complex photoinduced pathways.

An initial theoretical study was carried out for the solvated ApU dinucleoside [60], where each of the decay channels exploited by the monomeric units were investigated by means of CASSCF excited-state optimizations and conical intersection characterizations. Figure 16 shows the equilibrium geometries of the localized excited states, named here as ^1L_b for adenine and S(U) for uracil, as well as delocalized CT states as originated by excited-state geometry optimizations from the FC region, associated with a GS optimized geometry with vicinal interacting nucleobases (Fig. 16a). Transition states and conical intersections along the excited-state decay channels were also characterized, providing an estimate of the timescale of the relaxation times, as schematically reported in Fig. 16b. To disentangle among these relevant stationary points involved in the ApU photophysics, broadband 2DES spectra were computed on top of their structures, resulting in the simulated 2DES maps depicted in Fig. 16c. As can be seen, in the FC region (at zero or ultrashort t_2 waiting time, sub-100 fs), the main signals are expected to arise from the ^1L_a and S(U) traces, with peaks from the spectroscopic ^1L_a state of adenine rapidly depleted due to ultrafast deactivation and weak CT signals that are present only for those conformations where such non-localized states are bright. Due to the presence of a sizeable energy barrier for the deactivation of S(U) of uracil in ApU, in contrast to those recently computed for uridine in water [114], it was predicted that most spectral contributions in the sub-10-ps timescale would come from the intense S(U) trace. At longer waiting times, in the 10–100-ps range, the 2D maps would feature signals on the ^1L_b trace (at Ω_1 around 34,000–36,000 cm^{-1}) and a particularly intense fingerprint of the CT minimum, whose trace (at Ω_1 around 40,000 cm^{-1}) would be well separated from the ^1L_b trace, allowing quantification of the population of ApU conformations with accessible CT states. Thus, these simulations could help assign distinctive signals arising from different excited-state decay channels that might be recorded in upcoming 2DUV experiments, providing a state-specific characterization of the ApU excited-state dynamics. Here, with such an example, we demonstrated how 2DUV could be an extremely powerful tool for elucidating nucleic acid photophysics in model systems.

5 Outlook and Perspectives

In this review, we have outlined a route towards simulation of third-order nonlinear electronic spectroscopy with accuracy compatible with current experiments. The focus on 2D electronic spectroscopy is motivated by the huge impact that its application in the UV region could have on the study of molecular systems with absolute biological relevance, such as proteins and nucleic acids, the building blocks of life (Sect. 1). 2DUV experiments feature high spectral and temporal resolution, allowing us to follow the evolution of individual chromophores though their spectral fingerprints, providing direct signatures of electronic couplings, charge and energy transfers, and characteristic line shapes and dynamics, which provide information on PES topology and solvent reorganization timescales. The density matrix formalism (Sect. 2) provides an appropriate platform to describe the state-of-matter evolution in the Liouville space when interacting with weak electric fields, accessing the nonlinear response recorded in 2DES experiments.

We outlined a route towards simulation and prediction of 2DES spectra from first principles first by employing "drastic" (static) approximations to the working equations within the global eigenstates basis (SOS approach), which enables us to ascertain the two fundamental ingredients to be accurately computed, i.e. the electronic transition energies and dipole moments (TEs and TDMs), and then recovering the missing contributions that shape 2D electronic spectra, comprising the factors that determine the line shapes and the dynamic evolution of the nonlinear signals. This route is covered by illustrating some practical examples (Sect. 3), starting with the benchmark studies of excited-state manifolds of the target biomolecular systems (Sect. 3.1) by means of accurate electronic structure computations, providing reliable TEs and TDMs estimates. For accurate calculations of these fundamental ingredients, we rely on the multi-configurational wave functions combined with multi-reference perturbation theory energy corrections (the CASSCF//PT2 level of theory), which in our opinion offers the optimal compromise between accuracy/completeness and computational cost. We addressed its performance with respect to active space size, number and nature of simultaneously computed states, basis set and geometric parameters, and demonstrate its sensitivity to these parameters. CASSCF/CASPT2 enables adaptive protocols for obtaining converged energies and TDMs, reaching accuracy of 0.2–0.25 eV, which is considered state of the art in the field of photochemistry, yet corresponding to ~ 2000 cm^{-1}, a significant deviation from a spectroscopic point of view. Therefore, reaching this upper-bound limit of accuracy is essential for reliable theoretical spectroscopy. We have shown how such accuracy has been obtained for the TEs and TDMs of monomeric units of aromatic protein side chains and nucleobases in the gas phase, demonstrating how the (commonly used) full-valence active spaces yield large discrepancies relative to available experimental cross-sections, and large, as computationally quite expensive, approaches are required for quantitative comparisons. These high-level computations yielded important parameters that could be implemented in Hamiltonian models (within a quasi-particle representation) in order to extend the simulation to large scale, but become impracticable when working within an SOS approach, even for the smallest conceivable chromophore aggregates, i.e. the dimeric species. Computational recipes based on semi-empirical parameterization against our reliable reference data enabled accurate predictions of the excited-state manifold of dimeric species at a reasonable computational cost, paving the way to the first applications of 2DUV simulation protocols for estimating the effects of inter-chromophore interactions on the nonlinear response (Sect. 3.2).

By embedding multi-configurational/multi-reference computations within hybrid QM/MM schemes, it was possible to extend the SOS simulation (SOS//QM/MM) of 2D electronic spectra to realistic model systems of proteins and nucleic acids. This allowed us to demonstrate how 2DUV spectroscopy holds the potential to resolve inter-chromophore interactions occurring during GS dynamics of nucleic acid and protein systems. In particular, simulations of 2DUV spectra of the CFYC tetrapeptide and Trp-cage protein model revealed how both electronic (quartic) coupling shifts and charge-transfer states yield clear spectroscopic fingerprints (mainly associated with ESAs) in the one-color 2DUV–UV and two-color 2DUV–Vis spectra, respectively. This significant outcome prompted investigations on the potential of 2DUV spectroscopy for tracking the protein folding/unfolding processes of protein models with high temporal resolution and for the quantitative analysis of nucleobase stacking in nucleic acid models. In this regard, extensive studies on the unfolding dynamics of the CFYC tetrapeptide and on the conformational space characterization of the water-solvated ApA dinucleoside monophosphate indicated that the 2DUV technique represented a powerful alternative (or complement) to standard transient absorption experiments. These results highlight the importance of accurate electronic structure computations in 2DES simulation, to account for appropriate descriptions of the ESAs signals. However, the massive computational costs of ab initio computations restrict its application to small (still realistic) dimeric systems. Current developments in the field of multi-configurational, density-based and linear-response techniques [95, 115–122] would certainly provide beneficial tools for treating large (multi)chromophoric systems, provided they accomplish completeness in the description of excited-state manifolds (e.g. including doubly excited states) [123]. Efficient QM methods employed in hybrid QM/MM schemes would also make it possible to account for signal broadening due to solvent rearrangement, as shown in the ApA case.

Various types of dephasing-induced broadening (due to coupling to nuclear degrees of freedom and environment, finite excited-state lifetimes) can shape the 2D maps. When neglected, simulated signal line shapes and positions would differ significantly from their experimental counterparts. We moved a step towards a more realistic description of spectral line shapes through the inclusion of vibrational dynamics in the simulation (Sect. 3.3). The high temporal resolution featured in 2DUV experiments with ultrashort sub-10-fs pulses allows for the detection of vibrational features, which manifest in coherent oscillations of the spectral signatures. These features can be computed by combining quantum–classical excited-state dynamics simulations that account for discrete high-frequency intramolecular vibrational modes, with electronic structure calculations of the manifold of higher-lying excited states along the dynamics. By accounting for population transfer processes in the theoretical treatment, it became possible, for the first time, to directly compare theoretical predictions with experimental data, available from broadband 2DES spectra of the pyrene molecule, recorded along its excited-state relaxation process [16] (Sect. 3.4).

A comparison of theoretical and experimental 2DUV spectra revealed that the 2D map of pyrene recorded at waiting times on a timescale that allows dissipation of excess vibrational energy in the environment (i.e. when population is trapped in an excited state) is dominated by the spectroscopic signatures of the excited-state equilibrium geometry, as predicted by our computations. This outcome supports the extension of 2DUV spectra simulation for resolving the spectroscopic fingerprints of long-lived excited-state minima along the complex photoinduced decay pathways of DNA/RNA model systems. In particular, our predictions of state-specific fingerprints of excited-state dynamics in solvated ApU dinucleoside monophosphate exemplifies the potential impact of accurate simulations of 2DES spectra in revealing complex physicochemical properties of fundamental biological systems.

The accuracy of the theoretical treatments proposed here can be improved considerably on several fronts, from more efficient and reliable ab initio characterization of excited-state manifolds, to inclusion of non-Gaussian bath fluctuations, to trajectory-based approaches for handling bath fluctuations and population transfers concurring in the ultrafast timescale, just to mention a few developments that can be envisioned in this field.

Acknowledgements Ivan Rivalta acknowledges support from the French Agence National de la Recherche (FEMTO-2DNA, ANR-15-CE29-0010). Javier Segarra-Marti thanks Dr. Lara Martínez-Fernández for useful discussions. Marco Garavelli acknowledges support from the European Research Council STRATUS Advanced Grant (ERC-2011-AdG No. 291198). Shaul Mukamel gratefully acknowledges the support of the National Science Foundation (Grant CHE-1361516) and the Chemical Sciences, Geosciences, and Biosciences Division, Office of Basic Energy Sciences, Office of Science, U.S. Department of Energy.

Appendix: Retarded Green's function and third-order density matrix

In the absence of an external field ($\hat{H} = \hat{H}_0$), the free evolution of an unperturbed density matrix, $\hat{\rho}^{(0)}(t)$, is stated as

$$\hat{\rho}^{(0)}(t) = G(t)\hat{\rho}(0) = \Theta(t)e^{-\left(\frac{i}{\hbar}\right)\hat{H}_0 t}\hat{\rho}(0)^{\left(\frac{i}{\hbar}\right)\hat{H}_0 t} \tag{A.1}$$

where $\hat{\rho}(0) = |g\rangle\langle g|$ is the density matrix of the system in the GS equilibrium (g), and $\Theta(t) = \int_{-\infty}^{t} d\tau \delta(\tau)$ is the Heaviside step-function ensuring causality.

In the perturbation scheme described in Sect. 2.1 (Eq. 5), the third-order density matrix is stated as

$$\hat{\rho}^{(3)}(t) = G(t)\hat{\rho}(0) + \left(\frac{i}{\hbar}\right)^3 \int_0^t d\tau_3 \int_0^{\tau_3} d\tau_2 \cdots \int_0^{\tau_2} d\tau_1$$

$$G(t - \tau_3)[\mathrm{H}'(\tau_3)G(\tau_3 - \tau_2)[\mathrm{H}'(\tau_2)G(\tau_2 - \tau_1)[\mathrm{H}'(\tau_1)G(\tau_1)\rho(0)]]] \tag{A.2}$$

Lindblad equation and population transfer

In the CGF approach, population transfer is assumed to arise from fast (and thus memoryless) bath fluctuations, i.e. characterized by rapidly decaying correlation functions. Population transfers can be included phenomenologically by adding fluctuation and dissipation terms to the Liouville–von Neumann equation (Eq. 4), leading to the Lindblad equation

$$\dot{\rho} = \frac{i}{\hbar}[\hat{H}, \hat{\rho}] + \sum_\alpha \left(\hat{V}_\alpha \hat{\rho} \hat{V}_\alpha^\dagger - \frac{1}{1} \hat{V}_\alpha \overset{\dagger\wedge}{V}_\alpha \hat{\rho} - \frac{1}{2} \hat{\rho} \hat{V}_\alpha \overset{\dagger\wedge}{V}_\alpha \right) \tag{A.3}$$

where $\hat{V}_\alpha$ are operators describing system–bath couplings in the most general form. Applying the secular approximation to the Green's function, i.e. discarding the fast oscillating coherence terms that decay before population transfer is activated, results in the Pauli master equation describing the population relaxation

$$\frac{d\rho_{ee}(t)}{dt} = - \sum_{e'} K_{ee,e'e'} \rho_{e'e'}(t) \tag{A.4}$$

where K is the rate matrix, with elements $K_{ee,e'e'}$ depicting the rate of population transfer from state e into state e'. The solution of the differential equation is formally given by the population Green's function

$$\rho_{e'e'}(t) = - \sum_e G_{e'e',ee}(t) \rho_{ee}(0) \tag{A.5}$$

, and the elements of the matrix $G_{e'e',ee}(t)$ act as time-dependent weighting factors in Liouville pathways, where populations evolve in the excited state, like those of ESAs and SEs (in Eq. 9).

Phase functions and line shape

The phase functions can assume different forms depending on the level of sophistication applied to describe the vibrational dynamics of the system (an overview is given in Refs. [2, 47]). The main building block is the line shape function $g_{ij}(t)$, which is the integral transformation of the autocorrelation function of bath fluctuations

$$g_{ij}(t) = \frac{1}{2\pi} \int \frac{C_{ij}(\omega)}{\omega^2} \left[\coth \left(\frac{\hbar\omega}{2k_B T} \right) (1 - \cos \omega t) + i \sin \omega t - i\omega t \right] d\omega \tag{A.6}$$

It can be obtained from MD simulations or in closed-form expressions derived from different models. For example, the homogeneous (anti-diagonal) broadening of the spectral signals arising due to coupling to a continuum of fast-decaying low-frequency modes can be expressed by the line shape function of the semi-classical Brownian oscillator (OBO) [124]:

$$g_{ij}^{OBO}(t) = \frac{\lambda_{ij}}{\Lambda}\left(\frac{2k_BT}{\hbar\Lambda} - i\right)(e^{-\Lambda t} + \Lambda t - 1) \tag{A.7}$$

where λ_{ij} and Λ are the system–bath coupling strength and fluctuation timescale, respectively.

References

1. Aue WP, Bartholdi E, Ernst RR (1976) 2-dimensional spectroscopy - application to nuclear magnetic-resonance. J Chem Phys 64(5):2229–2246
2. Mukamel S (1995) Principles of nonlinear optical spectroscopy. O.U.P, New York
3. Zanni MT, Hochstrasser RM (2001) Two-dimensional infrared spectroscopy: a promising new method for the time resolution of structures. Curr Opin Struct Biol 11(5):516–522
4. Jonas DM (2003) Two-dimensional femtosecond spectroscopy. Annu Rev Phys Chem 54:425–463
5. Cowan ML, Ogilvie JP, Miller RJD (2004) Two-dimensional spectroscopy using diffractive optics based phased-locked photon echoes. Chem Phys Lett 386(1–3):184–189
6. Brixner T, Mancal T, Stiopkin IV, Fleming GR (2004) Phase-stabilized two-dimensional electronic spectroscopy. J Chem Phys 121(9):4221–4236
7. Brixner T, Stenger J, Vaswani HM, Cho M, Blankenship RE, Fleming GR (2005) Two-dimensional spectroscopy of electronic couplings in photosynthesis. Nature 434(7033):625–628
8. Collini E, Wong CY, Wilk KE, Curmi PMG, Brumer P, Scholes GD (2010) Coherently wired light-harvesting in photosynthetic marine algae at ambient temperature. Nature 463(7281):644–U669
9. Mukamel S, Abramavicius D, Yang L, Zhuang W, Schweigert IV, Voronine DV (2009) Coherent multidimensional optical probes for electron correlations and exciton dynamics: from nmr to x-rays. Acc Chem Res 42(4):553–562
10. Mukamel S, Bakker HJ (2015) Preface: special topic on multidimensional spectroscopy. J Chem Phys 142(21):212101
11. Fuller FD, Ogilvie JP (2015) Experimental implementations of two-dimensional Fourier transform electronic spectroscopy. Annu Rev Phys Chem 66(66):667–690
12. Selig U, Schleussner C-F, Foerster M, Langhojer F, Nuernberger P, Brixner T (2010) Coherent two-dimensional ultraviolet spectroscopy in fully noncollinear geometry. Opt Lett 35(24):4178–4180
13. C-h Tseng, Matsika S, Weinacht TC (2009) Two-dimensional ultrafast Fourier transform spectroscopy in the deep ultraviolet. Opt Express 17(21):18788–18793
14. Varillas RB, Candeo A, Viola D, Garavelli M, De Silvestri S, Cerullo G, Manzoni C (2014) Microjoule-level, tunable sub-10 fs UV pulses by broadband sum-frequency generation. Opt Lett 39(13):3849–3852
15. Borrego-Varillas R, Oriana A, Ganzer L, Trifonov A, Buchvarov I, Manzoni C, Cerullo G (2016) Two-dimensional electronic spectroscopy in the ultraviolet by a birefringent delay line. Opt Express 24(25):28491–28499
16. Krebs N, Pugliesi I, Hauer J, Riedle E (2013) Two-dimensional Fourier transform spectroscopy in the ultraviolet with sub-20 fs pump pulses and 250–720 nm supercontinuum probe. N J Phys 15(8):085016
17. Baum P, Lochbrunner S, Riedle E (2004) Tunable sub-10-fs ultraviolet pulses generated by achromatic frequency doubling. Opt Lett 29(14):1686–1688

18. Prokhorenko VI, Picchiotti A, Maneshi S, Miller RJD (2015) Broadband electronic two-dimensional spectroscopy in the deep UV. Ultrafast Phenom Xix 162:432–435
19. Tseng C-H, Sandor P, Kotur M, Weinacht TC, Matsika S (2012) Two-dimensional Fourier transform spectroscopy of adenine and uracil using shaped ultrafast laser pulses in the deep UV. J Phys Chem A 116(11):2654–2661
20. West BA, Womick JM, Moran AM (2011) Probing ultrafast dynamics in adenine with mid-UV four-wave mixing spectroscopies. J Phys Chem A 115(31):8630–8637
21. West BA, Moran AM (2012) Two-dimensional electronic spectroscopy in the ultraviolet wavelength range. J Phys Chem Lett 3(18):2575–2581
22. Prokhorenko VI, Picchiotti A, Pola M, Dijkstra AG, Miller RJD (2016) New insights into the photophysics of DNA nucleobases. J Phys Chem Lett 7(22):4445–4450
23. Jiang J, Mukamel S (2011) Two-dimensional near-ultraviolet spectroscopy of aromatic residues in amyloid fibrils: a first principles study. Phys Chem Chem Phys 13(6):2394–2400
24. Jiang J, Golchert KJ, Kingsley CN, Brubaker WD, Martin RW, Mukamel S (2013) Exploring the aggregation propensity of gamma s-crystallin protein variants using two-dimensional spectroscopic tools. J Phys Chem B 117(46):14294–14301
25. Oliver TAA, Lewis NHC, Fleming GR (2014) Correlating the motion of electrons and nuclei with two-dimensional electronic–vibrational spectroscopy. Proc Natl Acad Sci 111(28):10061–10066
26. Loukianov A, Niedringhaus A, Berg B, Pan J, Senlik SS, Ogilvie JP (2017) Two-dimensional electronic Stark spectroscopy. J Phys Chem Lett 8(3):679–683
27. Kowalewski M, Fingerhut BP, Dorfman KE, Bennett K, Mukamel S (2017) Simulating coherent multidimensional spectroscopy of nonadiabatic molecular processes: from the infrared to the x-ray regime. Chem Rev 117(19):12165–12226
28. Scholes GD, Fleming GR, Chen LX, Aspuru-Guzik A, Buchleitner A, Coker DF, Engel GS, van Grondelle R, Ishizaki A, Jonas DM, Lundeen JS, McCusker JK, Mukamel S, Ogilvie JP, Olaya-Castro A, Ratner MA, Spano FC, Whaley KB, Zhu XY (2017) Using coherence to enhance function in chemical and biophysical systems. Nature 543(7647):647–656
29. Mukamel S (2000) Multidimensional femtosecond correlation spectroscopies of electronic and vibrational excitations. Annu Rev Phys Chem 51:691–729
30. Brixner T, Stiopkin IV, Fleming GR (2004) Tunable two-dimensional femtosecond spectroscopy. Opt Lett 29(8):884–886
31. Tian P, Keusters D, Suzaki Y, Warren WS (2003) Femtosecond phase-coherent two-dimensional spectroscopy. Science 300(5625):1553–1555
32. Grumstrup EM, Shim S-H, Montgomery MA, Damrauer NH, Zanni MT (2007) Facile collection of two-dimensional electronic spectra using femtosecond pulse-shaping technology. Opt Express 15(25):16681–16689
33. Maiuri M, Brazard J (2018) Electronic couplings in (bio-) chemical processes. Top Curr Chem 376(2):10
34. Brańczyk AM, Turner DB, Scholes GD (2014) Crossing disciplines - a view on two-dimensional optical spectroscopy. Ann der Phys 526(1–2):31–49
35. Son M, Schlau-Cohen GS (2017) Ultrabroadband 2D electronic spectroscopy as a tool for direct visualization of pathways of energy flow. Proc SPIE 10(1117/1112):2273417
36. Abramavicius D, Palmieri B, Voronine DV, Sanda F, Mukamel S (2009) Coherent multidimensional optical spectroscopy of excitons in molecular aggregates; quasiparticle versus supermolecule perspectives. Chem Rev 109(6):2350–2408
37. Nenov A, Rivalta I, Cerullo G, Mukamel S, Garavelli M (2014) Disentangling peptide configurations via two-dimensional electronic spectroscopy: Ab initio simulations beyond the Frenkel exciton Hamiltonian. J Phys Chem Lett 5(4):767–771
38. Johnson PJM, Farag MH, Halpin A, Morizumi T, Prokhorenko VI, Knoester J, Jansen TLC, Ernst OP, Miller RJD (2017) The primary photochemistry of vision occurs at the molecular speed limit. J Phys Chem B 121(16):4040–4047
39. Bruggemann B, Persson P, Meyer HD, Maya V (2008) Frequency dispersed transient absorption spectra of dissolved perylene: A case study using the density matrix version of the MCTDH method. Chem Phys 347(1–3):152–165
40. Sanda F, Mukamel S (2008) Stochastic Liouville equations for coherent multidimensional spectroscopy of excitons. J Phys Chem B 112(45):14212–14220
41. Tanimura Y (2006) Stochastic Liouville, Langevin, Fokker–Planck, and master equation approaches to quantum dissipative systems. J Phys Soc Jpn 75(8):082001

42. Zimmermann T, Vanicek J (2014) Efficient on-the-fly ab initio semiclassical method for computing time-resolved nonadiabatic electronic spectra with surface hopping or Ehrenfest dynamics. J Chem Phys 141(13):134102
43. Tempelaar R, van der Vegte CP, Knoester J, Jansen TLC (2013) Surface hopping modeling of two-dimensional spectra. J Chem Phys 138(16):164106
44. Richter M, Fingerhut BP (2016) Simulation of multi-dimensional signals in the optical domain: quantum-classical feedback in nonlinear exciton propagation. J Chem Theory Comput 12(7):3284–3294
45. Petit AS, Subotnik JE (2014) Calculating time-resolved differential absorbance spectra for ultrafast pump-probe experiments with surface hopping trajectories. J Chem Phys 141(15):154108
46. Mukamel S (1983) Non-impact unified theory of 4-wave mixing and 2-photon processes. Phys Rev A 28(6):3480–3492
47. Abramavicius D, Valkunas L, Mukamel S (2007) Transport and correlated fluctuations in the non-linear optical response of excitons. Epl 80(1):17005
48. Jiang J, Mukamel S (2010) Two-dimensional ultraviolet (2DUV) spectroscopic tools for identifying fibrillation propensity of protein residue sequences. Angew Chem Int Ed 49(50):9666–9669
49. Stuhldreier MC, Temps F (2013) Ultrafast photo-initiated molecular quantum dynamics in the DNA dinucleotide d(apg) revealed by broadband transient absorption spectroscopy. Faraday Discuss 163:173–188; discussion 243–175
50. Ostroumov EE, Mulvaney RM, Cogdell RJ, Scholes GD (2013) Broadband 2D electronic spectroscopy reveals a carotenoid dark state in purple bacteria. Science 340(6128):52–56
51. Dean JC, Rafiq S, Oblinsky DG, Cassette E, Jumper CC, Scholes GD (2015) Broadband transient absorption and two-dimensional electronic spectroscopy of methylene blue. J Phys Chem A 119(34):9098–9108
52. Rivalta I, Nenov A, Cerullo G, Mukamel S, Garavelli M (2014) Ab initio simulations of two-dimensional electronic spectra: The SOS//QM/MM approach. Int J Quantum Chem 114(2):85–93
53. Nenov A, Rivalta I, Mukamel S, Garavelli M (2014) Bidimensional electronic spectroscopy on indole in gas phase and in water from first principles. Comput Theor Chem 1040:295–303
54. Nenov A, Beccara SA, Rivalta I, Cerullo G, Mukamel S, Garavelli M (2014) Tracking conformational dynamics of polypeptides by nonlinear electronic spectroscopy of aromatic residues: a first-principles simulation study. ChemPhysChem 15(15):3282–3290
55. Nenov A, Segarra-Marti J, Giussani A, Conti A, Rivalta I, Dumont E, Jaiswal VK, Altavilla SF, Mukamel S, Garavelli M (2015) Probing deactivation pathways of DNA nucleobases by two-dimensional electronic spectroscopy: First principles simulations. Faraday Discuss 177:345–362
56. Rivalta I, Nenov A, Weingart O, Cerullo G, Garavelli M, Mukamel S (2014) Modelling time-resolved two-dimensional electronic spectroscopy of the primary photoisomerization event in rhodopsin. J Phys Chem B 118(28):8396–8405
57. Nenov A, Giussani A, Segarra-Martí J, Jaiswal VK, Rivalta I, Cerullo G, Mukamel S, Garavelli M (2015) Modeling the high-energy electronic state manifold of adenine: calibration for nonlinear electronic spectroscopy. J Chem Phys 142(21):212443
58. Nenov A, Mukamel S, Garavelli M, Rivalta I (2015) Two-dimensional electronic spectroscopy of benzene, phenol, and their dimer: an efficient first-principles simulation protocol. J Chem Theory Comput 11(8):3755–3771
59. Giussani A, Segarra-Martí J, Nenov A, Rivalta I, Tolomelli A, Mukamel S, Garavelli M (2016) Spectroscopic fingerprints of DNA/RNA pyrimidine nucleobases in third-order nonlinear electronic spectra. Theor Chem Acc 135(5):1–18
60. Li Q, Giussani A, Segarra-Martí J, Nenov A, Rivalta I, Voityuk AA, Mukamel S, Roca-Sanjuán D, Garavelli M, Blancafort L (2016) Multiple decay mechanisms and 2D-UV spectroscopic fingerprints of singlet excited solvated adenine-uracil monophosphate. Chem—A Eur J 22(22):7497–7507
61. Segarra-Marti I, Jaiswal VK, Pepino AJ, Giussani A, Nenov A, Mukamel S, Garavelli M, Rivalta I (2018) Two-dimensional electronic spectroscopy as a tool for tracking molecular conformations in DNA/RNA aggregates. Faraday Discuss 207:233–250
62. Kim J, Mukamel S, Scholes GD (2009) Two-dimensional electronic double-quantum coherence spectroscopy. Acc Chem Res 42(9):1375–1384
63. Li Z, Abrarnavicius D, Mukamel S (2008) Probing electron correlations in molecules by two-dimensional coherent optical spectroscopy. J Am Chem Soc 130(11):3509–3515

 Springer

64. Roos BO (1987) The complete active space self-consistent field method and its applications in electronic structure calculations. Adv Chemical Phys. https://doi.org/10.1002/9780470142943. ch7
65. Andersson K, Malmqvist PA, Roos BO, Sadlej AJ, Wolinski K (1990) 2nd-order perturbation-theory with a casscf reference function. J Phys Chem 94(14):5483–5488
66. Roca-Sanjuán D, Aquilante F, Lindh R (2012) Multiconfiguration second-order perturbation theory approach to strong electron correlation in chemistry and photochemistry. Wiley Interdiscip Rev-Comput Mol Sci 2(4):585–603
67. Malmqvist PA, Rendell A, Roos BO (1990) The restricted active space self-consistent-field method, implemented with a split graph unitary-group approach. J Phys Chem 94(14):5477–5482
68. Aquilante F, Lindh R, Pedersen TB (2007) Unbiased auxiliary basis sets for accurate two-electron integral approximations. J Chem Phys 127(11):114107
69. Aquilante F, Autschbach J, Carlson R, Chibotaru L, Delcey MG, De Vico L, Fernández Galvan I, Ferré N, Frutos LM, Gagliardi L, Garavelli M, Giussani A, Hoyer C, Li Manni G, Lischka H, Ma D, Malmqvist PA, Müller T, Nenov A, Olivucci M, Pedersen TB, Peng D, Plasser F, Pritchard B, Reiher M, Rivalta I, Schapiro I, Segarra-Martí J, Stenrup M, Truhlar DG, Ungur L, Valentini A, Vancoillie S, Veryazov V, Vysotskiy V, Weingart O, Zapata F, Lindh R (2016) Molcas 8: new capabilities for multiconfigurational quantum chemical calculations across the periodic table. J Comput Chem 37(5):506–541
70. Avila Ferrer FJ, Cerezo J, Stendardo E, Improta R, Santoro F (2013) Insights for an accurate comparison of computational data to experimental absorption and emission spectra: beyond the vertical transition approximation. J Chem Theory Comput 9(4):2072–2082
71. Serrano-Andrés L, Merchán M, Nebot-Gil I, Lindh R, Roos BO (1993) Towards an accurate molecular-orbital theory for excited-states—ethene, butadiene, and hexatriene. J Chem Phys 98(4):3151–3162
72. Lorentzon J, Malmqvist P-Å, Fülscher M, Roos BO (1995) A caspt2 study of the valence and lowest rydberg electronic states of benzene and phenol. Theor Chim Acta 91(1):91–108
73. Serrano-Andrés L, Roos BO (1996) Theoretical study of the absorption and emission spectra of indole in the gas phase and in a solvent. J Am Chem Soc 118(1):185–195
74. Barbatti M, Aquino AJA, Lischka H (2010) The UV absorption of nucleobases: semi-classical ab initio spectra simulations. Phys Chem Chem Phys 12(19):4959–4967
75. Clark LB, Peschel GG, Tinoco I (1965) Vapor spectra and heats of vaporization of some purine and pyrimidine bases1. J Phys Chem 69(10):3615–3618
76. Voet D, Gratzer WB, Cox RA, Doty P (1963) Absorption spectra of nucleotides, polynucleotides, and nucleic acids in the far ultraviolet. Biopolymers 1(3):193–208
77. Yamada T, Fukutome H (1968) Vacuum ultraviolet absorption spectra of sublimed films of nucleic acid bases. Biopolymers 6(1):43–54
78. Abouaf R, Pommier J, Dunet H (2003) Electronic and vibrational excitation in gas phase thymine and 5-bromouracil by electron impact. Chem Phys Lett 381(3–4):486–494
79. Platt JR (1949) Classification of spectra of cata-condensed hydrocarbons. J Chem Phys 17(5):484–495
80. Roos BO, Andersson K, Fülscher MP, Serrano-Andrés L, Pierloot K, Merchán M, Molina V (1996) Applications of level shift corrected perturbation theory in electronic spectroscopy. J Mol Struct 388:257–276
81. Roos BO, Andersson K (1995) Multiconfigurational perturbation-theory with level shift - the cr-2 potential revisited. Chem Phys Lett 245(2–3):215–223
82. Giussani A, Marcheselli J, Mukamel S, Garavelli M, Nenov A (2017) On the simulation of two-dimensional electronic spectroscopy of indole-containing peptides. Photochem Photobiol 93(6):1368–1380
83. Hamm P, Zanni M (2011) Concepts and methods of 2D infrared spectroscopy. Cambridge University Press, Cambridge
84. Camilloni C, Broglia RA, Tiana G (2011) Hierarchy of folding and unfolding events of protein G, CI2, and ACBP from explicit-solvent simulations. J Chem Phys 134(4):045105
85. Ezra FS, Lee CH, Kondo NS, Danyluk SS, Sarma RH (1977) Conformational properties of purine-pyrimidine and pyrimidine-purine dinucleoside monophosphates. Biochemistry 16(9):1977–1987
86. Crespo-Hernandez CE, Cohen B, Hare PM, Kohler B (2004) Ultrafast excited-state dynamics in nucleic acids. Chem Rev 104(4):1977–2019

87. Takaya T, Su C, Harpe KdL, Crespo-Hernández CE, Kohler B (2008) UV excitation of single DNA and RNA strands produces high yields of exciplex states between two stacked bases. Proc Natl Acad Sci 105:10285–10290
88. Ruzicka P, Kral T (2013) Pyrene: chemical properties, biochemistry applications and toxic effects. Chemistry research and applications. Nova Science Publishers, New York
89. Reichardt C (1994) Solvatochromic dyes as solvent polarity indicators. Chem Rev 94(8):2319–2358
90. Kwon J, Park SK, Lee Y, Lee JS, Kim J (2017) Tailoring chemically converted graphenes using a water-soluble pyrene derivative with a zwitterionic arm for sensitive electrochemiluminescence-based analyses. Biosens Bioelectron 87:89–95
91. Figueira-Duarte TM, Mullen K (2011) Pyrene-based materials for organic electronics. Chem Rev 111(11):7260–7314
92. Niko Y, Didier P, Mely Y, Konishi G, Klymchenko AS (2016) Bright and photostable push-pull pyrene dye visualizes lipid order variation between plasma and intracellular membranes. Sci Rep 6:18870
93. Nenov A, Giussani A, Fingerhut BP, Rivalta I, Dumont E, Mukamel S, Garavelli M (2015) Spectral lineshape in nonlinear electronic spectroscopy. Phys Chem Chem Phys 17:30925–30936
94. Petit AS, Subotnik JE (2014) How to calculate linear absorption spectra with lifetime broadening using fewest switches surface hopping trajectories: a simple generalization of ground-state Kubo theory. J Chem Phys 141(1):014107
95. Nemeth A, Milota F, Mancal T, Pullerits T, Sperling J, Hauer J, Kauffmann HF, Christensson N (2010) Double-quantum two-dimensional electronic spectroscopy of a three-level system: experiments and simulations. J Chem Phys 133(9):094505
96. Butkus V, Zigmantas D, Valkunas L, Abramavicius D (2012) Vibrational vs. Electronic coherences in 2D spectrum of molecular systems. Chem Phys Lett 545:40–43
97. Butkus V, Valkunas L, Abramavicius D (2012) Molecular vibrations-induced quantum beats in two-dimensional electronic spectroscopy. J Chem Phys 137(4):044513
98. Raytchev M, Pandurski E, Buchvarov I, Modrakowski C, Fiebig T (2003) Bichromophoric interactions and time-dependent excited state mixing in pyrene derivatives. A femtosecond broad-band pump-probe study. J Phys Chem A 107(23):4592–4600
99. Krebs N (2013) New insights for femtosecond spectroscopy: from transient absorption to 2-dimensional spectroscopy in the UV spectral domain. PhD Dissertation, Faculty of Physics, Ludwig-Maximilians-University Munich
100. Zhang WM, Meier T, Chernyak V, Mukamel S (1998) Exciton-migration and three-pulse femtosecond optical spectroscopies of photosynthetic antenna complexes. J Chem Phys 108(18):7763–7774
101. Meier T, Chernyak V, Mukamel S (1997) Femtosecond photon echoes in molecular aggregates. J Chem Phys 107(21):8759–8780
102. Neuwahl FVR, Foggi P (1999) Direct observation of s2–s1 internal conversion in pyrene by femtosecond transient absorption. Laser Chem 19(1–4):375–379
103. Foggi P, Pettini L, Santa I, Righini R, Califano S (1995) Transient absorption and vibrational relaxation dynamics of the lowest excited singlet state of pyrene in solution. J Phys Chem 99(19):7439–7445
104. Chen J, Zhang Y, Kohler B (2015) Excited states in DNA strands investigated by ultrafast laser spectroscopy. In: Barbatti M, Borin AC, Ullrich S (eds) Photoinduced phenomena in nucleic acids ii. Topics in current chemistry. Springer International Publishing, Berlin, vol 356, pp 39–87. https://doi.org/10.1007/128_2014_570
105. Bucher DB, Kufner CL, Schlueter A, Carell T, Zinth W (2016) UV-induced charge transfer states in DNA promote sequence selective self-repair. J Am Chem Soc 138(1):186–190
106. Schreier WJ, Gilch P, Zinth W (2015) Early events of DNA photodamage. Annu Rev Phys Chem 66(1):497–519
107. Vayá I, Gustavsson T, Douki T, Berlin Y, Markovitsi D (2012) Electronic excitation energy transfer between nucleobases of natural DNA. J Am Chem Soc 134(28):11366–11368
108. Cadet J, Grand A, Douki T (2015) Solar UV radiation-induced DNA bipyrimidine photoproducts: Formation and mechanistic insights. In: Barbatti M, Borin AC, Ullrich S (eds) Photoinduced phenomena in nucleic acids ii. Topics in current chemistry. Springer International Publishing, Berlin, vol 356, pp 249–275. doi:10.1007/128_2014_553
109. Cadet J, Mouret S, Ravanat J-L, Douki T (2012) Photoinduced damage to cellular DNA: direct and photosensitized reactions†. Photochem Photobiol 88(5):1048–1065

110. Pecourt JML, Peon J, Kohler B (2001) DNA excited-state dynamics: ultrafast internal conversion and vibrational cooling in a series of nucleosides. J Am Chem Soc 123(42):10370–10378
111. Peon J, Zewail AH (2001) DNA/RNA nucleotides and nucleosides: Direct measurement of excited-state lifetimes by femtosecond fluorescence up-conversion. Chem Phys Lett 348(3–4):255–262
112. Onidas D, Markovitsi D, Marguet S, Sharonov A, Gustavsson T (2002) Fluorescence properties of DNA nucleosides and nucleotides: a refined steady-state and femtosecond investigation. J Phys Chem B 106(43):11367–11374
113. Hare PM, Crespo-Hernández CE, Kohler B (2007) Internal conversion to the electronic ground state occurs via two distinct pathways for pyrimidine bases in aqueous solution. Proc Natl Acad Sci 104:435–440
114. Pepino AJ, Segarra-Martí J, Nenov A, Improta R, Garavelli M (2017) Resolving ultrafast photoinduced deactivations in water-solvated pyrimidine nucleosides. J Phys Chem Lett 8(8):1777–1783
115. Segarra-Martí J, Garavelli M, Aquilante F (2015) Multiconfigurational second-order perturbation theory with frozen natural orbitals extended to the treatment of photochemical problems. J Chem Theory Comput 11(8):3772–3784
116. Vogiatzis KD, Li Manni G, Stoneburner SJ, Ma D, Gagliardi L (2015) Systematic expansion of active spaces beyond the casscf limit: a gasscf/splitgas benchmark study. J Chem Theory Comput 11(7):3010–3021
117. Li Manni G, Carlson RK, Luo S, Ma D, Olsen J, Truhlar DG, Gagliardi L (2014) Multiconfiguration pair-density functional theory. J Chem Theory Comput 10(9):3669–3680
118. Casida ME, Huix-Rotllant M (2012) Progress in time-dependent density-functional theory. Annu Rev Phys Chem 63:287–323
119. Dreuw A, Wormit M (2015) The algebraic diagrammatic construction scheme for the polarization propagator for the calculation of excited states. Wiley Interdiscip Rev 5(1):82–95
120. Sneskov K, Christiansen O (2012) Excited state coupled cluster methods. Wiley Interdiscip Rev 2(4):566–584
121. Chan GK-L, Sharma S (2011) The density matrix renormalization group in quantum chemistry. Annu Rev Phys Chem 62(1):465–481
122. Thomas RE, Sun Q, Alavi A, Booth GH (2015) Stochastic multiconfigurational self-consistent field theory. J Chem Theory Comput 11(11):5316–5325
123. Segarra-Martí J, Zvereva E, Marazzi M, Brazard J, Dumont E, Assfeld X, Haacke S, Garavelli M, Monari A, Léonard J, Rivalta I (2018) Resolving the singlet excited state manifold of benzophenone by first-principles simulations and ultrafast spectroscopy. J Chem Theory Comput 14(5):2570–2585
124. Li BL, Johnson AE, Mukamel S, Myers AB (1994) The Brownian oscillator model for solvation effects in spontaneous light-emission and their relationship to electron-transfer. J Am Chem Soc 116(24):11039–11047

 Springer

Top Curr Chem (Z) (2017) 375:86
https://doi.org/10.1007/s41061-017-0172-1

REVIEW

Ultrafast structural molecular dynamics investigated with 2D infrared spectroscopy methods

Jan Philip Kraack[1]

Received: 31 March 2017 / Accepted: 2 October 2017 / Published online: 25 October 2017
© Springer International Publishing AG 2017

Abstract Ultrafast, multi-dimensional infrared (IR) spectroscopy has been advanced in recent years to a versatile analytical tool with a broad range of applications to elucidate molecular structure on ultrafast timescales, and it can be used for samples in a many different environments. Following a short and general introduction on the benefits of 2D IR spectroscopy, the first part of this chapter contains a brief discussion on basic descriptions and conceptual considerations of 2D IR spectroscopy. Outstanding classical applications of 2D IR are used afterwards to highlight the strengths and basic applicability of the method. This includes the identification of vibrational coupling in molecules, characterization of spectral diffusion dynamics, chemical exchange of chemical bond formation and breaking, as well as dynamics of intra- and intermolecular energy transfer for molecules in bulk solution and thin films. In the second part, several important, recently developed variants and new applications of 2D IR spectroscopy are introduced. These methods focus on (i) applications to molecules under two- and three-dimensional confinement, (ii) the combination of 2D IR with electrochemistry, (iii) ultrafast 2D IR in conjunction with diffraction-limited microscopy, (iv) several variants of non-equilibrium 2D IR spectroscopy such as transient 2D IR and 3D IR, and (v) extensions of the pump and probe spectral regions for multi-dimensional vibrational spectroscopy towards mixed vibrational-electronic spectroscopies. In light of these examples, the important open scientific and conceptual questions with regard to intra- and intermolecular dynamics are highlighted. Such

Chapter 4 was originally published as Kraack, J. P. Top Curr Chem (Z) (2017) 375: 86. https://doi.org/10.1007/s41061-017-0172-1.

✉ Jan Philip Kraack
philip.kraack@chem.uzh.ch

1 Department of Chemistry, University of Zürich, Winterthurerstrasse 190, 8057 Zurich, Switzerland

  Springer

questions can be tackled with the existing arsenal of experimental variants of 2D IR spectroscopy to promote the understanding of fundamentally new aspects in chemistry, biology and materials science. The final part of the chapter introduces several concepts of currently performed technical developments, which aim at exploiting 2D IR spectroscopy as an analytical tool. Such developments embrace the combination of 2D IR spectroscopy and plasmonic spectroscopy for ultrasensitive analytics, merging 2D IR spectroscopy with ultra-high-resolution microscopy (nanoscopy), future variants of transient 2D IR methods, or 2D IR in conjunction with microfluidics. It is expected that these techniques will allow for groundbreaking research in many new areas of natural sciences.

Keywords Ultrafast 2D IR spectroscopy · Molecular structure · Spectral diffusion · Energy transfer · Chemical exchange · Vibrational coupling · Surface spectroscopy · Transient 2D IR spectroscopy · 2D IR microscopy · 2D IR electrochemistry · 2D IR Nanoscopy

1 Introduction

1.1 Ultrafast Infrared Vibrational Spectroscopy

Time-resolved infrared (IR) vibrational spectroscopy is a powerful method for tackling a large number of open scientific questions from chemical, biological and physical perspectives [1–11]. IR molecular spectroscopy has a high level of chemical specificity, which leads to the possibility of straightforward sample analysis based on characteristic absorption patterns from a broad range of functional groups (ca. 600–4000 cm^{-1}). Vibrational frequencies, peak intensities, band shapes and widths of IR transitions carry manifold information on molecular structure and intermolecular interactions, thus making IR spectroscopy an incredibly valuable tool for natural scientists. Adding a high time-resolution to IR spectroscopy even expands the benefits, since this allows resolving dynamic and kinetic changes in sample constitutions. Nowadays, a vast range of time scales can be investigated by pulsed IR spectroscopy, ranging from femtoseconds to milliseconds or longer. In particular, the field of ultrafast IR spectroscopy deals with vibrational dynamics predominately on the femto- to nanosecond timescale and has seen a considerable technological development in the last approximately 20 years. Consequently, methods from this field are now an almost routinely available analytical tool. Ultrafast IR spectroscopy allows the direct observation of dynamics on timescales that closely match molecular motions in real time. This makes it possible to observe directly the dynamics of vibrational relaxation, interactions of molecules with their environment (e.g. solvation), vibrational dephasing or the dynamics of vibrational echoes [8, 12–14]. An often invoked drawback of IR spectroscopy is, however, that the absorption coefficients of a large range of functional groups are rather low ($< 1000\ M^{-1}\ cm^{-1}$). Such weak absorbance values considerably limit the obtainable signal strengths. Despite that undeniable limitation, constant progress in experimental developments have made it possible to increase signal to noise ratios in a way that time-resolved IR spectroscopy can now even be applied to study monolayer thin samples at two dimensional interfaces of weakly absorbing chromophores [15]. To expand the capabilities of time-resolved IR spectroscopy, it was suggested approximately 20 years ago to advance ultrafast vibrational spectroscopy to a two-dimensional (2D) version [16]. This extension was designed to reveal even correlated dynamics of functional groups.

1.2 Multi-Dimensional IR Spectroscopy

Multi-dimensional optical molecular spectroscopy is generally widely known to exhibit the prominent benefit of resolving correlations and interactions between different resonances [1, 10, 17, 18]. That concept has been borrowed from well-established nuclear magnetic resonance (NMR) spectroscopy, where spreading the signals into two or more frequency dimensions allows the determination of coupled spins and the deconstruction of congested spectra [19–23]. Sequences of ultrashort laser pulses from different frequency ranges have been devised recently to allow establishing correlations of different types of molecular resonances, including also electronic and vibrational transitions in different types of samples. Up to now, the development of ultrafast multi-dimensional optical spectroscopy has led to a situation where a schematic frequency-frequency correlation map (Fig. 1) is almost completely covered by different versions of 2D spectroscopy. These methods range from the terahertz (2D THz, < 600 cm^{-1}) over the mid-IR (600–4000 cm^{-1}) up to the ultraviolet-visible (UV/VIS) spectral region ($< 45,000$ cm^{-1}) for 2D electronic spectroscopy (ES). It is noted in the context of Fig. 1 that the present chapter deals predominately with 2D IR spectroscopy. 2D IR can be applied nowadays over the entire spectral range that is generally used for analytical purposes to determine molecular structure (600–4000 cm^{-1}). Therefore, nearly every IR-active functional

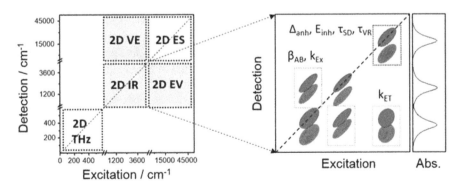

Fig. 1 Overview of currently available spectral ranges for ultrafast two-dimensional spectroscopy. 2D terahertz (THz), 2D infrared (IR) and 2D electronic spectroscopy (ES) all contain diagonal and off-diagonal contributions. Newer variants of mixed electronic-vibrational (EV) and vibrational-electronic (VE) spectroscopy contain exclusively cross peak signals. There are currently no mixed THz-IR/ES methods available. Note the breaks in the spectral axes. The spectral ranges covered in this chapter are shaded in gray

group can be investigated with this method. As an advantage of IR spectroscopy for structure determination, the observed bands are rather sharp and defined, whereas THz and electronic resonances often appear broad and featureless. Very recently also methods have been realized that allowed the investigation of correlated vibrational (V) and electronic (E) resonances. These techniques are consequently termed 2D EV and 2D VE spectroscopy, and they are especially promising for the elucidation of bio-molecular vibrational dynamics. Therefore, the spectral region of electronic excitations ($> 8000 \ cm^{-1}$) is also briefly covered in this chapter.

2D IR spectra are frequency-frequency correlation maps as schematically shown in the right part of Fig. 1. To construct the spectra, an initial excitation with IR light is used as a perturbation of a sample system, and its response is correlated to this initial event. The two corresponding frequency axes are, therefore, generally referred to as "excitation" and "detection frequencies" and the IR signals are spread in both dimensions. This spreading of the signals can generate complex shapes from the observed resonances (e.g. ellipses, circles or others) [10]. The shapes allow a more sophisticated interpretation of the properties of the resonance compared to linear absorption spectra. Many of the signals fall on the diagonal line in a 2D IR spectrum and are thus related to linear IR absorbance (right panel). The induced 2D IR signals come in pairs (blue/red) due to the anharmonicity of the vibrational potential under study (Δ_{anh}) and the induced nature of the differential signal (excitation). The ellipticity (E_{inh}) moreover reports on the degree of correlation in the resonance and an underlying spectral inhomogeneity. Importantly, also off-diagonal contributions can exist in a 2D IR spectrum, which may report on interactions between different resonances. Functional groups of molecules can interact in many different ways, e.g. via dipolar vibrational coupling (associated with a coupling constant β_{AB}). This coupling strongly depends on the distance of the interacting modes and is thus very sensitive to molecular structure (see Sect. 3.1.1). As a particularly powerful approach, a combination of experimental data and theoretical predictions can be used for almost quantitative structure modelling [24, 25].

Adding a time delay between excitation and detection events in 2D IR allows one to resolve different types of dynamics from the sample in a thermally equilibrated electronic ground state (see Sect. 2). The most straightforward information that can be obtained is the dynamics of vibrational relaxation (τ_{VR}), or the dynamics of spectral correlation via for instance spectral diffusion (τ_{SD}). Moreover, once excited, an excess vibrational energy in a vibrational bond can be transferred to another oscillator with a rate constant k_{ET}, which is characteristic of the sample system under study. Additionally, it is possible that the sample undergoes a chemical reaction following the initial excitation. Such a reaction can influence the vibrational properties of certain functional groups and the dynamics of so-called chemical exchange can be investigated by looking at the rate constants (k_{Ex}), with which off-diagonal signals evolve. These signals also strongly depend on the molecular structure of the sample and can, therefore, report on the dynamic evolution of sample structure. Finally, 2D IR spectroscopy can be combined with different types of pre-excitation methods (see Sect. 4.4). This allows one to address non-equilibrium dynamics, e.g. photo-induced dynamics of chemical bonds or dynamics from excited electronic states. It is clear that 2D IR spectroscopy can be used to obtain all the described information, which is often difficult to obtain by other experimental or analytical methods.

1.3 Scientific Impact of 2D IR Spectroscopy

2D IR spectroscopy has become a versatile analytical tool for bond-specific molecular structure determination on ultrafast timescales. Its inherent ability to resolve molecular dynamics of often less than a picosecond allows obtaining an unprecedented view of structural changes caused by direct rearrangements of chemical bonds. This advantage is based on the comparatively fast fluctuations that govern IR transitions, often resulting in dephasing times of about picoseconds or less. Similar to electronic spectroscopy of condensed phase systems, such fast dephasing times broaden the involved resonances at the expense of inherent spectral resolution. 2D IR can thus be viewed a complementary approach to the standard and widespread analytical techniques for structure determination such as X-ray crystallography and different forms of NMR spectroscopy. Especially the latter method exhibits dephasing times that are orders of magnitude slower (milliseconds) than those of IR transitions, which intrinsically limits the possible temporal resolution, at least regarding non-equilibrium dynamics.

To put 2D IR in a context for analytical chemistry methods, the technique can in principle be applied in two different ways. One variant uses a single 2D IR spectrum, generally at one very initial delay of vibrational relaxation, to correlate several vibrational resonances in a sample, just in the same way as different forms of 2D NMR do. Coupling between vibrational modes, evidenced by cross peaks in the spectrum, and the corresponding frequencies of diagonal and cross peaks are then the primary observables, from which molecular structure can be deduced. Couplings can exist as intra- and intermolecular interactions, which can occur "through bond" or "through space" [10]. One must keep in mind, however, that even in case of intramolecular couplings, these interactions are comparatively short-ranged, i.e. they occur on sub-nm distances. Although this is a typical value also often encountered for interactions in case 2D NMR spectroscopy, the difference is that in case of 2D IR cross peaks, the couplings drastically depend also on the strength transition dipole moments, which are generally low for IR transitions. Therefore, such couplings are often weak and difficult to resolve. In addition to measuring one isolated 2D IR spectrum, systematic variation of sample parameters at macroscopic time scales (milliseconds to hours or even days and weeks) then report on structural changes in the sample that are significantly slower compared to vibrational relaxation (picoseconds to nanoseconds).

In the second way of performing 2D IR spectroscopy for structure determination, a full series of 2D IR spectra is recorded throughout the vibrational lifetime of the sample. This variant fully exploits the striking advantage of laser-based optical spectroscopy methods in that they can operate on time scales down to the femtosecond regime. The ultrafast temporal resolution allows the direct observation of time-dependent structure and real time molecular dynamics, such as vibrations or the formation and breaking of a chemical bond. In that way, one determines the dynamic shapes of the peaks and amplitudes of diagonal and cross peaks to obtain structural information. It is important to note that such contributions do not exclusively stem from the sample molecules alone. Rather, the dynamics also originate from intermolecular interactions with its local environment, e.g. a solvent. In other words, 2D IR spectroscopy not only measures the dynamics of some sample molecules it also reports on the ultrafast dynamic fluctuations of the direct environment. If the sample molecule is too rigid to structurally fluctuate by itself on ultrafast time scales, then the 2D IR spectra can be interpreted as a snapshot of the dynamic environment. In that way 2D IR spectroscopy directly accesses an intermediate regime between purely homogeneous and purely heterogeneous systems by allowing one to observe a "time-dependent" heterogeneity of a sample down to the sub-picosecond timescale and thus gaining insight into line-broadening mechanisms. This approach is in contrast to most NMR methods, which generally probe time-averaged structure of molecules with an intrinsic temporal resolution of milliseconds. It is noteworthy, however, that different approaches exist also for NMR to go beyond that intrinsic temporal resolution, and to obtain information from time scales about picoseconds [26–28]. These methods are rather specialized and ultrafast laser spectroscopy constitutes the best and most direct way to access that temporal range and the associated dynamics.

2D IR spectroscopy exhibits a couple of other advantages over 2D NMR spectroscopy or X-ray crystallography regarding molecular structure determination. NMR spectroscopy often requires sample concentrations in the mM regime to obtain reasonable signals. However, many samples, and especially bio-molecules such as proteins, tend to aggregate under these conditions, thus making structure elucidation of the monomers challenging. Biomolecules are often also difficult to crystallize, what makes X-ray structure determination difficult in some cases. In addition, X-ray techniques do not resolve the molecular structure under fully solvated conditions. 2D IR spectroscopy allows circumventing such problems by its inherent ability to perform measurements at even very low concentrations (sub mM) of bulk solution samples [29]. As another advantage, isotope-labelling is widely known to provide rich structural information, especially in large molecules such as proteins [30]. That approach has become popular in order to overcome a comparatively low selectivity of IR spectroscopy (as compared to NMR) with respect to a decisive functional group if many residues exhibit similar transition frequencies, e.g. amide modes in proteins. Increasing the sensitivity even further and adding the possibility of spatial resolution, 2D IR spectroscopy can be combined with different forms of optical near field spectroscopy and micro- or even nanoscopy. This is done by coupling the incident light to plasmonically active nanostructures [31]. In a similar context, 2D IR spectroscopy can be applied to study a vast range of different samples. Variants exist that measure 2D IR spectra from small and large molecules in bulk solution, in solid-state samples such as amorphous powders or crystals, in biological membranes or from molecular monolayers at different types of interfaces.

Exploiting the properties of ultrashort pulses from different light sources, 2D IR can also optimally be extended by different other frequency ranges from optical spectroscopy. That includes the entire currently accessible range of laser light sources from the THz to the UV/VIS range (Fig. 1). Such an approach allows 2D IR to be performed in a transient manner on non-equilibrium dynamics, which significantly increases the temporal range of applicability by an initial light-induced perturbation. Using such light-triggered changes in the sample conditions in conjunction with synchronized laser systems thus bridges the gap between the quasi-static and the ultrafast 2D IR experiments and allows measurements on timescales over many orders of magnitude from femtoseconds to milliseconds and beyond.

Taking everything together, it is thus not astonishing that 2D IR has been developed to a point where it is considered as an advantageous method for molecular structural characterization, and many examples exist in the literature, where for instance secondary structure of proteins under fully solvated conditions has been determined [1, 24, 30].

1.4 Scope of this Chapter

Numerous overviews about 2D IR spectroscopy have been presented over the last approximately 18 years, each focusing either on selected aspects of the method, providing a brief overview, or giving a detailed theoretical description of the signals, as well as on technical aspects [10, 15, 17, 18, 30, 32–53]. In particular, detailed theoretical descriptions of time-resolved spectroscopy in general, and multi-dimensional optical spectroscopy in particular, are available elsewhere [1, 10, 18, 48, 54, 55]. This chapter attempts to (i) give the non-specialist reader a balanced summary of what is currently possible with 2D IR spectroscopy, (ii) indicate areas where shortcomings impose significant challenges for the technique to be applied and (iii) outline future research directions, which make the applicability of 2D IR spectroscopy even broader. The text uses some examples of "classic" 2D IR spectroscopy for ultrafast molecular structure determination, and in large parts additionally presents most recently reported highlights, where the method has been applied for addressing certain scientific questions, and to obtain molecular information that is difficult or even impossible to retrieve otherwise. Recent developments aim at making 2D IR spectroscopy available for analysis of molecules at surfaces, applications to samples in confined environments, combinations with microscopy, as well as different types of combinations of vibrational and electronic spectroscopy for the elucidation of non-equilibrium dynamics. From the status of routinely performable experiments, a full series of groundbreaking investigations can be envisioned, which will allow chemists, physicists and material scientists to obtain unprecedented insight into molecular structure from different perspectives. Some of these examples are indicated in the outlook of this chapter.

2 Essential Elements of 2D IR Spectroscopy

2D IR is a pump probe type of nonlinear spectroscopy, for which both the pump as well as the probe pulse are spectrally resolved to generate frequency-frequency correlation plots (Fig. 1) [16]. The signals are generated from a set of different

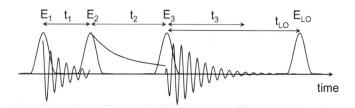

Fig. 2 Pulse sequence for third-order 2D IR spectroscopy. $E_{1,2,3}$ are excitation fields of a third-order nonlinear polarization that is emitted from the sample. Oscillatory lines correspond to vibrational coherences in the sample. E_{LO} is a local oscillator field used for heterodyne detection of the signal. t_{1-3} are time-delays of the successive light-matter interactions. Reprinted with permission from Ref. [15]. Copyright American Chemical Society (2016)

ultrashort laser pulses, which initially excite a molecular vibration and successively interrogate the evolution of the initial excitation by the help of a time-delayed probe pulse. Figure 2 shows a possible pulse sequence for 2D IR spectroscopy in a coherent variant, which exploits three laser pulses for excitation and probing (E_{1-3}), separated by variable time delays (t_{1-3}), as well as a fourth pulse, a so-called local oscillator (LO), which is used to resolve the signal field in its amplitude and phase, rather than its intensity (heterodyne detection) [10, 56, 57]. There exist several experimental implementations of 2D IR spectroscopy, which mainly differ in the number of laser beams that are exploited and the way in which the signal field is generated [10, 15, 42, 49, 58–60]. These methods exhibit generally different degrees of experimental complexity, data acquisition times and signal-to-noise levels. The pros and cons of the individual implementations have been discussed before in detail and the reader is referred to references [10, 15, 42, 49, 58–60]. All existing methods of 2D IR spectroscopy have, however, in common that three light-matter interactions (E_{1-3}) successively generate a third-order polarization in the sample ($P^{(3)}$), which is the source of a signal field that is emitted from the sample towards a detector.

Let the laser pulses act on a hypothetical sample with a molecular vibration that is described by a qualitative ground state potential as depicted in Fig. 3a. The colored arrows indicate possible transitions between the different vibrational levels

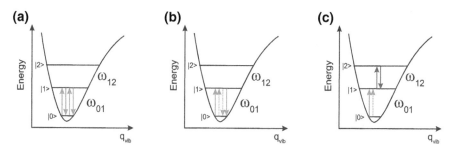

Fig. 3 Simplified electronic ground state potential relevant for ultrafast 2D IR spectroscopy and the most important signal contributions. **a** Ground-state bleach signal, **b** stimulated emission signal, and **c** excited state absorption signal

upon successive excitation and the color-coding will become clear in what follows. In the pulse sequence shown in Fig. 2, the first pulse (E_1) acts on the sample and induces a vibrational coherence between the $|0\rangle$ and $|1\rangle$ levels (oscillatory line). Followed by a certain time t_1, the so-called coherence time, a second laser pulse (E_2) converts this coherence into a population state, e.g. in $|1\rangle$, which afterwards decays during vibrational relaxation taking place during the waiting time t_2, the so-called population time. The evolution of the system is interrogated by the help of a third field (E_3), which again induces a vibrational coherence that oscillates during the time t_3 and that may be generated between levels $|0\rangle$ and $|1\rangle$, or $|1\rangle$ and $|2\rangle$. The local oscillator (E_{LO}) does not interact with the sample and is only used for heterodyne detection.

To generate two frequency axes for the 2D IR data, two Fourier transformations are necessary for obtaining the "pump" and the "probe" axis. The first Fourier transformation is generally performed by a detector, on which the signal light is spectrally resolved. This generates the probe spectral axis, just as in ordinary transient absorption spectroscopy. To obtain the pump spectral axis, the delay t_1 between the first two pulses is successively scanned, thereby mapping out the evolution of the coherence generated by the first field interaction. If the fields E_1 and E_2 have a fixed phase relation, such a scan results in an oscillatory signal along the t_1 axis, which, after Fourier transformation yields the pump spectral axis [10, 49, 61].

In a more basic physical interpretation, the third-order time-domain polarization $P^{(3)}(t)$ induced in the sample is the source of the signal light and can be described by a convolution of the external electric fields with the samples third-order response function $R^{(3)}(t)$ (Eqs. 1 and 2).

$$P^{(3)}(t) \propto \int_0^\infty dt_3 \int_0^\infty dt_2 \int_0^\infty dt_1 \sum_n R_n^{(3)}(t_3, t_2, t_1) E_3(t - t_3)$$

$$\cdot E_2(t - t_3 - t_2) E_1(t - t_3 - t_2 - t_1) \tag{1}$$

with

$$R^{(3)}(t) = \sum_n R_n^{(3)}(t_3, t_2, t_1) \propto -i\langle \hat{\mu}(t_3 + t_2 + t_1)[\hat{\mu}(t_2 + t_1), [\hat{\mu}(t_1), [\hat{\mu}(0), \rho(-\infty)]]]\rangle \tag{2}$$

That third-order time-domain response function is the quantity that contains all the relevant information about the sample, and which one is ultimately interested in. In brief, its properties are based on the temporal evolution of the density matrix, starting from thermal equilibrium, $\rho(-\infty)$, after successive interactions with the external fields and the dipole operators $\hat{\mu}_i$ [10, 62]. Taking into account a vast range of possible environments and a distribution of influences, the ensemble average is taken into account by evaluating the trace over the density matrix $\langle...\rangle$. After expansion of the three commutators, the response function contains a series of elements, which account for the total signal [10, 62, 63]. Here, exemplary energy level diagrams are used to visualize these contributions (Fig. *3*). Detailed descriptions of such diagrams can be found in Refs. [10, 62–64].

In brief, the four arrows represent four light-matter interactions in a hypothetical vibrational three-level system of an anharmonic mode (q_{vib}) in an electronic ground state potential. The blue/red color-coding of the arrows is intended to visualize the interactions between different vibrational levels. Some of the elements of the response function involve only transitions between the ground state $|0\rangle$ and the first excited state $|1\rangle$ and are termed ground state bleach (Fig. 3a) and stimulated emission (Fig. 3b). In addition, elements exist that involve transitions from the first to the second excited state $|2\rangle$. These elements are therefore referred to excited state absorption (Fig. 3c). Note that the exemplarily depicted energy level diagrams in Fig. 3 only represent parts of the total amount of signal contributions. Generally, the total response contains contributions, for which either a vibrational echo is emitted from the sample (rephasing pathways) or not (non-rephasing pathways). The latter group is also sometimes denoted as "virtual echo" contributions [65]. A final 2D IR spectrum, generally referred to as a "fully absorptive" spectrum, will contain all different contributions. Several reports on 2D IR spectroscopy exist, where rephasing and non-rephasing contributions have been plotted separately; however, these representations are not widely used anymore in the recent literature [10, 66, 67]. An important aspect for 2D IR spectroscopy is the involvement of four different light-matter interactions (Eq. 2), each of which is associated with a transition-dipole moment (the fourth and last interaction is the emission of the signal light). As a result, the signal scales with the absorption coefficient squared as opposed to linear (e.g. FT IR) spectroscopy, which contains only two light matter interactions [10, 68]. This property of nonlinear spectroscopy can be used to enhance the contrast of 2D IR spectroscopy over linear methods and to help interpreting congested spectra (see Sects. 3.1.2 and 4.2).

2.1 Molecular Information from Diagonal and Off-Diagonal Peaks in 2D IR Spectra

2.1.1 Diagonal Peaks

2D IR spectra contain significantly more information about the sample compared to linear FT IR spectroscopy. Besides the aspect of ultrafast temporal resolution, valuable information is already contained in the observation of pairs of signals for each vibrational transition that is excited. These pairs are associated with ground state bleach/stimulated emission (blue, Fig. 3a) and excited state absorption (red, Fig. 3a), respectively. The origin of the two contributions is the anharmonicity of the electronic ground state potential, which gives rise to frequency-shifted, higher-lying transitions (Fig. 4a). Thus, the frequency separation of the blue and red peaks in a 2D IR spectrum can be used to retrieve the value of the anharmonicity. However, the frequency separation of the two bands does not automatically give the anharmonicity value. If the width of the band is larger than the anharmonic shift, positive and negative contributions add in the overlapping region and the anharmonicity has to be obtained by fitting lineshape functions to the bands.

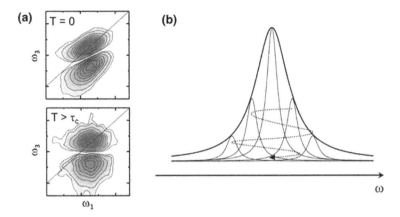

Fig. 4 **a** Examples of 2D IR spectra for a bulk solution sample of KSCN in water for different population delays. Correlation is lost for population delays long compared to the intrinsic correlation time τ_c. **b** The principle of spectral diffusion in 2D IR spectroscopy, which explains the change of the shapes of the peaks in a 2D IR spectra in **a** as a function of the population delay

As one of the most powerful aspects, 2D IR spectra at a series of population delays can report on the dynamics of the sample. Variations of the sample conditions (e.g., solvent, temperature, pressure) can then be used to elucidate the influence of the environment on the ultrafast dynamics. The most obvious information that can be obtained from the intensity of the peaks is the dynamics of vibrational relaxation. After pumping the vibrational transition, vibrational relaxation diminishes the signals over time due to repopulation of the ground state $|0\rangle$ from the excited state $|1\rangle$. However, as shown in Fig. 4a, also the shapes of the peaks evolve with increasing population delays. Starting from fairly elongated signals in the direction of the diagonal line at very initial delays, the spectra become rounder over time, and their widths along the anti-diagonal line changes. Under certain assumptions, the anti-diagonal linewidth at initial population delays reports on the homogeneous linewidth of the transition, whereas the diagonal linewidth relates to the total linewidth that is observed in an FT IR experiment after correcting for the different dependencies on the transition dipole moments [10]. The temporal change of the 2D IR lineshapes is termed spectral diffusion and is based on the dynamic interconversion of different environments in the sample. Spectral diffusion can report on the origin and dynamics of line broadening of IR transitions. In a simple picture, the elongated lineshapes in 2D IR spectra resemble correlation between pump and probe frequencies in the transitions. Completely elongated lines along the diagonal, therefore, indicate full correlation, which means that the frequency resolved pump interaction effectively selects a certain subset of oscillators under the broadened transition (origin of the dashed arrow in Fig. 4b). Over the course of vibrational relaxation, the initially pumped molecules lose their memory of the initial frequency and interconvert to other possible frequencies under the envelope of the IR transition. As this holds for all different combinations of excited molecules, the shapes of the signals become round and uncorrelated at delays that are longer than a characteristic correlation time (τ_c). The correlation is generally expressed in terms of a two time-point time-dependent frequency-frequency correlation function (FFCF, $C_2(t)$, Eq. 3), which in most cases is taken to follow the form of a sum of exponential functions, weighted with amplitudes Δ_i. However, it is important to note that this assumption is not rigorously valid for all samples and much more complicated dynamics can exist.

$$C_2(t_2) = \langle \delta\omega_{01}(t_2)\delta\omega_{01}(0)\rangle \propto \sum_i \Delta_i \exp\left(-t_2/\tau_{c,i}\right) \tag{3}$$

Different parameters of 2D IR signals can be used to obtain a measure that is directly proportional to the FFCF, e.g. the nodal slope between the excited state absorption and ground state bleach signals, the ellipticity of the signals, the center-line slope (CLS) [10, 69–71]. Which method is the best to characterize the dynamics should be evaluated on a case-by-case basis to be as accurate as possible [69–71].

Spectral diffusion can take place on time scales of a few tens of femtoseconds up to hundreds of picoseconds and is thus directly addressable with 2D IR spectroscopy. Next to the frequently observed exponential behavior of the FFCF, sometimes quasi-static contributions are determined experimentally, that is, contributions which have correlation times much longer than the experimentally accessible temporal range. From a molecular point of view, different mechanisms are responsible for the observation of spectral diffusion in 2D IR signals. Thermal motion of molecules in their environment causes collisions of the IR active functional groups with solvents. The persistent changes in the environmental conditions (re-orientation, solvation, hydrogen bonding) influence the frequency of the functional groups and cause dynamic transitions between the frequencies. Other mechanisms can be structural fluctuations of the sample molecule itself (i.e. conformational changes, bond rotation/bending) [37]. Finally, also intermolecular interactions between different functional groups can cause spectral diffusion such as energy transfer between the same type of oscillators [72]. Different mechanisms are often active at the same time and contribute to the ultrafast response. Therefore, a combined approach of experiments and theory, mostly based on molecular dynamics simulations and density-functional theory geometry optimization, is very helpful in identifying which of the contributions is most dominant and what the observed time scales tell about molecular dynamics and properties. In this chapter, different examples for the use of spectral diffusion for interpreting molecular dynamics are discussed, which focus on molecules in different environment, i.e. molecules in bulk solution environment, as well as under different types of dimensional confinement. These examples are discussed in Sects. 3 and 4

2.1.2 Off-Diagonal Peaks

The real strength of 2D IR spectroscopy is reflected by the possibility to resolve interactions between IR-active functional groups, and in particular by the possibility to follow such interactions on the sub-picosecond timescale. Such interactions show up as cross peaks between different diagonal peaks in a 2D IR spectrum, just as in analytical data from 2D NMR spectroscopy [10]. Cross peaks are very sensitive to molecular structure and geometry and are, therefore, very useful to determine distances and angles between transition dipoles [10]. However, it is important to specify on a case-by-case basis what one refers to as an interaction, since there are multiple possibilities from a molecular point of view how cross peaks can be generated. Distinguishing these cases is important, since the origin is very meaningful for a correct interpretation of the spectra and the dynamics. The most important contributions are based on coupled oscillators, which can give rise to a spatially delocalized excitation in the molecule (a vibrational exciton, in analogy to the well-known Frenkel exciton). Vibrational coupling may not only be restricted to intramolecular cases, since functional groups of different molecules can be coupled as well, e.g. in closely packed aggregates. Alternatively, even in the case of largely localized excitations and weak coupling, again both in intra- and intermolecular cases, the excitation energy can be transferred between different oscillators, when a donor and an acceptor are spatially close enough and exhibit a proper orientation with respect to each other. It is, therefore, well established that the intensity and the shape of the cross peaks reflects the strength of the coupling between two functional groups, and this directly relates to molecular structure (Sect. 3.1.1). Strong coupling and energy transfer are not the exclusive mechanisms by which cross peaks can be generated in a 2D IR spectrum. Moreover, once an ultrashort laser pulse excites a molecular vibration, the molecule can undergo a chemical reaction within the lifetime of the vibrational excitation, thereby possibly influencing the vibrational frequency of the initially excited bond. In addition, changes in the molecular properties can give rise to cross peaks and the whole concept is generally referred to as "chemical exchange". There are even more possibilities for cross peaks to appear in a 2D IR spectrum, e.g. Fermi resonances [10, 73]. These cases are rather special and are not further discussed here.

In a more detailed physical description, there are several ways in which two coupled oscillators can be described, and these have been outlined in detail in Ref. [10]. The distinction is made based on the coupling mechanism (e.g. electrostatic, electrodynamic, or mechanical), as well as their relative contributions, since multiple mechanisms can be active at the same time. Mechanical coupling exists in many molecules, is a very efficient mechanism for energy delocalization in a molecule, is generally interpreted with the analogy of interacting springs ("through-bond") and has been well characterized with 2D IR spectroscopy [1, 10, 74, 75]. In contrast, "through space" electrostatic coupling between two oscillators does not require chemical connectivity, but can be very strong for molecules with very large transition dipole moments. However, it is short-ranged due to the strongly non-linear dependence of the coupling terms on the inter-group distance (Sect. 3.1.1) and often only the nearest neighbors need to be considered. In addition to this, the representation of oscillators as dipoles is often a rough approximation. Higher-order multipoles can contribute to the coupling as well. Moreover, orbitals are often strongly delocalized over an entire molecule. This establishes a charge density distribution that responds to the vibration of certain oscillators. Such charge-flow effects are not included in the simple picture of transition dipole coupling and thus impose limitations to the applicability. However, simple transition dipole coupling, without mechanical, electrodynamic or other quantum contributions, is still possibly the most widely used model to interpret electrostatic coupling in 2D IR spectra and often yields acceptable results for the description of molecular signals (Sect. 3.1.1) [10].

To illustrate the strength of cross peaks in 2D IR spectra, Fig. 5 shows two prominent cases for their appearance, i.e. (a) vibrational coupling and (b) chemical exchange. Figure 5a shows a sketch of a 2D IR spectrum that would be qualitatively expected from two coupled oscillators in the regime of strong coupling, i.e. when the magnitude of the coupling is stronger than the frequency difference of the two oscillators. Such coupling between vibrational modes is usually described in analogy to the Frenkel exciton picture [10]. Additionally given is an exemplary energy level diagram for the interpretation of the spectrum and its transition from the local mode picture to the exciton picture. In this transition, coupling between the two modes influences the relative energetic position of the contributing vibrational

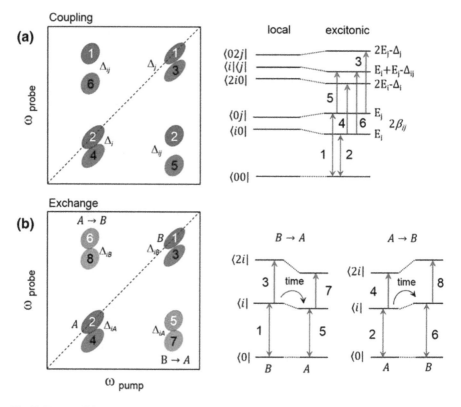

Fig. 5 Two possible sources of cross peaks in a 2D IR spectrum. **a** Schematic 2D IR spectrum (left) of two coupled modes i and j together with a corresponding energy level diagram (right). Coupling generates instantaneous cross peaks in the 2D IR spectrum, which are located in the upper and lower diagonal region. Note that the separations of excited state absorption and ground state bleach signals (Δ_i, Δ_j, and Δ_{ij}) in the diagonal and off-diagonal regions is different. **b** Schematic 2D IR spectrum for chemical exchange (left) and corresponding energy level diagrams (right). Arrows indicate possible transitions between the different levels

levels, which gives rise to distinct frequency separations in the 2D IR spectra. In addition to the expected single (e.g. $\langle 0j|$) and doubly (e.g. $\langle 02j|$) excited states of each mode exist in this description, which give rise to the pairs of diagonal peaks (1/3 and 2/4), a combination state ($\langle i|\langle j|$) also contributes, which influences the spectral positions for the cross peaks (2/5 and 1/6). Note that the coupling directly manifests itself by the existence of a common ground state of the two modes, which gives rise to ground state bleach signals at cross peak positions for both modes simultaneously, even if only one of them is initially excited (1/2). The 2D IR spectrum contains information not only on the frequencies of the two modes, but also the frequency differences between ground state bleach (blue) and excited state absorption (red) transitions, which relate to the diagonal anharmonicities (Δ_i, Δ_j). Additional information is contained in the off-diagonal anharmonicity (Δ_{ij}), which is directly related to the magnitude of the coupling constant β_{ij}. Especially for the latter parameter, strong and weak coupling regimes can be directly distinguished [10]. This is because in the case of weak coupling, the frequency-separation of the oppositely signed signals (2/5 and 1/6) is much smaller than in the strong coupling case, which leads to mutual partial cancellation. Thus, the intensity of the cross peaks is a direct measure of the strength of coupling. Moreover, the magnitude of the coupling constant contains information about the distance and angles of the contributing modes [1, 10, 18, 42, 76], i.e. the molecular geometry. These quantities may be retrieved from polarization-resolved 2D IR spectra using an appropriate model to account for the coupling (Sect. 3.1.1). In fact, it is noted that the coupling also slightly changes the appearance of the diagonal peaks with respect to an uncoupled case, due to its influence on the shape of the potential energy surface. The described picture is static and does not include any time-dependence of the coupling constant. As a consequence, the cross peaks can already be expected directly after excitation at zero population delay, but may show some dynamics according to the molecular system under study. Moreover, as the samples molecular structure generally fluctuates in solution, also the coupling constant changes its magnitude. This results in dynamic effects regarding the cross peak shapes and amplitudes, which has been characterized experimentally and theoretically in detail [10, 37, 77].

The situation for the observation of cross peaks is different if the signals arise from other processes in the sample, e.g. chemical exchange (Fig. 5b). The term chemical exchange generally refers to a reaction that takes place in the sample, which can be either intra- or intermolecular. This reaction can involve either intramolecular conformational changes of the sample, as well as formation and breaking of chemical bonds. The details of the mechanism and the dynamics can in fact be quite complex, and different examples for chemical exchange have been reported for 2D IR spectroscopy (Sect. 3.1.4). The power of 2D IR spectroscopy is that the dynamics of the reaction can be followed in real time, i.e. on the femto- to picosecond timescale. In general, one considers a reaction, which contains two species that exist in chemical equilibrium, here: A and B. That is, A can convert to B and vice versa. Necessary conditions for chemical exchange to be observed with 2D IR are again the spectral separation of the modes that contribute to the species A and B, along with the IR-activity, a non-zero anharmonicity for each mode

($\Delta_{iA, iB}$) and a reaction rate that is fast enough for the conversion to take place within the vibrational lifetime of the sample. In an energy scheme representation of a chemical exchange between A and B transitions occur between the different levels for each species. At initial population delays, when the molecules did not have enough time for the reaction to take place, one observes ground state bleach, stimulated emission and excited state absorption transitions on the diagonal for both A (2/4) and B (1/3). These peaks are due to direct excitation of the sample before the reaction. When the population delay is successively increased, cross peaks will show up in the upper and lower diagonal region due to the reaction taking place, i.e. A converting to B (6/8) or vice versa (5/7, shaded blue and red). This means that the sample molecules change their characteristic vibrational frequencies due to the reaction on the picosecond timescale. Therefore, the cross peaks grow in with characteristic time constants that are governed by the intrinsic dynamics of the sample, i.e. the reaction rate constant as well as the rate of vibrational and rotational relaxation. The applicability of chemical exchange 2D IR spectroscopy is very broad and has been used to shed light on different chemical reactions. Moreover, temperature-dependent measurements of the reaction rates have even been used to derive thermodynamic properties of different samples, thus making the method very attractive from a physico-chemical point of view.

3 Applications of 2D IR Spectroscopy for Molecular Structure Determination

3.1 2D IR for Chemistry and Biology

3.1.1 3D Molecular Structure from Vibrational Couplings in 2D IR Spectra

Although 2D IR spectroscopy is intrinsically designed as a time-resolved technique that exploits sequences of femtosecond pulses, single "quasi-static" 2D IR spectra already contain a lot of information and have been used extensively since the advent of the method to elucidate molecular structure of samples in bulk solution [1, 10, 30]. By its inherent sensitivity to the different mechanisms of coupling within a molecule or between different molecules, distances between functional groups, angles between transition dipole moments and coupling strengths can report on the detailed structure of even large molecules such as proteins [10, 30]. Comparing experimental results and theoretical predictions, detailed information about molecular structure can be obtained. Still, the possibility to time-resolve the structural information and to elucidate fluctuations is an important advantage of 2D IR spectroscopy over other quasi-stationary methods such as crystallography and NMR [77].

A very instructive example of how 2D IR spectroscopy can be used to determine the three-dimensional structure of molecules is given in Fig. 6 for a solution phase experiment on a small and cyclic peptide [25]. The goal of that experiment was to demonstrate how the coupling matrix can be uniquely determined, from which

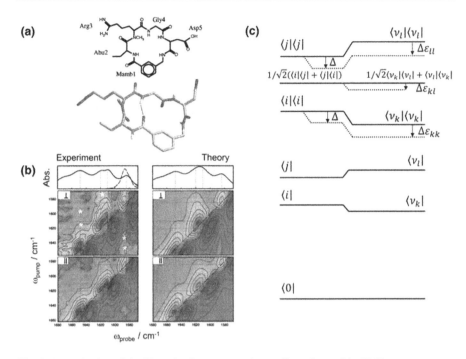

Fig. 6 Determination of the 3D molecular structure via couplings observed in 2D IR spectroscopy. **a** Sketch of the chemical structure of the penta-peptide along with a known 3D model based on the crystal structure of the compound. **b** Experimental and simulated 2D IR spectra in the amide-I spectral range together with linear FT IR spectra. Most intense cross peaks are marked with asterisks. Ground state bleach signals are blue whereas excited state absorption signals are red. **c** Simplified energy scheme of two coupled oscillators, which is used to describe the mutual coupling of the amide-I modes within the peptide chain. For details, see the text. Adapted with permission from Ref. [25]. Copyright National Academy of Sciences (1999)

molecular structure can be retrieved. Figure 6a shows a sketch of the chemical structure of the investigated penta-peptide (cyclo-Mamb–Abu–Arg–Gly–Asp) along with its crystal structure [25]. The structure is stabilized by an intramolecular hydrogen bond (dashed line) and thus forms a so-called beta-turn at the Abu-Arg units. As an essential feature of the application of 2D IR in that report, all amide-I modes of the different units were at least partially spectrally resolved (Fig. 6b, left top panel), making it possible to address each site separately. Polarization-resolved 2D IR spectroscopy (Fig. 6b, left column, population delay 800 fs) in combination with detailed comparison to a theoretical model (Fig. 6b, right column) could then be used to unambiguously determine the solution phase structure. As the important measure of structural information in the experiment, cross peaks between the different amide-I bands were observed, which most clearly showed up in the configuration of perpendicular pump and probe pulse polarizations (central panel, white asterisks) [25, 78, 79]. These cross peaks indicated the coupling between the different amide-I modes. The coupling could be analyzed by considering an energy level scheme shown in Fig. 6c that is similar to Fig. 5. In that scheme coupling manifests itself by a common ground state of two modes, as well as allowed transitions from one-exciton states to two-exciton states of mixed character ($\langle i | \langle j |$). Starting from harmonic potentials for uncoupled oscillators (i and j), the local anharmonicity Δ was taken into account as a perturbation and shifts the doubly excited states to lower frequencies compared to the harmonic 0–1 transitions. The positive (red) and negative-going (blue) signal contributions in the 2D IR spectra at diagonal and cross peak positions can thus be identified as normal mode transitions to the single ($\langle v_k |$) and doubly excited states ($\langle v_k | \langle v_k |$ and $1/\sqrt{2}[\langle v_k | \langle v_l | + \langle v_l | \langle v_k |]$), respectively, which exhibit associated diagonal as well as off-diagonal anharmonicity $\Delta \varepsilon_{kk}$ and $\Delta \varepsilon_{kl}$. The values for these quantities as well as the associated intensities of the cross peaks were directly obtained from the experimental spectra. Together with the anisotropy of the cross peaks, which relates to the relative orientation of the modes in the limiting cases of rather weak coupling, along with a model coupling Hamiltonian [derived from the crystal structure in Fig. 6a], the experimental data could be approximated by least-square fits [right column in Fig. 6b], which agreed remarkably well. The coupling in this particular case was assumed to originate predominately from electrostatic interactions together with a set of transition charges, which couple neighboring groups by "through bond" interactions.

As seen in the present example, the identification and quantification of vibrational coupling makes 2D IR spectroscopy of the widespread amide-I modes sensitive to secondary structure of large molecules. Numerous other examples have been reported, in which vibrational coupling could be used to elucidate the structure of large molecules. Cross peaks between backbone amide groups in proteins have been used to determine different types of structural motifs [30, 80–82]. The crucial step in the procedure described is the determination of the coupling Hamiltonian, which, in the considered case, involved significant information from the already known crystal structure. As an additionally important point, the quality of the structural information significantly depends on the level to which the coupling mechanism is treated theoretically [10, 30]. Sophisticated approaches are often needed to include effects of mechanical coupling, calculations of molecular orbitals, or charge-flow effects. Ultimately, the goal would be to directly obtain the structure of the molecule solely from 2D IR experiments and theoretical fitting, or modelling approaches. Such approaches involve significant and in parallel applications of molecular modelling, molecular dynamics and density-functional theory calculations to derive a complete understanding of the solution phase molecular structure [24, 83]. That approach can be expected as straightforward for rather small systems [84, 85], but becomes increasingly complicated for molecules as large as a protein. This is even more of an issue when the samples crystal structure is not known a priori. Additional experimental support for structure determination can often be obtained from isotope-labelling experiments or shifts upon solvation. This is, however, costly and time-consuming. It is therefore advantageous in many cases to merge three methods for structure determination of molecules, i.e. NMR, crystal structure and 2D IR spectroscopy to derive a complete understanding of the sample.

3.1.2 Bio-Molecular Structure from Quasi-Static 2D IR Spectra

2D IR spectroscopy can be used to elucidate detailed molecular structure also for molecules as large as proteins. Such systems exhibit dynamics that can range from picoseconds up to milliseconds and longer. Therefore, the method can shed light on all of the dynamics of a protein itself, as well as on its dynamic environment such as biological water molecules. The latter is believed to be a crucial aspect in developing an understanding of a relationship between bio-molecular structure and biological function [30]. The inherent capability of resolving lineshapes, couplings and dynamics thereof makes 2D IR spectroscopy unique among the many existing methods for structural biology. As a particularly powerful combination, 2D IR is often used in conjunction with isotope-labelling and mutational approaches to study protein structure [30, 47, 86, 87]. Such studies generally look at the amide-I region of the proteins (about 1700–1600 cm^{-1}), which is largely composed of CO-stretch (and some additional NH-wagging contributions). In many cases, the amide-I modes are delocalized over many residues, thus giving a detailed structural fingerprint of the protein backbone. Less common is the investigation of the amide-II region (1600–1500 cm^{-1}) for this purpose. Alternatively, also other vibrational labels can be introduced by help of unnatural amino acids or protein-bound ligands [88, 89]. 2D IR in combination with FT IR spectra for comparison can be advantageously used to determine particularly the secondary structure of proteins (helices, sheets, coils, turns), due to the quadratic scaling of the signal intensity on the absorption coefficient (Sect. 2). This effect enhances the spectral contrast over FT IR signals. Moreover, structural information is contained in the resolution of cross peaks between different residues that originate from vibrational excitons (Sect. 3.1.1). Such couplings can be particularly well resolved by applying specific combinations of pump and probe polarizations [25, 78]. Finally, experimental data can often be compared to detailed structural simulations, thereby helping to clarify even faint details of bio-macromolecules.

Of particular interest with regard to protein structure is the elucidation of misfolding. The reason for this is that misfolded protein structures are often responsible for diseases in living organisms. The identification of molecular origin of the misfolding is expected to reveal routes to potential drugs and therapies. 2D IR spectroscopy has been demonstrated as an ideal analytical tool for studying structural defects in proteins, such as partial unfolding or aggregations [30, 86]. Of particular interest is the formation of so-called amyloid fibrils, which have been identified as origins for several distinct human diseases. Zanni et al. have shown how to use 2D IR spectroscopy in combination with isotope-labelling to identify specific intermediates in amyloid formation and thereby getting access to the mechanism of the process [90]. That group investigated the kinetics and the mechanism by which the human islet amyloid polypeptide (hIAPP) forms fibrils via an intermediate that exhibits a parallel beta-sheet structure. In particular, the authors exploited the characteristic features of beta-sheets and disordered peptides in 2D IR spectra to uncover the backbone structure [86].

Figure 7a shows the sequence, as well as a solid-state NMR model structure of the investigated hIAPP with 37 residues. The colored letters refer to amino acids

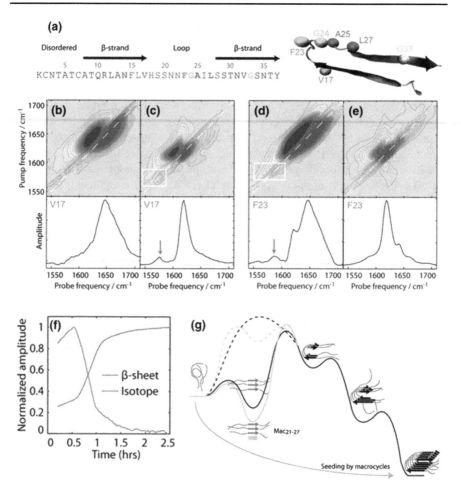

Fig. 7 Fibril formation in the human islet amyloid polypeptide (hIAPP). **a** Sequence and solid-state NMR model. **b–e** 2D IR spectra and diagonal slices of hIAPP during fibril formation. V17 and F23 indicate the $^{13}C^{18}O$ isotope-labelled part of the sample (boxes and red arrows). **f** Kinetics of the unlabeled beta-sheet and isotope-labelled F23 during fibril formation. **g** Energy scheme for the hIAPP aggregation. Adapted with permission from Ref. [90]. Copyright National Academy of Science (2013)

that have been $^{13}C^{18}O$ isotope-labeled to obtain residue-specific structural information and to break the vibrational coupling of the amide-I modes along the protein backbone. Figure 7b–e shows the corresponding 2D IR spectra and diagonal slices of exemplarily-labeled peptides of the so-called lag-phase [(b)/(d)] and the equilibrated phase [(c)/(e)] during fibril formation. The white boxes and red arrows indicate the spectral position of the isotope-labeled peptides. The distinct structural changes are observed for V17 by going from broad unstructured patterns during the lag-phase [random coil, (b)] to sharp peaks [beta-sheet, (c)] upon equilibrated fibril formation. Contrasting to that, labeling F23 in the backbone resulted during the lag-phase, (d), in a combination of broad random-coil peaks and a sharp isotope-label peak (red arrow, < 1600 cm^{-1}). These features evolved to a broad and featureless peak (label) and a sharp beta-sheet feature [(e), > 1600 cm^{-1}], respectively. This evolution indicated the transition from a beta-sheet to a random-coil motif in the F23 region, exactly opposite to what was observed for V17. The kinetics for the evolution of this feature for the F23 sample are shown in (f) and occur on a timescale of several hours. Importantly, the F23 peak increases and afterwards decreases, typical for what is expected for an intermediate in the formation process. This intermediate was observed by isotope labeling throughout the entire FGAIL region (residues 23–27), what highlights the importance of this beta-sheet type structure over an extended region.

To test the importance of this region for fibril formation, a macrocycle binding approach was adopted and the kinetics of the fibril formation upon region-specific binding to the hIAPP were thoroughly screened. As a result, it was found that only targeting the FGAIL region with a macrocycle (Mac$_{21-27}$) slowed down the kinetics markedly. It was thus inferred that the macrocyclic recognition of the FGAIL region stabilizes an intermediate on the potential energy curve along fibril formation (Fig. 7g) with parallel beta-sheet arrangement (parallel red arrows). Overall, this extensive study of fibril formation nicely demonstrated the value of 2D IR spectroscopy in real time and in combination with isotope labelling for the elucidation of intermediates in the process of protein misfolding.

Similar examples exist that have supported the value of 2D IR spectroscopy regarding structural biology. Zanni's group has continuously refined and expanded the model of fibril formation in hIAPP over several years, which eventually led to the conclusive picture presented above [91–93]. However, labelling of a single residue limits this approach to rather small proteins due the spectral resolution of the label in the congested spectra, and additionally requires the preparation of a large amount of labelled variants, as demonstrated. Other approaches have also been shown to allow the investigation of much larger proteins. Specific regions from the amino acid sequences in proteins as large as 173 residues have been resolved by use of segmental isotope-labelling [94]. This approach on the one hand significantly increases the number of absorbing isotope-substituted residues and is equally applicable to smaller proteins [81], but on the other hand might be limited to proteins, which can be expressed in different parts.

In other studies, the same group recently also used 2D IR to investigate in detail ion configurations in the selectivity filter of a potassium ion-channel [95]. The authors exploited 2D IR in combination with MD simulations to reveal ion distributions in the channel and to compare the results with proposed mechanisms for ion-permeation. The investigations were made possible via an experimental approach termed "semi-synthesis", in which the particular protein under study is assembled from different parts, i.e. a synthetic peptide that can be modified on demand, and other recombinant peptides that constitute the remaining protein parts. Such an approach therefore significantly increases the synthetic flexibility for sample preparation. Together with the possibility to incorporate unnatural amino acids [96, 97]. with tailored IR-labels, these methods constitute now a significantly broadened applicability of 2D IR for the corresponding structural investigations of proteins.

Much earlier even, 2D IR has been applied to investigate structural properties of DNA by the resolution of inter- and intra-strand coupling between bases [79, 98]. In other examples, Tokmakoff's group has extensively investigated the secondary structure of globular proteins by use of 2D IR [87]. The group has built a whole library of 16 proteins and assigned fractions of residues to alpha helices, beta sheets and random structures by help of a singular-value-decomposition analysis of 2D IR data. That way, a remarkable agreement between 2D IR data and crystal structure information could be proven. Moreover, the same group also investigated the dissociation of insulin dimers and unfolding with 2D IR [99]. They used the observed changes in the spectra to obtain binding constants and to parametrize a full thermodynamic model for dimerization.

Biomolecular structure and its relation to different types of diseases is a particularly active field of 2D IR research. Starting from the demonstrations over the last couple of years, it will be interesting to see if 2D IR has the potential to be ultimately established as an analytical tool for life sciences. A dream would for instance be to construct a 2D IR spectrometer, which yields structural information from biological tissues or could even be applied to in vivo investigations.

The results regarding structural properties of bio-molecular samples that have been presented in these sections are based mostly on quasi-static 2D IR spectra, from which structural interpretations have been deduced from observations of either spectral positions, vibrational couplings, or slow dynamical evolutions thereof. However, 2D IR has the important advantage to yield structural information from an ultrafast timescales as fast as picoseconds and below. To illustrate how this is exploited for molecular structure determination, the following sections will concentrate on changes in 2D IR spectra that occur between successively delayed pump and probe pulses during the vibrational lifetime of a particular IR label.

3.1.3 Structural Dynamics Resolved from Spectral Diffusion in 2D IR Spectroscopy

To start with, consider the example of spectral diffusion (Sect. 2), which yields information on interconversion of the frequencies of oscillators under the envelope of a broadened vibrational transition. 2D IR spectroscopy is an ideal method for investigating such dynamics since spectral diffusion can be easily extracted from 2D IR spectra. Therefore, the method can provide unique insight into the molecular origin of what is frequently termed "dynamic heterogeneity". Various methods for data analysis have been developed such as the central line slope method, the nodal slope or peak ellipticity, all of which can be applied to study spectral diffusion [69, 70, 100]. Such methods have been tested even in the case of non-Gaussian dynamics [101]. In addition, theoretical approaches such as structure optimization on the DFT level or molecular dynamics simulations can be used to directly compare the experimental results to theoretical predictions and, therefore, lead to an understanding of the molecular origin of the encountered dynamics [77, 84, 102, 103]. Consequently spectral diffusion has been extensively used by many groups to study dynamic structure and environmental interactions in various samples.

A good example of the value of the informational content obtainable from spectral diffusion has been reported by Hamm et al. for the case of the origin of inhomogeneous broadening in peptides [102]. By comparing spectral diffusion in N-methylacetamide (NMA) and trialanine in water, the origin of spectral broadening of the absorption bands could be revealed. This was done by a combined experimental and theoretical approach that led to the overall most detailed understanding of structural dynamics in peptides at the time of the experiments [77, 84, 102–104]. Looking at the amide-I band in the two samples, it could be determined that the respective transition is significantly inhomogeneously broadened in trialanine, but not in NMA. This holds despite the fact that the FTIR data exhibit very similar shapes. To illustrate this, Fig. 8 shows FT IR, as well as 2D IR data of both samples at different population delays. From the 2D IR data it is immediately apparent that the spectral elongation of, e.g. the ground state bleach signals (blue) is different for the two samples with stronger inhomogeneity for trialanine. Moreover, the changes in the spectral elongation are different for increasing population delays with much slower dynamics for trialanine. Specifically, complete homogeneous broadening is observed on the 4 ps timescale in case of NMA, while inhomogeneity still persists on the same timescale in trialanine. Quantification of spectral diffusion dynamics was obtained from theoretical fits to

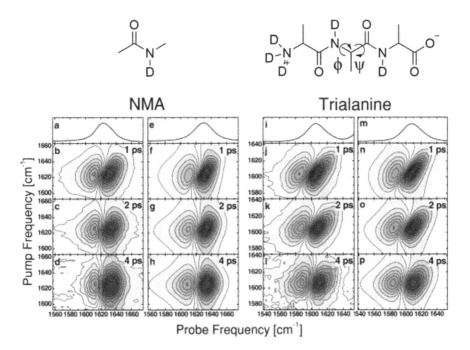

Fig. 8 Linear IR spectra and 2D IR signals for spectral diffusion of deuterated NMA (**a–h**) and deuterated, ^{13}C-substituted trialanine (Ala-Ala*-Ala) (**i–p**) in D$_2$O. **e–d** and **i–l** represent experimental spectra whereas **e–h** and **m–p** represent computational simulations further described in the text. Adapted with permission from Ref. [102]. Copyright American Institute of Physics (2002)

the 2D IR data based on the response function formalism for both systems ((f) − (h) and (n) − (p)), which matched the experimental data almost perfectly. A simple Kubo-picture [62] was sufficient to achieve close matching between experimental and theoretical results, with only one component in the case of NMA, but two components in the case of trialanine. Concomitantly performed MD simulations indicated that the spectral diffusion in NMA is nearly entirely caused by hydrogen bonding to the surrounding water and that the same timescale is observed for trialanine. This showed that the long-time difference in the elongation of the 2D IR spectra must originate from intramolecular structural fluctuations. Further, DFT calculations on the sample systems indicated that there exist two distinct conformations for trialanine, which differ in their dihedral angles of the central peptide bond (ϕ, ψ), and which interconvert on timescales much longer ($\gg$4 ps) than the vibrational lifetime of the amide-I mode. Combining the results from the experiment with the insights from MD and DFT calculations, a detailed understanding of the structural dynamics alongside the intermolecular hydrogen bonding dynamics of small peptides could be obtained and reliably traced back to the spectroscopic observables that are directly extractable from 2D IR data. Since then, that approach has been proven as very valuable regarding the general ultrafast structural elucidation of molecules in the condensed phase [10].

For a broad range of samples, the observation of spectral diffusion has led to a detailed interpretation of ultrafast molecular dynamics such as solvent-solute interactions, hydrogen bonding excitation energy transfer, or molecular reorientation [105–108]. Spectral diffusion has been observed for even molecular samples as large as proteins by employing different types of IR-labels [30, 32, 40, 44, 96, 109, 110]. Such studies can sense the conformational dynamics of the protein itself as well as the solvation environment, and these are generally slowed down as compared to the spectral dynamics of isolated IR-labels. Extremely slow spectral diffusion dynamics have been observed in liquid crystal samples [111, 112]. In such cases, the dynamics can take place on timescales as slow as hundreds of picoseconds and are likely to originate from density fluctuations rather than from conformational re-orientations. In many cases, spectral diffusion can be even quiet complex and originate from different types of interactions of a molecule with its environment. Non- or multi-exponential spectral diffusion has been observed in glass-forming liquids near the glass transition temperature [113], in proteins [30, 109], hydrogen bond dynamics in liquid water [114], or IR-labels in ionic liquids [115–117]. The origin of the various timescales is often the existence of different molecular ensembles (e.g. aggregates), re-orientational and solvation-induced spectral diffusion, or structural spectral diffusion.

Overall, spectral diffusion is an important and readily extractable observable from 2D IR spectra and its application is significantly broader than solely bulk solution dynamics. Recently, a series of reports have appeared where 2D IR spectroscopy has been used in conjunction with non-isotropic environments, e.g. confined environments such as interfaces, metal-organic-frameworks and nanostructures. Examples from such applications will be considered in Sect. 3.2.

3.1.4 Structural Dynamics Resolved with 2D IR Chemical Exchange Spectroscopy

The formation and breaking of chemical bonds in molecules often occurs on the femto- to picosecond timescale. It is relatively easy to observe such a process with conventional time-resolved pump-probe type variants of ultrafast spectroscopy when a starting material and a photoproduct exhibit spectrally distinct features and the starting material is in a pure form [118–128]. The situation is much less straightforward if one considers a chemical equilibrium between such spectroscopically distinct species. In this case, a single spectral axis is not sufficient to characterize the dynamic aspects of bond formation and breaking in chemical equilibrium. 2D IR spectroscopy considerably helps in this respect by resolving correlated spectral features, just as in case of 2D NMR spectroscopy, albeit with a significantly higher temporal resolution down to sub-picoseconds [10, 20, 23, 25, 129, 130].

One of the classic examples where 2D IR spectroscopy was used to study dynamics of continuous bond formation and breaking in chemical equilibrium is shown in Fig. 9 for the case of hydrogen bond dynamics in an unusual phenol-benzene complex that forms in strongly non-interacting solvents such as CCl_4 [50, 131, 132]. In this complex, a benzene molecule acts as a π-base for the OD functional group of the phenol (Fig. 9a), forming a comparatively weak hydrogen bond. The OD-stretch vibration in such a complex is spectrally well separated

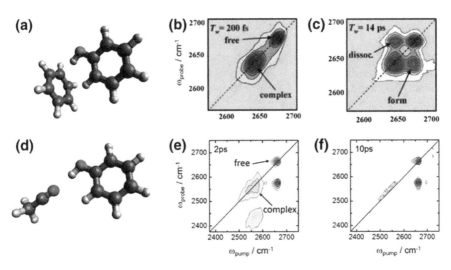

Fig. 9 Ultrafast chemical exchange spectroscopy for the observation of hydrogen bonding complexation in benzene/phenol(OD) solutions in CCl_4. **a** DFT-calculated structure of the hydrogen bonded phenol/benzene complex for which formation and dissociation is observed. **b** and **c** Representative, normalized 2D IR spectra at early and late population delays in the spectral region of the OD-stretch vibration. Complex formation/dissociation is observed via the appearance of cross peaks. **d** DFT-calculated structure of the hydrogen bonded acetonitrile/phenol complex for which no chemical exchange is observed during the vibrational lifetime. **e** and **f** Representative, normalized 2D IR spectra at early and late population delays in the spectral region of the OD-stretch vibration. Adapted with permission from Ref. [50]. Copyright American Chemical Society (2007)

$(2625\ \text{cm}^{-1})$ from the frequency of the free phenol $(2660\ \text{cm}^{-1})$ form, both resulting in ground state bleach signals at early waiting time (200 fs), located on the diagonal line of a 2D IR spectrum (red, Fig. 9b). The associated excited state absorption signals (outside the observation window) are additionally spectrally strongly red-shifted for the bleach contributions, which, for this particular system, makes the characterization of chemical exchange advantageous due to essentially non-existing spectral overlap between oppositely signed ground- and excited-state absorption features. At initial delays (200 fs), only diagonal peaks exist, which indicates that both forms have not yet interconverted. However, when the population waiting time is increased to several picoseconds [e.g. 14 ps (Fig. 9c)], cross peaks start to appear between the complex and free phenol bands for both the ground state, as well as the excited state absorption bands. A detailed kinetic analysis of a full population time series of spectra based on an equilibrium model for the complex and the free form of the phenol yielded a dissociation time constant of about 8 ps, which is well resolvable within the vibrational lifetime of the complex and the free phenol (ca. 10–13 ps) [131]. By the observation of the same exchange kinetics for the ground state and the excited state absorption signals, it was furthermore revealed that the excitation and probing of the complex by the help of the 2D IR pulse sequence does not perturb the equilibrium and thus reveals a direct insight into the dynamics behind the chemical equilibrium. Such expectations might not always be encountered, especially when highly excited vibrational levels are populated (see Sect. 5.2).

The formation of hydrogen bonds is a good example of the applicability of 2D IR spectroscopy to ultrafast dynamics also from another perspective: Once the hydrogen bond gets much stronger, with consequently slower exchange rates, chemical exchange can no longer be observed within the vibrational lifetime of the sample. This is illustrated in Fig. 9d–f, for the case of hydrogen bonding between phenol-OD and acetonitrile. The nitrile functional group acts as a comparatively strong hydrogen bond acceptor that down shifts the complex IR absorption band by almost $100\ \text{cm}^{-1}$. For this particular system, the hydrogen bond is too strong and does not break on the timescale of the vibrational lifetime as demonstrated by the absence of cross peaks in the 2D IR spectra.

The benzene-phenol complex is also a good example for studying influences of the electronic structure of the hydrogen bonding partners. Further investigations by Fayer's group have looked at the impact of electron-donating/withdrawing functional groups [132, 133], solvent compositions [134], or the presence of multiple acceptor sites in the π-base [135]. As one key result, it could be established that the dissociation rates of the complexes are strongly correlated with the corresponding formation enthalpies, what was argued to be describable in an Arrhenius-like manner [132].

The example from Fig. 9 has served as a paradigm for chemical exchange dynamics during solute-solvent complexation. Further examples have been investigated, which demonstrated bond formation/breaking in the context of chemical exchange also in simpler, as well as much more complicated sample systems. Hydrogen bonding was further studied between methanol and organic nitriles [68, 136], methanol and N-methylacetamide [137], methanol and different esters

[138], between water and small ions at high concentrations [139], or intramolecular hydrogen bonding in diols [140]. All the obtained results highlighted the importance of picosecond formation/dissociation of intermolecular hydrogen bonds and discussed its impact on systems that are more complex. The concept of chemical exchange is, however, much more general, and since its advent, many examples other than hydrogen bonding have been reported. Fayer et al. have demonstrated picosecond exchange in carbon-carbon single-bond isomerization [141]. For a particular system, isomerization time constants around 43 ps have been determined and estimations have been made regarding non-substituted ethane and *n*-butane. These experiments thus give direct insight into the barrier height of intramolecular bond rotation. Other experiments looked at conformational switching in mutants of myoglobin and demonstrated that ultrafast chemical exchange dynamics play a role even for such complex bio-molecular systems [142]. Harris et al. achieved to determine transition state geometries by 2D IR chemical exchange spectroscopy in fluxional rearrangements of $Fe(CO)_5$. They were able to demonstrate a pseudo-rotation mechanism for the rearrangement and to rule out IVR between the different involved modes, which is another possibility for growing cross peaks in 2D IR spectra (Sect. 3.1.5). Finally, Gaffney et al. studied ligand exchange dynamics in the coordination sphere of small cations [143, 144]. They demonstrated the preferential contact ion-pair formation between soft Lewis acids and hard Lewis bases, which cannot be easily explained by Pearson's acid-base concept. Such experiments are highly relevant for biological systems, since insights into solvation and coordination-shell dynamics by ultrafast exchange dynamics with naturally abundant ions are of pivotal relevance for instance in biological signal transduction.

3.1.5 Intermolecular and Intramolecular Vibrational Energy Transfer Observed with 2D IR

Another possibility for cross peaks to show up in a 2D IR spectrum is from vibrational energy transfer between different oscillators in the sample [10]. Ultrafast 2D IR spectroscopy is, therefore, the ideal method to characterize such processes since the underlying dynamics intrinsically occur on the femto- to picosecond timescale. As an important point, vibrational energy transfer is strongly sensitive to molecular structure, thereby allowing obtaining direct information about the sample constitution. Moreover, energy transfer can take place in different ways, e.g. in an inter- or intramolecular manner. In the former case, the donor and acceptor molecules need not be of the same type, i.e. they can be a solvent and a solute molecule. Similarly, in the latter case, the donor and acceptor modes need not be the same functional groups, but can be any two groups within a sample molecule. As such, the dynamics of intramolecular cross peaks from energy transfer in a 2D IR spectrum can under some circumstances be taken as a measure for intramolecular vibrational energy redistribution (IVR). Energy transfer can moreover occur in either a resonant or a non-resonant way. The access energy of a certain mode after excitation can be transferred to another oscillator that at least partly spectrally overlaps with the transition frequency of the donor. In such a case spectral diffusion of the donor and the acceptor bands plays an important role since the transition frequencies can fluctuate under the envelope of the vibrational bands. In the case of non-resonant energy transfer, the existing energy and momentum difference between donor and acceptor modes is compensated by environmental degrees of freedom.

Energy transfer is an important feature in 2D IR spectroscopy since clear physical interpretations about the underlying mechanisms have been developed for different cases [1, 10, 43, 52]. From a comparison between experimental results and theoretical predictions detailed conclusions can then be drawn with respect to the molecular structure. As an example, energy transfer rates strongly depend on the distance between donor and acceptor modes, as well as the relative orientations between the transition dipole moments [10, 145–147]. Energy transfer has consequently been used as a "molecular ruler" to determine distances of molecules in condensed phase samples [43]. Typically, intermolecular energy transfer is relevant for distances of functional groups of less than a nanometer [148, 149]. In contrast, intramolecular energy transfer has been observed up to distances of 6 nm [150]. If one considers condensed phase systems, the concentration of a sample molecule in bulk solution is in many cases fairly low ($\sim$ mM). If aggregation of the molecules can be neglected, the average distance between two particles of the same type is then often too large for intermolecular energy transfer to be observed experimentally and intermolecular energy transfer predominately takes place between the solute and the solvent. However, once the molecules are brought closer together, there exist a couple of cases for which energy transfer can play a significant role for the ultrafast dynamics. Intermolecular energy transfer has been observed in concentrated ($\sim$M) solutions of ions in liquids [151]. A special case is encountered for bond-mediated energy transfer in cases where two molecules exhibit chemical interactions such as hydrogen bonds [152–154]. Besides bulk environments, other examples have been reported, where thin films of molecules exhibited vibrational energy transfer between chemically identical molecules [155, 156].

3.1.5.1 Intermolecular Energy Transfer To indicate how detailed molecular information can be obtained by the observation of energy transfer from 2D IR data, Fig. 10 shows a recently reported example of in-depth analyzed energy transfer in thin films ($\sim$ 1 μm thickness) of Pentaerythritol tetranitrate (PETN), a model system to investigate intermolecular interactions in explosives [157]. PETN exhibits four nitrate ester functional groups (Fig. 10a), which, in 2D IR spectra, give rise to an intense asymmetric stretch transition at about 1660 cm^{-1} (A_1) with a weak shoulder at 1685 cm^{-1} (A_2), when the molecule is dissolved in bulk solution (Fig. 10b and c). The second band stems from distinct conformational heterogeneity of the molecule. These two peaks become considerably broadened for molecules in PETN films (B_1, B_2, Fig. 10e and f). Looking at the time-dependence of the 2D IR signals moreover reveals distinct differences between the bulk solution and thin-film samples, i.e. the appearance of cross peaks between the two bands (B_{12}), only in the case of the thin film. This growing-in of cross peaks, which was determined to happen with a time constant of about 2 ps for the PETN film, is a typical signature of energy transfer in 2D IR measurements [10, 75, 148].

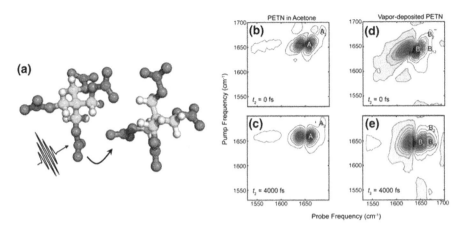

Fig. 10 Energy transfer in films of PETN. **a** Structure of two PETN molecules from the crystal structure of the sample. **b** and **c** 2D IR spectra of PETN in bulk Acetone. A_1 and A_2 indicate diagonal peaks of the asymmetric stretch vibration of a nitrate-ester functional group in bulk solution PETN. **d, e** 2D IR spectra of PETN as a vapor-deposited film. B_1 and B_2 indicate diagonal peaks of the asymmetric stretch vibration of a nitrate-ester functional group in a PETN film and B_{12} indicates a cross peak between the two signals. Adapted with permission from Ref. [157]. Copyright American Chemical Society (2016)

Alternative explanations for the origin of the peaks, e.g. chemical exchange between different conformations, could directly be ruled out for the PETN film by several control experiments. The first support came from the absence of comparable cross peaks in bulk solution, where the molecules exhibit significantly larger structural flexibility. The second support came from time-dependent anisotropy measurements of both samples, which resulted in faster anisotropy decay for the films as compared to the solution. Both points supported the assignment of the cross peaks to intermolecular energy transfer, since the PETN molecules exhibit a significantly shorter distance in the film. From the experimental results, as well as theoretical predictions based on the crystal structure of PETN, a detailed interpretation of the energy transfer could be derived. The two most prominent exciton transitions were calculated to exhibit a mutual angle of 83°, which agreed remarkably well with the experimentally determined values from anisotropy measurements (80°). Using a transition dipole coupling model and the mutual distances as well as orientations of the nitrate ester groups in the crystal, it could be shown that the intermolecular coupling constants between adjacent nitrate ester groups on different molecules are as large as, and partially even larger than the intramolecular coupling constants. The transition dipole coupling includes pure electrostatic effects between different units and calculates the coupling constant β_{ij} between two local modes i and j based on the transition dipole moments $\mu_{i,j}$ and the distance vectors between the modes $r_{i,j}$ (Eq. 4). Through these quantities, the coupling is directly sensitive to intermolecular distances and relative orientations (κ is an orientational factor). Similar as in Sect. 3.1.1, the coupling constant is used to obtain the normal mode frequencies and eigenvectors from a diagonalization of a one-exciton Hamiltonian, and to finally relate the 2D IR signals to molecular structure.

$$\beta_{ij} = \frac{1}{4\pi\varepsilon_0}\left[\frac{\vec{\mu}_i \cdot \vec{\mu}_j}{r_{ij}^3} - 3\frac{(\vec{r}_{ij} \cdot \vec{\mu}_i) \cdot (\vec{r}_{ij} \cdot \vec{\mu}_j)}{r_{ij}^5}\right] \equiv \frac{1}{4\pi\varepsilon_0}\frac{\kappa|\vec{\mu}_i||\vec{\mu}_j|}{r_{ij}^3} \tag{4}$$

Under the assumption of Fermi's golden rule, the transfer rate k_{ij} is then proportional to the square of the coupling elements, which results in a strongly nonlinear distance dependence of the energy transfer rate

$$k_{ij} \propto \frac{|\vec{\mu}_i|^2|\vec{\mu}_j|^2}{r_{ij}^6}. \tag{5}$$

The transition dipole coupling suggested that the vibrations in the crystal are excitonically delocalized over several molecules. From these observations and the experimentally determined transition dipole strengths of the nitrate esters in the thin films, one derives that the vibrational coupling extends to up to 30 PETN molecules in the crystal. More accurate coupling models, which additionally include charge flow, higher-order multipoles and mechanical coupling, may be applied in future studies on such systems to refine the interpretation of the coupling between the molecules in the film. Given the delocalized nature of the high-frequency modes in the PETN films, it will be interesting to see how such possibilities for energy transfer and energy propagation affect the initiation of explosions in these type of materials.

Similar to the discussed thin films of explosives, coupling and vibrational energy transfer between chromophores has recently been identified in immobilized transition metal complexes at semiconductor surfaces [155, 156, 158]. For such systems 2D IR data even demonstrated the coherent nature of the delocalized excitations based on oscillating cross peaks between different modes [156]. Depending on the preparation conditions of such samples under two-dimensional confinement, the coupling between the chromophores can also be too weak to result in strongly delocalized vibrational states. Under such circumstances of weak coupling between adsorbates, energy transfer has been identified by the temporal evolution of cross peaks in isotope-labelled metal-carbonyl monolayers [158]. That energy transfer has been demonstrated to occur on the timescale of up to 100 ps, which is significantly slower as most inter- and intramolecular dynamics of the IR-label, even under conditions, where the molecules come as close as their van der Waals distances.

In a more general discussion of such effects, immobilized molecules and thin films obtain increasing importance in several fields of research and applications such as solar cells [159, 160], catalysis [161–163] and organic electronics [164–166]. It can, therefore, be expected that intermolecular interactions are persistent in such fields as well and may influence for example the decisive performance of working devices. In a similar manner, the concept of delocalized excitations can easily be extended towards other transitions such as electronic coupling, also because the transition dipole moments for electronic transitions are generally much larger as for vibrational transitions.

3.1.5.2 Intramolecular Energy Transfer The flow of energy within a given sample molecule after vibrational excitation is another important mechanism, which can be used to elucidate molecular structure, dynamics and intramolecular interactions [10]. This is of particular importance for complex organic molecules in a solution environment with a large range of functional groups as well as inter-group distances. The observation of energy flow can help to understand in detail the mechanisms, which underlie conformational fluctuations, vibrational relaxation and IVR. As in the case of spectral diffusion (Sect. 3.1.3), such processes influence time-dependent contributions in the Hamiltonian of the system. The typical timescale for the fluctuations that determine these processes is approximately picoseconds or faster and is thus very well accessible with ultrafast 2D IR spectroscopy.

As stated in the introduction, 2D IR spectroscopy is often invoked as being the optical analogue of 2D NMR spectroscopy, where intramolecular distances and fluctuations are measured using for instance the nuclear Overhauser effect (NOESY) [10, 25, 46, 66, 74, 77, 167]. Indeed, 2D IR spectroscopy has been used to demonstrate incoherent energy transfer between amide-I modes in small peptides (e.g. trialanine) by the observation of dynamic evolution of cross peak intensities together with a detailed comparison of experimental results with MD simulations [77]. As a consequence of population transfer from an initially excited oscillator to an acceptor mode, both of which may be coupled to some extent, the intensity of the observable cross peaks relative to the diagonal peaks grows with increasing population delays, similar as in Fig. 10. Importantly, such relaxation rates depend on the energy difference between the coupled states and slow down with increasing energy separation. The determination of such variations in cross relaxation rates due to increased spectral separation based on isotope-substituted samples, along with polarization-dependent 2D IR measurements that reveal the time-averaged relative orientations and coupling strengths of different vibrational modes, have overall allowed revealing a detailed picture of structural fluctuations in such short peptide chains in the amide-I spectral range [77].

As demonstrated in Sects. 3.1.2 and 3.1.3 the amide-I band is a very useful marker for the determination of molecular structure; however, its occurrence is largely restricted to peptides and proteins. It is highly desirable to follow energy transport in molecules over a much larger spectral range, also covering different other types of functional groups with varying absorption coefficients and varying vibrational lifetimes. To resolve such intramolecular energy transport in different types of molecules and to study the energy transport based on different mechanisms (e.g. transition-dipole coupling, anharmonic coupling, mechanical coupling, or heat diffusion) dependent on distances of functional groups, Rubtsov et al. have developed a multi-color approach for 2D IR spectroscopy [74, 75, 168–170]. In their so-called relaxation-assisted (RA) 2D IR method, a high frequency mode (a tag, Fig. 11a) is initially excited and the influence of the initial excitation on other

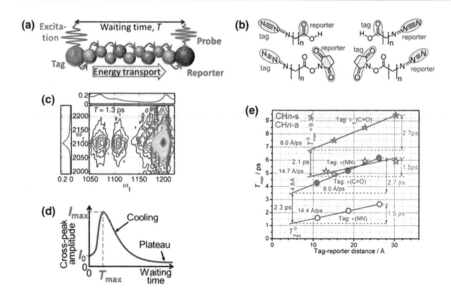

Fig. 11 Energy transport in molecules investigated with relaxation-assisted (RA) 2D IR spectroscopy. **a** Mechanism of RA 2D IR spectroscopy. **b** Collection of sample molecules investigated with RA 2D IR. **c** A typical magnitude RA 2D IR spectrum showing the cross peak region between high- and low-frequency modes (for details, see the text). **d** Schematic representation of encountered dynamics in RA 2D IR. **e** Distance-dependence of the arrival time (T_{max}) of excitation energy from a tag to reporter. The plot reveals the transfer speed via the slope of the lines. Adapted with permission from Ref [170]. Copyright American Chemical Society (2015)

modes (reporters) over a broad spectral range is determined for a series of time delays. Several types of tags and reporters have been investigated and the speed with which the excitation propagates through the sample has been determined in detail. In general, energy deposited in a certain high-frequency mode of a molecule may propagate towards other functional groups in a "through-bond" manner via mechanical or anharmonic coupling either in a diffusive or a ballistic way [75, 170]. Indeed, both types of propagation have been observed experimentally and show clear trends with respect to the chemical nature of the molecules. That is, ballistic energy transfer was observed for molecules with linear chains of decisive repeating units such as alkanes, poly(ethyleneglycoles), or perfluoroalkanes. Ballistic energy transport occurs with a faster speed than diffusive type transport and has been suggested to take place via the excitation of wave packets from delocalized chains states. In contrast, diffusive energy transport is generally described by the physics of thermal conduction in materials. Ballistic transport has been demonstrated extending up to distances of 60 Å between tags and reporters and is suggested to be limited by the mean free path length of associated phonon modes [150]. In contrast, a diffusive type of transport was determined for molecules without such repeating units, thus termed disordered.

Figure 11b depicts a collection of representative examples of molecules that have been studied with RA 2D IR. These molecules consist of linear alkyl chains with different chain-length (n) and tags, as well as reporter groups attached to the ends of the chain. Of particular use proved the azide asymmetric stretch vibration and CO stretch vibrations of several types of carbonyl functional groups [75, 170]. Figure 11c represents a typical magnitude RA 2D IR spectrum of a sample from the lower left class of molecules with $n = 2$, where the azide group has been excited and several lower frequency groups of the succinimide are probed. Note that there exists no diagonal line since the signals are located exclusively in the cross peak region. Following the evolution of the single cross peak signals over time for each peak yields kinetics as schematically drawn in (d), which exhibit clear maxima of the intensity that report on the arrival time of the excitation energy at the specific mode, followed by an exponential decay due to cooling of the sample. When varying the chain length of the sample molecule, the intensity maxima of the cross peaks are found to be continuously delayed. This observation represents the propagation speed of the energy in the molecule via the slopes of the plots, (e).

Several interesting aspects are revealed by the experimental data: First, the speed of propagation within the sample is constant over large distances, which already implies a ballistic energy propagation mechanism. Second, different propagation speeds are observed, depending on the chemical nature of the tag and the reporter units. Similar results can also be observed by varying the chemical nature of the chain [170]. Third, the speed in extremely high, reaching values of about 1.5 nm/ps, which exceeds even the speed of sound in some metals [171]. The reason for the different speeds was identified to originate from a wave packet formation after initial excitation: For the varying samples, different types modes from the chain of repeating units contribute to the wave packet propagation. Depending on the actual folding structure of the molecules in solution phase, the propagation of the wave packet can be significantly affected, thereby allowing obtaining information of the molecular conformations. Overall, the clear linear behavior shows that chain as well as tag/reporter properties control the energy propagation in the sample and excludes the involvement of "through space" energy transfer between the functional groups, as this would result in a strongly nonlinear distance dependence (Eq. 5). The possibility of controlling the energy transport in materials through rational design of the molecular structure is an interesting aspect for material science and technological applications. Next to furthering the fundamental understanding of the materials, the optimization of these effects may allow designing energy dissipation materials in electronic devices.

3.2 Structural Dynamics of Molecules in Confined Environments

3.2.1 2D IR from Molecules Confined in Two Dimensions: Surfaces and Interfaces

In recent years, ultrafast molecular spectroscopy has undergone a significant development towards surface-related investigations. This evolution was mainly triggered by progress in fields of energy science, heterogeneous catalysis and chemical sensing [15]. 2D IR spectroscopy allows obtaining direct information of how molecules under different forms of confinement behave and how confinement influences molecular structure and properties. As many of the important aspects in the mentioned fields are based on the structural and dynamical properties of molecules, 2D IR can be considered the perfect method for analytical purposes. Spectroscopy of molecules at two-dimensional interfaces is particularly challenging because there is only a low number of contributing molecules in the focal region of the laser spot. Moreover, in the IR spectral range ultrafast molecular spectroscopy is often considered difficult due to often very low absorption coefficients of functional groups (< 1000 cm^{-1}). However, to date a series of methods, which are able to resolve ultrafast, multi-dimensional IR spectra from two-dimensional interfaces, has been developed by different groups. All of these methods exhibit distinct strengths and weaknesses that have been thoroughly compared before, both from an experimental and theoretical signal point of view [15]. In this section, an overview is presented on the demonstration of recent insights from the sample perspective.

The first demonstration that ultrafast 2D IR spectroscopy can be sensitive enough to obtain signals from only monolayer thin samples of organic molecules at two-dimensional interfaces was reported by Bredenbeck et al. by use of homodyne-detected 2D sum-frequency generation (SFG) spectroscopy [172]. Although limited structural and dynamic information was revealed in that study, the report triggered subsequent ground-breaking developments of surface 2D IR spectroscopy. Detailed investigations employed the high sensitivity of 2D IR in transmission and external reflection BOXCARs geometry and focused on strongly absorbing metal-carbonyl headgroups in self-assembled organic monolayers at solid-liquid/gas interfaces. These experiments looked at mainly spectral diffusion dynamics in dependence of different experimental parameters such as solvent properties, surface morphology or monolayer structural defects [173–178]. The high absorption coefficients of metal-carbonyl complexes (> 1000 M^{-1} cm^{-1}) [173] strongly facilitated the generation of the nonlinear signals from the monolayers. The application of plasmonic substrates together with surface-enhanced 2D attenuated total reflectance (ATR) IR spectroscopy [179] recently then allowed pushing the sensitivity to molecular monolayers equipped with azide and nitrile functional groups that exhibit absorption coefficients as low as 550 to < 200 M^{-1} cm^{-1} [180, 181]. Figure 12 shows 2D ATR IR spectra of the asymmetric stretch vibration from azide functional groups, attached to linear alkyl chains of short [2-azidoethanthiol (2-N3), (a) and (b)] and long [11-azidoundecanethiol (11-N3), (c) and (d)] on ultrathin gold (Au) layers (average thickness 1 nm). Spectral diffusion was characterized for these systems by the help of the CLS method (white lines), and the values were found to drastically depend on the distance between the Au surface and the functional group. The interpretation is that a close proximity to the rough surface results in a high degree of structural heterogeneity and slower spectral diffusion for the short-chain monolayers. This difference in the dynamics was rationalized with a lower quality of packing between the monolayer constituents of the short-chain versus the long-chain sample molecules (2-N3 vs. 11-N3) based on the varying degrees of inter-chain hydrophobic interaction.

In the context of the formation of self-assembled monolayers, important questions often exist regarding the interaction between the monolayer constituents. If molecules are tightly packed in well-defined aggregates, the intermolecular distance is generally less than 1 nm, which is sufficiently close proximity for the

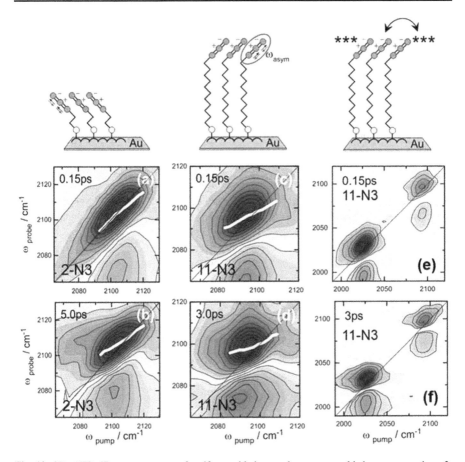

Fig. 12 2D ATR IR spectroscopy of self-assembled monolayers on gold layers. **a** and **a** 2-azidoethanthiol (2-N3) at early (0.15 ps) and late (5 ps) population delays. **c** and **d** 11-azidoundecanthiol (11-N3) at early (0.15 ps) and late (3 ps) population delays. **e** and **f** isotope-dilution experiments in 11-N3 (^{15}N$_3$ (2025 cm^{-1}) and ^{14}N$_3$ (2105 cm^{-1})) samples to resolve possible intermolecular interactions such as energy transfer and vibrational coupling. Adapted with permission from Ref. [180]. Copyright American Institute of Physics (2015)

molecules either to establish delocalized vibrational states or exchange excitation energy via dipole-dipole interactions (Sect. 3.1.5). Surprisingly, 2D ATR IR spectroscopy revealed the absence of cross peaks in monolayer structures by the help of isotope-labelling of functional groups (Fig. 12e and f). Monolayers of 11-N3, partially labelled with all-^{15}N-azide groups (ca 2025 cm^{-1}) have been co-adsorbed with all-^{14}N-azide moieties thus generating sets of spectrally clearly separated bands. Scanning the population delay for multiples of the vibrational lifetime of the azide group ($\sim$ 1 ps), cross peaks are absent for all delays [180]. This clearly indicates the absence of vibrational coupling or energy transfer between the molecules. Existing strong intermolecular vibrational coupling should result in the immediate appearance of cross peaks, which also oscillate dependent on the population time [66]. In contrast, energy transfer would have been expected by a gradual growth of cross peaks, followed by a subsequent decay due to vibrational relaxation [157]. Taking into account the dependence of the energy transfer rate [77, 145, 182, 183] on $\mu^4 r^{-6}$ (Eq. 5), two possible reasons have been proposed, which may explain the missing cross peaks. First, the azide functional groups exhibit a comparatively low extinction coefficient ($\sim 550\ \mathrm{M}^{-1}\mathrm{cm}^{-1}$), thus keeping the energy transfer rate intrinsically low. Second, the even stronger dependence on the inter-particle distance further reduces the transfer rate, thus possibly making the process even more unlikely. Moreover, the azide group has a comparatively short vibrational lifetime (< 2 ps) and any energy transfer significantly exceeding the lifetime will be just too difficult to be observed experimentally. This interpretation is consistent with the observation that vibrational coupling and intermolecular energy transfer have been resolved in aggregated or structurally more rigid samples with much stronger absorption coefficients such as metal-carbonyl samples [156, 158].

Further investigations of 2D IR at interfaces concerned the orientation of molecules and in particular orientation of functional groups with respect to the interface. The interfacial orientation of functional groups has important consequences for specific applications, e.g. for coordination-chemistry compounds, for molecular recognition or chemical reactivity of catalysts. In addition to simple considerations regarding chemical accessibility, interfacial orientation can also have an impact on substrate-adsorbate interactions. Xiong et al. have demonstrated how to exploit the properties of the fourth-order nonlinear susceptibility of a sample from immobilized molecules in 2D SFG spectroscopy to obtain insight into molecular orientation and substrate-adsorbate interactions [184–186]. Being based on the even-order susceptibility terms, SFG methods (stationary as well as time-resolved) can beneficially determine the average orientation of functional groups by the sign of signal [15, 187]. This advantage is based on the fact that the susceptibilities are in general complex-valued quantities with real and imaginary parts, which may have both positive and negative contributions [188–191]. Note that in contrast to even-order methods, odd-order methods (e.g. 2D IR) result in the same signal signs for parallel and anti-parallel orientation of transition dipole moments with respect to the surface normal and thus cannot differentiate the two cases of orientation. Figure 13 (a) shows an exemplary heterodyne-detected (HD) 2D SFG spectrum of a model compound, Re(diCN-bpy)(CO)$_3$Cl, on an Au surface. Such tricarbonyl complexes exhibit three strongly IR-active normal modes in the spectral range of investigation [A′1 ($> 2000\ \mathrm{cm}^{-1}$) as well as A′2 and A″ ($< 2000\ \mathrm{cm}^{-1}$), respective directions of the vibrational modes are indicated by the blue sticks in Fig. 13c]. Figure 13b shows for comparison a third-order 2D IR spectrum of the same complex dissolved in bulk solution (DMSO). The 2D SFG spectrum shows that upon immobilization, the A′1 and A′2/A″ modes experience extensive line broadening as compared to the bulk solution sample and furthermore exhibit opposite signs in the ground state bleach and excited state absorption signals (yellow/red vs. blue/green, respectively). The opposite signs indicate opposite orientations of the modes with respect to the surface. However, this information was concluded to be insufficient to determine the *absolute* orientation of the modes. To gain detailed insight into absolute

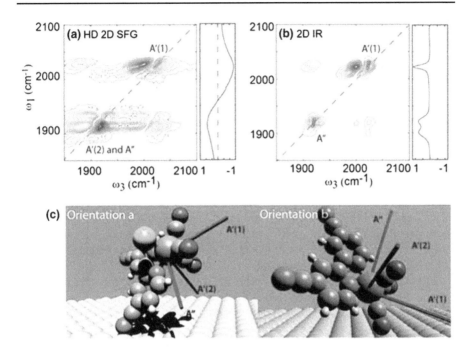

Fig. 13 2D SFG spectros-copy of a CO_2-reduction catalyst $(Re(diCN-bpy)(CO)_3Cl)$ on an Au surface. **b** 2D SFG signal of a monolayer sample on Au. **b** 2D IR signal of a bulk solution sample of the complex for comparison. **c** Simulated structures of the complex on Au using DFT. Orientation a is suggested to be preferred. Adapted with permission from Ref. [184]. Copyright American Chemical Society (2015)

orientations, the signals were analyzed by help of the CLS method of both the diagonal peaks, as well as the intramolecular cross peaks. This analysis revealed that $A'2/A''$ modes experience extra contributions from homogeneous broadening in the sample, which is not present in the $A'1$ modes. This additional contribution has been proposed to originate from interactions of the sample with image-dipoles in the metal layer [184]. Based on this information and by help of DFT calculations of the adsorbed molecule, it could be concluded that the preferred adsorption geometry of the model complex must be close to the structure shown as "Orientation a" in Fig. 13c.

The full power of 2D SFG spectroscopy to determine molecular orientation has been demonstrated in a series of other reports including theoretical considerations [192], as well as experimental studies on different sample systems, for instance on DNA strands [193] and peptides [194]. Using the intrinsic properties of the 2D SFG method revealed that in poly(thymine) monolayers on Au surfaces [193] different CO stretching modes exhibit different orientation with respect to the surface. Furthermore, based on intra- and inter-base cross peaks, as well as by combining DFT-calculations and experimental results in previously unresolved inter-base coupling between bright and dark T1 modes of neighboring thymine bases was observed. In similar experiments on peptides on Au surfaces [194] the secondary structure of molecules, as well as their relative orientation with respect to the surface could be determined. As a furthermore interesting observation, modes from random coil motifs that are intrinsically invisible to SFG spectroscopy due to orientational averaging were shown to be resolvable in 2D SFG data through cross peak features with SFG-active vibrations. This experimental observation was taken as a confirmation of earlier theoretical predictions [192].

2D SFG spectroscopy is not exclusively applicable to immobilized molecules at interfaces. This is due to the intrinsic surface-specificity of the method, which stems from the fourth-order nonlinear susceptibility that is exploited for signal generation [15]. Thus, the possibility to measure non- or only weakly adsorbed molecules is the big advantage of 2D SFG spectroscopy over third-order methods. In that way, a series of different interfaces can be addressed by 2D SFG, i.e. liquid-liquid or gas-liquid, all of which are currently not possible to study in a similarly straightforward manner with third-order methods. In fact, 2D SFG in different experimental implementations has been used extensively to study orientation, dynamics and interactions of molecules at liquid-gas interfaces, in particular for aqueous systems [182, 195–202]. Figure 14 shows stationary as well as ultrafast 2D SFG spectra in the OH stretching region of water (HOD) molecules at charged liquid-gas interfaces. Isotope-diluted water was used instead of pure H_2O to avoid impacts from Fermi-resonances or coupling contributions. Positive and negative surface charges have been induced in that study by interfacial monolayers from surfactants with cationic (DPTAP) and anionic (DPPG) headgroups, respectively. It was demonstrated that the surface water molecules are oppositely oriented in the two different cases (Fig. 14a): positive surface charge (DPTAP) induces orientations with H-atoms pointing away from the surface (down), whereas negative surface charge induces

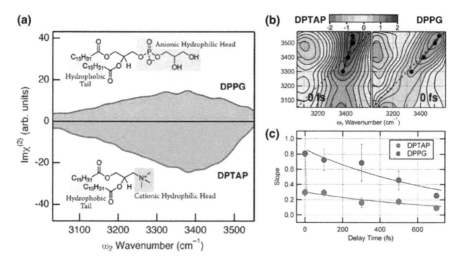

Fig. 14 Heterodyne-detected 2D SFG spectroscopy of water (HOD) at charged liquid-gas interfaces. **a** Stationary SFG spectra of water at negatively (with DPPG, blue) and positively (with DPTAP, red) charged interfaces. The different signal signs indicate different orientation of the water molecules. **b** 2D SFG spectra of water from the OH-stretching region at the DPTAP and DPPG interfaces at a population time of 0 fs. **c** Spectral diffusion dynamics of the 2D SFG signals. Adapted with permission from Ref. [198] Copyright Wiley VCH (2016)

orientations with H-atoms pointing towards the surface (up). While the frequency and width of both features is similar, which indicates similar strengths of hydrogen bonding, a close analysis of the CLS from 2D SFG spectra at a series of population delays revealed that the hydrogen bond dynamics in the HOD molecules near the charged interfaces drastically depend on the surface charge (Fig. 14b and c). That is, initial CLS values (0.8 for DPPG and 0.3 for DPTAP) strongly depend on the surface charge, but both exponentially decay with very similar time constants of about 750 fs [198]. The strong difference in the initial CLS values was argued to result from a hidden underlying process of spectral diffusion only in case of positively charged DPTAP, which is too fast to be observed with the given temporal resolution ($\sim$ 200 fs). Such presumably sub-100 fs dynamics were argued to be similar to dynamics that have been observed in the bulk phase of water [203]. As a consequence, it was inferred that in case of negative surface-charge, the water molecules form hydrogen bonds to the charged phosphate headgroup of DPPG, which inhibit the bulk-like behavior in the ultrafast spectral diffusion. Such hydrogen bond acceptor sites do not exist in the case of DPTAP, which exhibits a positively charged ammonium headgroup. The observed orientation and dynamics of water molecules can in principle have a strong impact in chemistry and biology. Many lipid membranes are surrounded by water molecules and the hydrogen bonding, as well as the orientation might have a profound influence on the properties of e.g. cell membranes.

Ultrafast dynamics of molecules at liquid-gas interfaces is currently a strongly active field of research, and 2D SFG spectroscopy has been used to reveal further detailed understanding of the mechanisms, which dictate the underlying molecular dynamics. Other negatively charged surfactants have been used at liquid-gas interfaces to identify different types of water molecules at the interface by the observation of distinct vibrational bands in the OD stretch region [199]. The isolated bands have been argued to originate from localized vibrations near the surfactant and delocalized modes, which are shared between different OD-bonds, respectively. Resolving the energy transfer between the sub-ensembles via cross peak dynamics, it could be proven that the different vibrations strongly interact. The prominence of energy transfer at water-gas interfaces was highlighted even earlier with 2D SFG studies that employed homodyne detection [182, 195]. Such experiments addressed ultrafast energy transfer between water molecules at uncharged water-air interfaces. Using isotope-dilution, it was determined that intermolecular energy transfer is the process, which overall dominates the dynamics of spectral diffusion at the interface. The slowdown of spectral diffusion at the interface as compared to water molecules in bulk solution environments was discussed to stem from the reduced number of acceptor molecules at the interface, which alters the probability of energy transfer.

The strong contributions of energy transfer that dominate the ultrafast dynamics of small molecules at liquid-gas interfaces are remarkable if compared to the missing energy transfer between small molecules [204, 205] (e.g. CO and CN$^-$) at ultrathin (few nanometers) metal layers and also to much bulkier and much more flexible monolayer molecules at solid-liquid and solid-gas interfaces [173, 177, 178, 180]. The observation of missing energy transfer in the latter systems (e.g. Fig. 12) suggests that the distance between the headgroups of the constituents in functionalized monolayers is most likely too large for intermolecular energy transfer to occur or that the relaxation rates are too fast compared to a possible transfer rate. The factors, which determine the energy transfer rates, are the distance and orientation between the molecules, as well as the magnitudes of the transition dipole moments of donor and acceptor molecules (Eq. 5). Taken into account that the often employed carbonyl stretch vibrations exhibit transition dipole moments [173, 206] that by far exceed that of OH stretches from liquid water [207], it is very likely that a dominant factor is the r^{-6} dependence of transfer rate (Eq. 5), which inhibits the observation of cross peaks in the 2D IR experiments.

3.2.2 2D IR Spectroscopy of Three-Dimensionally Confined Molecules

Next to the two-dimensional confinement discussed in Sect. 3.2.1, additional important cases exist in chemistry and biology where a confinement of molecules in three dimensions plays a crucial role for the dynamics and molecular interactions. Consider water molecules that take part in the hydration of minerals [208], protein hydration and molecules in micelles [209, 210], or molecules and ions in channels across biological membranes [95, 211]. Other important cases are organic molecules in crystals of functional materials [212], molecules at interfaces of layered heterogeneous catalysts [213], or molecules caged in defined synthetic environments such as metal-organic frameworks [214–216]. In such systems, molecular motions are considerably restricted and the electrostatic environment around a local probe can be significantly different as compared to an isotropic case. This can have a profound impact on the stability and the particular function of materials. It is, therefore, important to understand in detail, by which mechanisms three-dimensional confinement influences molecular dynamics. 2D IR spectroscopy has been shown to yield manifold information about this effect.

To demonstrate how 2D IR spectroscopy can to this, Fig. 15 shows an example of the ultrafast dynamics of confined water in structurally different environments. In that particular case, two minerals, gypsum and basanite, have been chosen for the investigations due to their similar chemical constitution ($CaSO_4 \cdot n\, H_2O$), but different three-dimensional structure (top panels in Fig. 15a and b, respectively). The number of water molecules thus plays an important role in the structural framework of the crystals. As a key point regarding the dynamics, the containing water molecules do not form a similarly branched hydrogen bonding network as in bulk water, but rather interact with the ions in the crystal structure. Thus, their orientational, structural and hydrogen bond dynamics and contributions from energy transfer can be expected to drastically differ from, e.g. bulk water.

Looking at the OD-stretch in isotope-diluted water (HOD), Fayer et al. were able to observe directly the structural dynamics of water molecules intercalated in inorganic crystals. The authors chose gypsum and basanite crystals as the samples, which exhibit similar chemical constitution, only differing by the amount of intercalated water molecules (Fig. 15) [208]. As in many studies on water dynamics, HOD was used instead of pure H_2O or D_2O to avoid intra- and intermolecular energy transfer. Measuring 2D IR signals of gypsum from freshly prepared and annealed samples (middle and lower panel in Fig. 15a, respectively, at $T = 0.5$ ps), it could be shown

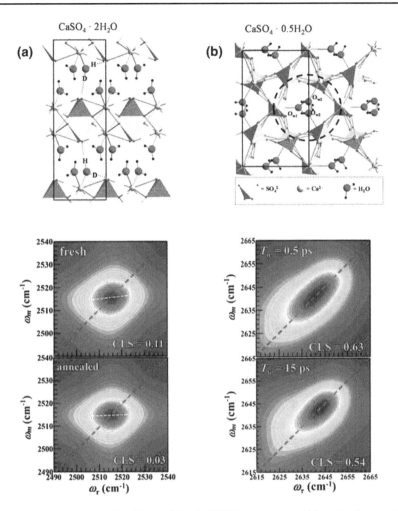

Fig. 15 2D IR spectroscopy of the OD stretch band of HOD in **a** gypsum and **b** basanite. Top panels give the crystal structure with the color-coding given in (**b**). Middle and lower panels give 2D IR spectra of **a** fresh and annealed gypsum at $T = 0.5$ ps, and **b** indicated population delays at a temperature of 298 K. Adapted with permission from Ref. [208]. Copyright American Chemical Society (2016)

that only very limited structural heterogeneity exists in that sample. Thermal annealing can cause drastic structural changes in the crystals thereby permanently influencing the molecular dynamics. For the gypsum sample, annealing essentially resulted in a nearly completely homogeneously broadened band (lower panel). Time-dependent measurements additionally demonstrated that spectral diffusion is practically inexistent in the samples within a temporal observation window of about 15 ps. In contrast to gypsum, water in basanite's tubular channels shows completely different spectroscopic behavior; that is, the data exhibit strong and dynamic disorder on the picosecond timescale (Fig. 15b middle and lower panel). Depending on the temperature, this partial sampling of the inhomogeneous lineshape was found to be as slow as 30 ps. On much longer timescales considerable quasi-static inhomogeneous contributions remain. Additional temperature dependent measurements of the spectral diffusion dynamics indicated that the structural fluctuations accelerate at elevated temperatures and that portions of the lineshape become even motionally narrowed. Finally, permanent structural rearrangements in the crystals by annealing was demonstrated to be absent in basanite since repeated cycles of temperature-dependent experiments yielded identical spectral diffusion for the different temperatures. These experiments thus overall demonstrated that water molecules in such mineral structures are not static, but sensitively respond to the three-dimensional arrangement of the crystal constituents. The observed dynamic contributions may have a considerable influence on the stability of three-dimensional structures also in other natural crystals. Establishing a systematic understanding of the origin of structural fluctuations in different naturally abundant minerals might furthermore allow establishing a firm view of the thermodynamic as well as kinetic properties of these materials.

In a similar manner, 2D IR has already been used to study molecular dynamics in other systems under three-dimensional confinement. Also, Fayer's group investigated structural fluctuations of molecules in synthetic metal-organic frameworks (MOFs) [215]. Of particular interest in these experiments were the structural flexibility of the MOF itself, which is invoked as a key property for the several aspects of functionality [217]. Attaching metal-carbonyl local probes to the structural backbone of the MOF, it could be shown that structural fluctuations of the framework are complex, bi-exponential and occur to large extents on timescales longer than 0.5 ns, what is considerable slower as dynamics in isotropic environments and even at surfaces and interfaces (Sect. 3.2.1) [15]. Importantly, contrasting with an intuitive expectation, the dynamics become even slower (> 2 ns) when the pores of the MOF are filled with solvent molecules. This was attributed to significantly restricted dynamics of the solvent molecules in the pores, as well as severe restrictions of the MOF to undergo structural deformations.

It is, however, important to note that structural fluctuations of molecules under three-dimensional confinement do not always occur on the timescale of multiple tens of picoseconds or longer. Organic cations in 3D lead-iodide perovskite crystals are a good example of relatively fast molecular motions on timescales of a few hundreds of femtoseconds to picoseconds, while slower dynamics are only of very minor importance [212]. These structural fluctuations have been identified as "wobbling-in-a-cone", as well as jump-like reorientation of the organic molecules in this type of material. It may be expected that such ultrafast orientational dynamics can be of hallmark importance for the dielectric response of the sample and, therefore, for opto-electronic properties of these novel materials.

4 Extensions of 2D IR Spectroscopy

Recent extensions of 2D IR spectroscopy involve detailed applications, as well as technological advancements in the field of spectro-electrochemistry, microscopy, the combination with electronic spectroscopy and efforts to merge electronic and vibrational spectroscopy. In the following sections, highlights from these recent research directions are presented and discussed in the context of obtaining dynamic structural information of samples that is difficult to obtain by other methods.

4.1 2D IR Electrochemistry: Vibrational Stark-Shift and Redox Spectroscopy

Developments in the field of surface-sensitive 2D IR spectroscopy (Sect. 3.2.1) have evoked interest recently to expand the capabilities of time-resolved spectroscopy towards purposely charged solid-liquid interfaces. Such interfaces play a prominent role in the fields of electro-catalysis, fuel cells, or battery assemblies. Important questions to be addressed are re-orientational dynamics of molecules in electric fields, dynamics of charge-transfer processes between an electrode and a solute, or the molecular basis of vibrational Stark-shifts of adsorbates. Additional points to be tackled are dynamics of intermediates in electrocytalytic systems, the molecular origin of overpotentials, or transient responses of molecules to potential-jumps at an electrode interface, only to mention a few important aspects. 2D IR spectroscopy is again an ideal tool for these type of investigations since it directly allows looking at the molecular vibrations. Frequencies, bandwidths and dynamics are very sensitive reporters for interfacial electric fields and thus allow precise measurements under working electrochemical conditions.

An important step towards addressing and answering these questions has been recently made by combining 2D IR spectroscopy with surface-sensitive spectro-electrochemistry by the development of 2D ATR IR electrochemistry [218]. In this variant of 2D ATR IR, the reflecting interface of an ATR substrate is made conductive and can thus work as an electrode at which molecules can be immobilized and studied with ultrafast time resolution (Fig. 16a). The application of the ATR arrangement is particularly beneficial in the IR spectral range, where often strongly IR-absorbing electrolyte solutions are employed (such as water). In the initial demonstration of 2D ATR IR electrochemistry, the ultrafast dynamics of

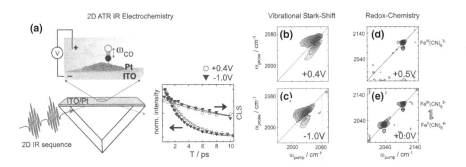

Fig. 16 a Schematic principle of 2D ATR IR electrochemistry together with examples of the ultrafast vibrational relaxation dynamics (norm. intensity, left scale) and spectral diffusion dynamics (CLS, right scale) of CO adsorbed to ITO/Pt electrodes. **b** and **c** Examples of vibrational Stark-shift 2D ATR IR spectra of CO adsorbed to Pt-coated ITO electrodes. Adapted from Ref. [218] with permission. Copyright American Chemical Society (2016)

immobilized carbon monoxide (CO) at the interface between aqueous electrolytes and indium-tin-oxide (ITO)/platinum (Pt) electrodes has been investigated dependent on the applied electrochemical potential to the ATR surface [218]. The CO molecule sensitively responds to the applied potential through the vibrational Stark effect [219–222], which influences the CO stretching frequency by two distinct mechanisms. In a first mechanism, application of the potential modifies the electron density in the CO bond, thus strengthening or weakening it with varying potential. In a second mechanism, the relative energetic positions of ground- and excited vibrational states in the anharmonic electronic ground state potential of the molecule are modified by the interfacial electric field [221, 223]. This originates from slightly larger average bond lengths in the vibrationally excited state as compared to the ground state, which results in different values for the dipole moment, depending on the degree of vibrational excitation. Experimentally, the CO frequency was thus modulated (Fig. 16b and c) in a potential range between about -1.0 to 0.4 V (against an Ag/AgCl reference electrode). Interestingly, despite a more than 30 cm^{-1} shift in the vibrational frequency in this potential range, the vibrational dynamics of the adsorbed CO are only slightly influenced by the application of the interfacial electric field. Vibrational relaxation (Fig. 16 (a), left scale) becomes only slightly slower (ca. 20%) for positive potentials (0.4 V, open circles) as compared to negative potentials. In a similar manner, spectral diffusion, as determined from the CLS method (Fig. 16a, right scale), becomes only slightly faster (ca. 10%) for positive potentials (0.4 V, open circles). Such comparatively mild variations in the ultrafast dynamics indicate that the interfacial electric field is not strong enough to result in structural changes in the hydrogen bond network of water molecules at the interfaces, which interacts with the adsorbed CO [204, 218].

Although 2D ATR IR has been so far mainly used to investigate ultrafast dynamics of immobilized molecules at solid-liquid/gas interfaces, the technique is intrinsically sensitive also to molecules from an isotropic environment above the interface, e.g. the bulk electrolyte solution. This is due to the generation of the evanescent wave at the ATR interface, which penetrates up to several micrometers into the medium of lower refractive index [224]. Therefore spectro-electrochemistry can also be performed for isotropic samples such as redox-active metal complexes. This has been demonstrated by electrochemical switching between stable redox species of ferro- and ferricyanide in aqueous solution ($Fe^{III/II}(CN)_6^{3-/4-}$, Fig. 16d and e). Starting from the pure Fe^{III} species (2113 cm^{-1}) at positive potentials ($+$ 0.5 V against Ag/AgCl), a decrease of the electrochemical potential to 0 V starts reducing the metal center, thus partially creating Fe^{II} species at lower vibrational frequencies (2040 cm^{-1}). With these two demonstrations spectro-electrochemistry, i.e. vibrational Stark-shift spectroscopy and redox-chemistry, 2D ATR IR has set the stage for ultrafast investigations directly at electrode-electrolyte interfaces. This rapidly growing field of research is expected to reveal important insight into structural dynamics under electrochemical conditions, which are relevant in the fields of catalysis or energy research. It is noteworthy that alternative approaches for measuring ultrafast dynamics under electrochemical conditions have also been presented, which can partly be applied for analogous investigations. External reflection 2D IR spectroscopy has been introduced by Bredenbeck et al. to measure redox-active samples in bulk solution under electrochemical conditions [225, 226]. Although not yet demonstrated, this method should also be applicable to immobilized molecules at the electrode-electrolyte interface [227]. Furthermore, surface-specific SFG spectroscopy can also be used under electrochemical conditions [228–230], however, has not yet been implemented as a multi-dimensional variant for measuring ultrafast dynamics.

4.2 2D IR Microscopy

Conventional 2D IR spectroscopy in transmission or reflection mode uses laser beams, which are focused onto a sample in spatial dimensions of several tens to hundreds of micrometers. The signal thus contains spatially averaged contributions over extended regions of the sample. While this is generally not problematic for isotropic samples such as molecules in liquids, spatial averaging can be challenging for samples that are immobilized on surfaces, or confined in three-dimensional structures. This is because local structures of the molecules can be considerably different since orientations with respect to surfaces or local solvent environments can be drastically different. 2D IR is an ideal method for resolving such spatial differences in ultrafast vibrational dynamics and to obtain chemically specific temporally and spatially resolved information of samples. Consequently, recent experimental developments have focused on the combination of the method with microscopy [59, 60]. In an initial variant of 2D IR microscopy a tight focusing method with reflective optics in combination with a single-beam implementation of pump-probe-type 2D IR spectroscopy was employed to achieve a nearly diffraction-limited spatial resolution [59]. Chemical contrast between molecules dissolved in a bulk and strongly interacting liquid against adsorbed molecules on small, non-interacting polystyrene beads could then be achieved by measuring different vibrational lifetimes.

The essential challenge in the combination of 2D IR spectroscopy and microscopy is to acquire data for a sufficiently large number of different spots with similar data quality compared to a spatially integrated signal on a reasonable timescale. Depending on the required spatial resolution and the area of interest, this can easily result in a few thousand-fold increase of measurement time. To overcome this problem, Zanni et al. have recently developed a 2D IR wide-field microscope, using a dual acousto-optic modulator (AOM) pulse shaper and a focal-plane array detector (FPA). This way the authors were able to simultaneously acquire 16,000 2D IR spectra, thus avoiding scanning the sample "step-by-step". Figure 17a shows a schematic overview of the experimental setup, in which the output of a single OPA is split into two portions that are fed into a dual AOM to generate two pairs of polarization-orthogonal pulses for double time-domain data acquisition [231]. The two beams are afterwards recombined to irradiate a spot of about 100 μm in diameter of the sample. Behind the sample the pump and probe pulses are separated by a polarizer and the probe light is imaged onto the FPA. Under the chosen conditions, the microscope records for each pixel about 1.1 μm separated spots from the sample. The performance of the microscope was characterized by measuring 2D IR spectra of samples containing polystyrene beads

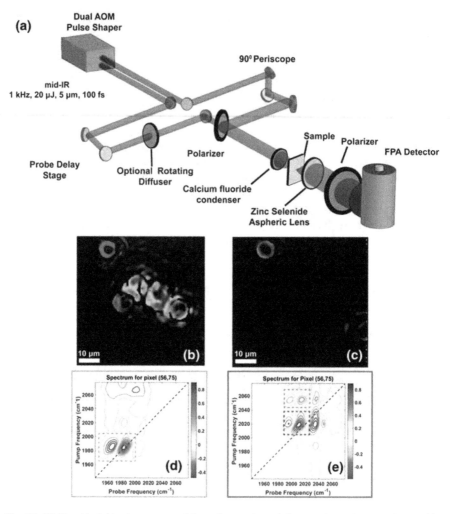

Fig. 17 2D IR wide-field microscopy. **a** Schematic overview of the experimental setup using a wide-field microscope and a dual AOM pulse shaper. **b** and **c** Chemical images with color coding as blue for $W(CO)_6$ (WHC) and red/orange for $Mn_2(CO)_{10}$ (DMDC) diagonal/(intramolecular) cross peak, respectively. **d** and **e** Representative single pixel fully-absorptive 2D IR spectra of WHC and DMDC, respectively. Adapted from Ref. [60]. Copyright American Chemical Society (2016)

impregnated with different metal-carbonyl compounds, i.e. $W(CO)_6$ (WHC) and $Mn_2(CO)_{10}$ (DMDC) (Fig. 17b and c, respectively). These compounds yield diagonal and diagonal plus intramolecular cross peaks, respectively (Fig. 17d and e), which can afterwards be used to construct chemically encoded images (Fig. 17b and c). Mixing beads with the different carbonyls, the independently labelled particles can be identified in the mixture based on their diagonal and cross peak vibrational characteristics. That is, the blue dashed square (Fig. 17d) is a representative spectrum of the blue region in Fig. 17b, and the red/orange dashed squares (Fig. 17e) are representative contributions of the red/orange regions in

Fig. 17b and c. An additional important feature of nonlinear spectroscopy in combination with microscopy is the dependence of the signals strength on the fourth power of the transition dipole moment (Sect. 2) [10]. This enhances the contrast in the images over comparable data from FT IR microscopes due to the square-dependence of the stationary signal in the latter case [59, 60]. Although up to now no temporal resolution has been demonstrated in combination with imaging, such experiments will be very useful in biological applications, tissue analysis, or for studying chemical exchange and energy transfer in heterogeneous systems, as well as combinations with transient 2D IR spectroscopy methods (see Sect. 4.4).

4.3 3D IR Spectroscopy

2D IR spectroscopy has been incredibly valuable in characterizing inter- and intramolecular dynamics of high-frequency vibrational modes from molecules in liquids and in different other environments. However, there exist a couple of important aspects and situations, which cannot easily be probed via third-order nonlinear spectroscopy. These cases thus require higher-order methods, such as so-called 3D IR spectroscopy and transient 2D IR spectroscopy, and these are considered in this and the upcoming section.

We start with considering the 3D IR technique as an extension of 2D IR spectroscopy. Although any type of time-resolved spectroscopy that involves multiple combinations of temporal and frequency axes (> 2) can be considered as at least "three-dimensional" [185], it is important to classify what exactly is meant by 3D IR spectroscopy in the context discussed here. Conventional 3D IR spectroscopy is thought of as being limited to only those cases where the sample evolves in coherent states during at least three time-variables ($t_{1/3/5}$) that are distributed between five excitation and detection fields in combination with a local oscillator field (Fig. 18a) [232]. Time-variables $t_{2/4}$ are then responsible for the dynamics of for instance population and inter-state coherence relaxation of the sample. In contrast to other methods, which can in principle be used to report partly similar information (e.g. 2D Raman spectroscopy [233–236]), 3D IR involves only fully resonant interactions in the electronic ground state potential of the sample molecule.

Just as 2D IR spectroscopy has been introduced as a pump-probe experiment with frequency-resolved pump and probe pulses, 3D IR spectroscopy has been proposed as a "hole-burning experiment with 2D IR detection", for which pump-frequency dependent lineshape dynamics such as spectral diffusion or chemical exchange, energy transfer and couplings can be investigated [232, 237]. Considering the five contributing excitation and detection fields, a much larger set of energy level diagrams as for 2D IR spectroscopy can be expected to take part in the signal generation process [10, 62]. Figure 18b–d gives only a few examples, which are relevant in a three-level system that is generally considered for ground state bleach and excited state absorption pathways. Figure 18b shows a contribution that is an extension of the ground state bleach signal from third-order 2D IR spectroscopy and an analogous diagram can be drawn for stimulated emission (not shown explicitly) [238, 239]. In contrast, Fig. 18c and d shows contributions that involve also excited state absorption transitions (1–2) and measure dynamics of higher-lying vibrational

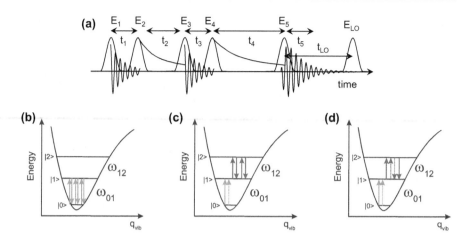

Fig. 18 b Pulse sequence for 3D IR spectroscopy and associated temporal delays involved in the experiment. The sample is in coherent states during t_1, t_3 and t_5 and undergoes for instance vibrational relaxation during $t_{2/4}$. $E_{1\text{-}5}$ are excitation/detection fields whereas E_{LO} is the local oscillator. **b–d** Exemplary energy level diagrams of fifth-order 3D IR spectroscopy. Color codes for the arrows refer to the electronic ground state potential given in Fig. 3. Examples of the full set of fifth-order diagrams have been given elsewhere [244]

levels. The diagram in Fig. 18c can be thought as a transient bleach signal of the $v = 1$ level after initial excitation. Similarly, the diagram in Fig. 18d represents a transient stimulated emission from the $v = 2$ level. Higher vibrational ($v > 2$) levels can in principle contribute as well, but have so far only rarely been considered [240, 241]. Such contributions will be considered in a different context in a subsequent Sect. 5.1 [242, 243].

3D IR spectroscopy has been proposed to yield valuable information concerning mainly two types of dynamics. The first one is not very obvious and concerns the dynamics of intermolecular low-frequency modes [232]. Although third-order 2D IR spectroscopy is already sensitive to such molecular contributions by observing spectral dynamics of some high-frequency "spectator-modes", the method is only linear with respect to the intermolecular low-frequency degrees of freedom. 2D IR probes the two-point FFCF (Sect. 2), and extensions are needed to obtain information beyond that level. This means in general that one has to consider higher-order pulses sequences compared to the ones used for 2D IR. 3D IR spectroscopy has been designed to do exactly that, i.e. to spectrally label the sample system at an additional time point, from which fluctuations can be probed. This allows one to obtain information about higher-order correlations by investigating the characteristics of a three-point correlation function ($C_3(t_4, t_2)$).

$$C_3(t_4, t_2) = \langle \delta\omega_{jk}(t_4)\delta\omega_{ij}(t_2)\delta\omega_{01}(0)\rangle \tag{6}$$

Note that in principle the frequencies can be different due to the involvement of higher-lying excited vibrational levels. Looking at this type of correlation function, a distinction can be made between homogeneous vs. heterogeneous dynamics by introducing additional spectrally resolved pump interactions. It is important to note in this context that only in the case of non-Gaussian statistics of the transition frequency fluctuations $\delta\omega(t)$ one would obtain new physical information about the system [232]. Because the local interactions that often dominate vibrational dynamics, it can be expected that non-Gaussian statistics may play a significant role in vibrational dynamics of several types of samples, especially in non-isotropic environments, where asymmetric lineshapes are observed in the samples. However, demonstrations of 3D IR spectroscopy in this context are rare [232]. and the discussion will be limited to the other essential strength of 3D IR.

The second important example that has been proposed for 3D IR spectroscopy concerns so-called (non-)Markovian dynamics in a chemical reaction. Broadly speaking, the Markovian assumption implies that the temporal evolution of a molecular system depends only on its current state, but not on previous events. That is, the sample system has no "memory" about earlier events. In such a case, correlation functions such as Eq. (6) would simply factorize in two lower-order correlation functions (Eq. 7). As an example for the breakdown of this assumption, it has been suggested that chemical reactions, which occur as fast as the solvent relaxation around it, can become non-Markovian [245].

$$C_3(t_4, t_2) = \langle \delta\omega_{jk}(t_4)\delta\omega_{ij}(t_2)\delta\omega_{01}(0)\rangle \equiv \langle \delta\omega_{jk}(t_4)\delta\omega_{ij}(0)\rangle\langle \delta\omega_{ij}(t_2)\delta\omega_{01}(0)\rangle \quad (7)$$

The potential of 3D IR spectroscopy for resolving non-Markovian dynamics in chemical reactions has recently been explored by Hamm et al. [245]. As a model reaction, the authors re-investigated the chemical exchange of hydrogen bond formation/dissociation between phenol-OD/benzene in CCl_4 (see also Sect. 3.1.4) from the perspective of Markovianity. The aim of this experiment was to test if the OD-stretch vibration performs as a Markovian coordinate or not, i.e. if the complexes associated/dissociated during the vibrational lifetime after initial excitation perform spectroscopically differently as the ones that did not associate/dissociate. As an essential result, it was found that the OD-stretch vibration is indeed a non-Markovian coordinate. Using molecular dynamics simulations to back up that interpretation, it could furthermore be revealed that the non-Markovian dynamics originate from the heterogeneous structure of the mixed solvent (benzene/CCl_4), which forms clusters due to the different chemical nature of the molecules [246]. In this situation, the dynamics can be fairly complex since hydrogen bond formation/dissociation can depend on the position of the phenol molecule in the cluster.

The 3D IR signals of the phenol-OD/benzene system depicted in Fig. 19 have been obtained at the indicated waiting times, where the first value is associated with the t_2 delay and the second value with t_4. As described in Sect. 3.1.4 the signal at 2660 cm^{-1} is associated with the free phenol-OD (F) whereas the broader signal at 2630 cm^{-1} is associated with the complex (C). Note that only the ground state bleach/stimulated emission signal contributes in the considered spectral range, i.e. no energy level diagrams of the form of Fig. 18c and d contribute to the signals. In extension to 2D IR spectroscopy, diagonal peaks show up, but now on the body-diagonal line, if the waiting times are (close to) zero (upper left spectrum). In contrast to 2D IR spectroscopy, now six possibilities exist for cross peaks to show

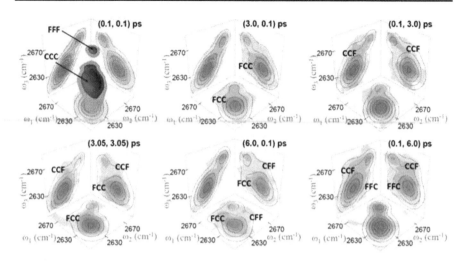

Fig. 19 3D IR spectra of chemical exchange in the hydrogen bond formation/breaking reaction between phenol-OD/benzene [245]. Population times t_2 and t_4 are indicated for each spectrum. Only the 0-1 ground state bleach/stimulated emission signal is resolved in the considered spectral range. Adapted with permission from Ref. [245]. Copyright National Academy of Sciences (2014)

up in a 3D IR spectrum if chemical exchange occurs between C and F states of the sample for increasing waiting times. For a better visualization of the cross peaks, all other 3D IR spectra are shown as 2D projections onto the different combinations of frequency axes. For an increase of only the first population time, exchange can occur along t_2, but not along t_4. A free phenol (F) can therefore turn into a complexed (C) molecule, but subsequently C has too little time during t_4 to fall apart (FCC, upper middle). The counterpart is the CCF cross peak, for which the complex has no time to dissociate during t_2, but dissociates during t_4 (upper right). If both waiting times are scanned (lower left), then dissociation/formation of C and F can both occur, which leads to the different cross peaks CCF and FCC. Unfortunately, the most interesting cross peaks, for which formation/dissociation is observed two times (CFC and FCF), could not be resolved within the signal-to-noise ratio. From the experimental data, however, Markovianity of the exchange reaction could still be tested by evaluating probability ratios of the different contributions, e.g. p_{FCC}/p_{CCC} [245]. The basis for this test is that the in case of a two-state, Markovian process, a three-point correlation function expressed in terms of probabilities rather than frequencies and which is equally accessible from 3D IR factorizes into two two-point probability functions, just as in Eq. (7). Any deviation from an expected two-state behavior then indicates non-Markovian contributions, what was indeed observed both experimentally, as well as from MD results [245].

A much simpler test for non-Markovian behavior of a system was proposed to be based on experimentally observed non-exponential dynamics in cases where strictly exponential responses are expected from kinetic considerations. However, it was argued that experimental noise, as well as other overlaying kinetic processes such as relaxation can have a too strong impact on the kinetics that non-exponential dynamics are very hard to identify reliably. In that sense, 3D IR spectroscopy is a very powerful method for rigorously testing basic assumptions about chemical dynamics.

The concept of (non-)Markovian dynamics is in fact much broader as in the described comparatively simple case of hydrogen bond formation/dissociation between two small molecules in a heterogeneous solvent. Markov-state models are very relevant for protein dynamics and the timescales for structural changes in such complex systems cover many orders of magnitude [247–249]. However, the intrinsic dependence of the 3D IR signal on the lifetime of the vibrationally excited levels is likely to limit the application of the method for comparatively slow dynamics of vibrational relaxationn. Only in certain cases vibrational labels are long-lived enough to sample dynamic processes on the timescale of several hundreds of picoseconds or even nanoseconds [15, 111, 112].

3D IR spectroscopy has also been successfully applied to other sample systems. Also Hamm et al. have applied the concept to study the dynamics of highly vibrationally excited states of isotope-diluted ice (HOD in H_2O) [239]. It was found that the higher lying vibrational states in the ground state potential decay extremely fast, that is, on the order of 200 fs. By theoretical considerations, it was concluded that this ultrashort lifetime is based on mode mixing of the OD-stretch coordinate with lattice degrees of freedom of the ice crystal. For this particular system, the authors, therefore, identified a crossover behavior between adiabatic and non-adiabatic approximations, which occurs in comparatively low-lying energetic regions of an electronic potential due to the strong anharmonicity of the OD-stretch vibration of water.

Unfortunately, 3D IR spectroscopy is a complex experimental method and the contributing fifth-order response functions scale in a cubic manner with the absorption coefficients and so does the signal. Contrary to initial hopes and expectations [10], it will thus be very challenging to apply 3D IR to a broad range of sample systems and particularly cases such as surfaces and interfaces, where the method bears a great potential to investigate heterogeneous dynamics.

4.4 Transient 2D IR Spectroscopy

While 3D IR spectroscopy is able to resolve non-equilibrium dynamics in the electronic ground state of a sample, considerably more information can be obtained by the help of a couple of additional variants of transient 2D IR spectroscopy. In transient 2D IR methods, the sample is brought to a non-equilibrium state, and subsequently 2D IR spectra are measured either in dependence of the delay of initial perturbation, or in dependence of the conventional IR population delay. The amount of information that can be obtained is manifold, including dynamics from excited electronic states [39, 250–257], photo-induced chemical reactions [38, 258, 259], or different types of non-equilibrium sample conditions such as temperature variations [260–264]. Most importantly, this concept allows an extension of experimentally available timescales far beyond vibrational relaxation. This is because the only thing that is needed is a spectral shift, intensity change or broadening of a vibrational transition upon perturbation, which, depending on the system under investigation, may principally last forever. The discussion here is focused on essentially the above mentioned three types of non-equilibrium states of preparation to demonstrate the power of transient 2D IR spectroscopy.

The different variants of transient 2D IR spectroscopy are generally realized by adding more pulses to the standard 2D IR pulse sequence and extending the spectral range of investigations (Fig. 20). The additional pulses can be thought of as triggers for the generation of a non-equilibrium state of the sample. In many cases, the additional pulses directly act on the sample molecules and promote them to, e.g., an excited electronic state. In other variants, the initial perturbation pulses can be used to raise the temperature of the sample, or change the pH of the solution environment. A standard pulse sequence for transient 2D IR spectroscopy is depicted in Fig. 20a. Black pulses and oscillatory lines have the same meaning as in a standard 2D IR experiment (Fig. 2), i.e. excitation and detection fields as well as vibrational free-induction decays. The delays $t_{1/3}$ still indicate temporal axes with respect to which Fourier-transformations are carried out to derive a 2D IR spectrum ($\omega_{1/3}$) and t_2 is the IR population delay. In addition, excitation with a UV/VIS field ($E_{UV/VIS}$) precedes the 2D IR sequence and induces the non-equilibrium state. The $E_{UV/VIS}$ field is schematically split into two replica since in most cases considered here that pulse contributes two light-matter interactions. The temporal resolution on the axis of UV/VIS excitation that can be achieved with this method is approximately about the pulse duration, i.e. generally between 10 and 100 fs. Note that the definition of the pre-excitation delay is used in different ways in the literature, depending on the actually performed experiments. Variants exist, in which the delay T is defined with respect to the second light-matter interaction of the IR sequence (upper short arrow) [39, 251, 252, 255, 256]. Other experiments

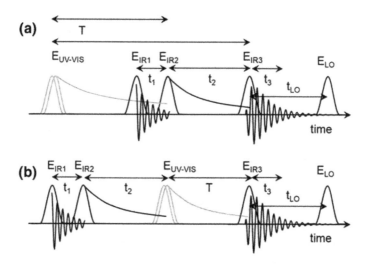

Fig. 20 Two important pulse sequences for transient 2D IR spectroscopy. **a** Conventional transient 2D IR. A UV/VIS pulse precedes the 2D IR sequence and perturbs the sample in a given way (see text) to result in 2D IR spectra in dependence of the initial perturbation. **b** Triggered exchange transient 2D IR spectroscopy. Following the initial interaction with the two IR pump fields, the perturbation pulse intercepts the samples evolution during t_2. The probe pulse interrogates this newly perturbed sample following the UV/VIS interaction

have defined the delay T with respect to the third light-matter interaction of the IR fields (lower long arrow) [250]. In fact, most of the reported transient 2D IR experiments have utilized 2D IR spectra generation using a Fabry–Perot filter [252, 265], where the initial two IR interactions both take place under the envelope of the same excitation pulse. This avoids complications in the data analysis based on the necessity to scan additional delays between $E_{IR1,2}$, which may partially overlap with the T delay to some extent (e.g. at very short T delays). If the delay between the initial excitation and the IR excitation is sufficiently long, i.e. when excited state vibrational equilibration and solvation has completed, then transient 2D IR spectra can be interpreted as a third-order 2D IR spectrum from an excited electronic state. Experiments of that sort have been used to study structural dynamics in biomolecules or bio-molecular sub-units [38, 258, 259, 266], solvation dynamics in excited electronic states [254, 267], or electronic relaxation dynamics and associated vibrational signals of excited electronic states. Other examples exist regarding structural dynamics in large organic molecules [255, 256], charge-injection dynamics between adsorbates and mesoporous semiconductors [250], or spectral shifts in charge-transfer states of inorganic complexes [251].

Other implementations of transient 2D IR spectroscopy exist, which utilize the UV excitation pulse after the initial interaction with the two IR fields (Fig. 20b). In that particular case, the UV/VIS pulse intercepts the relaxation during the IR population delay and promotes the system to an additional non-equilibrium state, e.g. again an excited electronic state or a photochemical reaction channel. Following this interception, the probe pulse interrogates the total perturbation of the system, which consists of IR, as well as UV/VIS excitation. That variant is commonly referred to as "triggered exchange transient 2D IR spectroscopy" [51, 252, 253, 268]. A possible interpretation of this type of transient 2D IR is that the ground state IR spectrum is correlated with the excited state IR spectrum, or even a starting material to a photo-product [250, 252]. Such experiments have revealed spectral signatures of metal-to-ligand charge transfer in carbonyl complexes [252], photo-induced ligand migration in proteins [269], or different electron-transfer efficiencies of adsorbates at semiconductor surfaces [250]. Selected highlight examples of both of the transient 2D IR implementations are discussed further below in this section.

A final note is due on the interpretation of the spectra. As the IR sequence is resonant with IR transitions from the sample molecules that have undergone pre-excitation, as well as those that did not, a subtraction of the two types of 2D IR spectra is often needed to derive a transient 2D IR spectrum, i.e. with and without the pre-excitation interaction.

The first example to be discussed here is the observation of transient hydrogen bond dynamics in a beta-turn of a cyclic peptide, cyclo(Boc–Cys–Pro–Aib–Cys–OMe) (Fig. 21a). Hamm et al. demonstrated that transient 2D IR can be used to follow in real time the weakening of an intramolecular hydrogen bond and the subsequent opening of a beta-turn upon UV-photolysis of a disulfide bridge [258]. The considered sample exhibits several distinct carbonyl functional groups (color-coded in (a)), which can be observed in the amide-I region of the IR spectrum (color code in Fig. 21b). Following 266 nm photolysis of the disulfide-bridge, the carbonyl

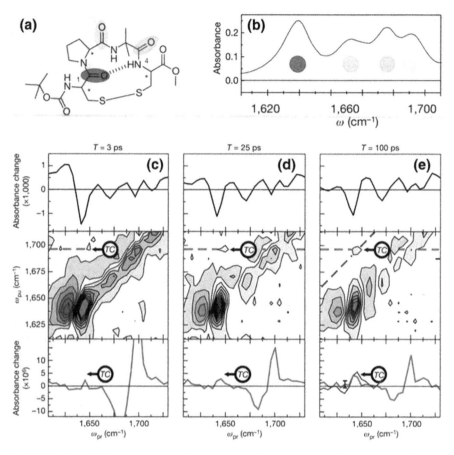

Fig. 21 Transient 2D IR spectroscopy of hydrogen bond dynamics in a cyclic beta-turn. **a** Chemical structure of the employed sample molecule cyclo(Boc–CPUC–OMe). The dashed line indicates the intramolecular hydrogen bond. **b** Stationary IR absorption spectrum of the sample in the amide-I region, color-coded to indicate the chemical bonds in (**a**). **c–e** Transient 1D (top row) and transient 2D IR difference spectra (center row) after photolysis of the disulfide bridge at indicated delays together with cuts along the probe spectral axis (lower row). TC refers to the transient cross peak. Adapted with permission from Ref. [258]. Copyright Nature publishing group (2006)

bands are considerably affected in their spectral positions. One effect that contributes to an almost instantaneous redshift of all bands is the transient formation of heat in the molecule. The photolysis excites low-frequency modes in the molecule, which are anharmonically coupled to the amide-I high-frequency modes. A spectral redshift due to this heat generation can be seen in both the 1D and 2D transient spectra as positive/negative going features (Fig. 21c–e). These signals are located along the diagonal line in the transient 2D IR spectra. The important feature that is seen in the transient 2D IR spectra is the growth of a transient cross peak (TC) between Cys(1) and Aib (positions color-coded in red and blue in Fig. 21a and spectral positions in Fig. 21b). The growth of that cross peak can be best visualized in cuts along the probe axis [red dashed line, lower row in (c)–(d)].

The two carbonyl groups are located at the two positions in the molecule, which are most sensitive to the intramolecular hydrogen bond. Additional equilibrium 2D IR spectra were used to show that these two amide-I modes are coupled in the molecule and thus changes in the 3D structure of the molecule concerning the hydrogen bond strongly affects this coupling. The time constant of that process was determined as about 160 ps, which is considerably faster than predicted speed limits for contact formation between side chains in related molecules.

Transient 2D IR spectroscopy has additionally been used to study properties of molecules in excited electronic states and in particular of adsorbates at semiconductor interfaces. Figure 22 shows spectra demonstrating that different binding configurations of a dye on a nano-crystalline and sensitized TiO_2 substrate exhibit different electron-transfer rates and efficiencies [250]. An equilibrium 2D IR spectrum was used to show that the dye molecules adopt different binding configurations on the substrate (Fig. 22a), visible through different IR transitions on the diagonal line (2010–2040 cm^{-1}). The different binding configurations were furthermore invoked to be uncoupled by the observation of exclusively diagonal peaks in the 2D IR spectrum without any hint for cross peaks. Adding a 400 nm pump pulse that precedes the 2D IR sequence by 20 ps, the authors showed by the observation of clear new features in the spectral region > 2050 cm^{-1} that photo-induced electron-injection indeed takes place at the dye-semiconductor interface (Fig. 22b). The bleach signals in the transient 2D IR spectrum shows that all three binding configurations absorb the UV light, but surprisingly only one new feature is generated. This suggests that the different binding configurations exhibit drastically varying electron-injection rates and efficiencies. To identify the binding configuration, which predominately contributes to the charge-injection process, the UV pump pulse was placed between the IR pump and IR probing pulse, i.e. a triggered exchange sequence. The obtained triggered exchange 2D IR spectrum (Fig. 22c) again exhibits the diagonal peaks from the different binding configurations, as well as a broad cross peak for the two binding configurations with the lowest frequencies (< 2030 cm^{-1}). That cross peak indicates that only these two binding configurations

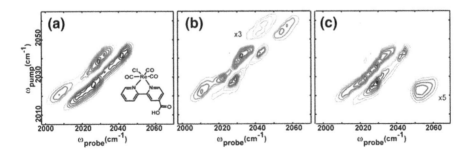

Fig. 22 Transient 2D IR spectroscopy for studying charge-injection dynamics and efficiencies of Re-carbonyl adsorbates at TiO_2 surfaces. **a** Equilibrium 2D IR spectrum of the sketched Re-carbonyl dye adsorbed at nano-crystalline TiO_2 surfaces at a population delay of 0 ps. **b** transient 2D IR spectra of the dye obtained with a UV pulse preceding the full 2D IR sequence by 20 ps. **c** Triggered exchange 2D IR spectrum with an IR-UV delay of 0.3 ps and an IR delay of 0.8 ps. Reprinted with permission from Ref. [250]. Copyright American Chemical Society (2009)

contribute to the charge-injection process, whereas the injection for the third configuration is negligible. It can be envisioned that future applications of transient 2D IR spectroscopy to similar systems will lead to valuable insights with regard to the performance of heterogeneous photo-catalysts, or several forms of dye-sensitized solar cells [159].

Kubarych et al. have established transient 2D IR spectroscopy to also allow studying differences in solvation dynamics between ground and excited electronic states in metal complexes [254]. The authors investigated solvation dynamics of the thermally equilibrated triplet state of Re-carbonyl complexes, which is populated after UV excitation. It could be determined that the spectral diffusion dynamics in the triplet state are three times slower as compared to the electronic ground state. These differences were attributed to the markedly different charge distributions in the molecules in the different electronic states. Future studies can be envisioned, which do not consider thermally equilibrated excited state systems, but which look at differences in spectral diffusion dynamics of vibrationally excited species as compared to thermally relaxed molecules. Such experiments could, therefore, be used to shed additional light on relaxation dynamics in excited electronic states.

Other versions of transient 2D IR spectroscopy investigated electronic relaxation dynamics, as well as excited state vibrational features in biological molecules such as carbonyl carotenoids [255, 256]. Excitation wavelength, as well as solvent-dependent spectral signatures and distinct cross peaks could be resolved in these studies, which originated from coupled stretching modes of the conjugated chain in electronic states of different symmetry. Besides revealing detailed insight into the special dynamics of these samples, it should be noted that these studies are of the few, which demonstrate that transient 2D IR spectroscopy can be applied to other sample systems besides strongly IR-absorbing metal-carbonyl complexes. As such, the method is likely to find in future much broader application than demonstrated so far.

4.4.1 Extending the Observation Window for 2D IR Spectroscopy

One particular issue in ultrafast vibrational spectroscopy is the comparatively short lifetime of the excited vibrational levels. Strong coupling of the sample to the environment, as well as rapid IVR often leads to signal decay on the timescale of a few picoseconds. This limits the temporal observation window and makes in intrinsically difficult to observe processes that occur on longer timescales. Recently, Bredenbeck et al. have proposed a method that can circumvent that lifetime-limitation by combining VIS and IR excitation, similar to triggered exchange spectroscopy. In their method of vibrationally promoted electronic resonance (VIPER) 2D IR spectroscopy a vibrationally resonant 2D IR sequence is combined with an additional UV/VIS excitation pulse that is for itself *non-resonant* with any electronic transition of the sample (Fig. 23b) [270]. The UV/VIS excitation pulse is arranged to arrive after the IR excitation pulse at the sample position. Only the combination of IR and UV/VIS excitation can promote the system to an excited electronic state. By this method the applicability of 2D IR spectroscopy can be extended to timescales that are much longer compared to the vibrational relaxation

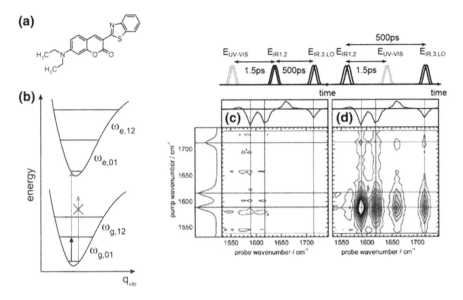

Fig. 23 VIPER 2D IR spectroscopy applied to coumarin 6. **a** Chemical structure of Coumarin 6. **b** Schematic illustration of the VIPER excitation mechanism, which involves resonant IR-excitation (black) and non-resonant VIS-excitation (green). Only the combination of IR and UV/VIS pulses can excite the sample to the excited electronic state. **c** and **d** 2D VIPER signals with the non-resonant VIS pulse preceding the 2D IR sequence in between the IR pump and probe pulses, respectively. Only the latter combination results in a measureable 2D VIPER signal. Adapted with permission from Ref. [270]. Copyright Wiley VCH (2014)

in the electronic ground state. This was demonstrated by applying VIPER 2D IR spectroscopy to the coumarin 6 dye (Fig. 23a), which exhibits ring-modes and CO stretching modes in the spectral range 1550–1800 cm^{-1} (Fig. 23c left panel, red). It could be shown that the VIPER 2D IR combination with a preceding UV/VIS pump pulse leads to negligible signal intensity on the timescale of hundreds of picoseconds due to much faster vibrational relaxation and only non-resonant UV/VIS excitation (Fig. 23c). However, once the non-resonant UV/VIS pulse was placed between IR excitation and probing pulses, the VIPER signals with extended lifetime could clearly be measured (Fig. 23d). The application of VIPER 2D IR to other systems may thus allow in future studies to record solvation dynamics, chemical exchange or energy transfer far beyond the vibrational lifetime of sample molecules in the electronic ground state. It is, however, noted that the selective excitation combination with IR and UV/VIS pulses requires fairly steep edges of the UV-VIS absorption spectrum of the sample. If the UV-VIS is too spread out, then non-resonant electronic excitation will hardly be possible.

4.4.2 Temperature-Jump Transient 2D IR Spectroscopy

All examples discussed so far involve direct interaction of the additional UV/VIS pulse with the sample molecules to induce the non-equilibrium state, which is probed by the 2D IR sequence. Additional implementations have been realized, in which the non-equilibrium state is induced indirectly, e.g. by interaction of the trigger pulse with the solvent. One such realization that has been used extensively is the method of temperature-jump (T-jump) 2D IR spectroscopy [260, 262, 264, 271]. In this variant, a high-energy nanosecond (1–10 ns) pump pulse is tuned to a resonant excitation of the solvent molecules, i.e. water in most cases. During that excitation pulse, large amounts of thermal energy are deposited in the irradiated sample volume, which raises the temperature of the solvent molecules. T-jumps as large as 10 K have been demonstrated experimentally [262]. The heat is generally assumed as evenly distributed in the focus volume following the ns-interaction and the sample temperature remains approximately constant on timescales up to the ms-regime, i.e. when the heat diffuses out of the focal spot. T-jump 2D IR spectroscopy can thus probe structural dynamics on the timescale from ns to ms and has been used to predominately study bio-molecular systems and in particular protein-unfolding reactions [260, 262], as well as tautomerization reactions [264].

Figure 24 showcases the application of T-jump 2D IR spectroscopy to unravel the unfolding dynamics of ubiquitin in water (Fig. 24a) [260]. A T-jump pulse was used in this case by Tokmakoff et al. to study the unfolding phases of the protein

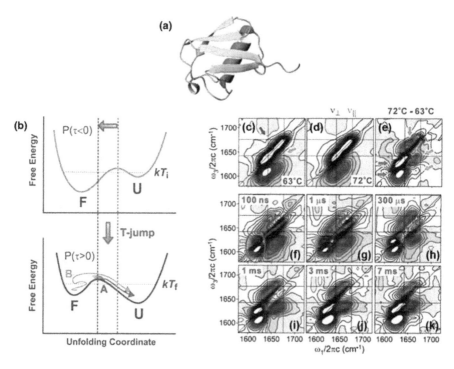

Fig. 24 a Crystal structure representation of ubiquitin. **b** Mechanism of a T-jump experiment. **c–k** Steady-state 2D IR and temperature-jump transient 2D IR spectroscopy of ubiquitin unfolding. **c** and **d** Temperature-dependent equilibrium 2D IR spectra of the amide-I region of ubiquitin at indicated temperatures, and **e** the difference between **c** and **d**. **f–k** Time-dependence of transient 2D IR difference spectra at indicated T-jump delays. Adapted with permission from [260]. Copyright National Academy of Sciences (2007)

that is close to unfolding transition state (Fig. 24b). To be able to reach the unfolding transition state, the sample needs to be pre-heated to a temperature of 63°C. This pre-heating of the sample is necessary since comparable T-jumps do not lead to unfolding of the protein at lower temperatures. The T-jump induces a barrier shift towards the folded state (F), leading to a downhill unfolding (U) of a part of the sample (A in Fig. 24b). Afterwards, a remaining sub-ensemble can unfold by overcoming the shifted energy barrier on a longer timescale (B in Fig. 24b). Figure 24c shows the corresponding 2D IR spectrum at elevated temperature in the amide-I spectral range, which is characterized by intense diagonal peaks and weak cross peaks (arrows). The two beta-sheet modes lie at positions of 1642 cm^{-1} ($v_\perp$) and blue 1676 cm^{-1} ($v_\parallel$), respectively. A change in the stationary sample temperature (Fig. 24d) leads to disruption of the beta-sheet of ubiquitin. This is observed by a blue-shift of the $v_\perp$ mode, which can be best visualized in a difference 2D IR signal between the two temperatures (red arrows in Fig. 24e). The blue and green arrows indicate an increase in the random-coil structure and a depletion of the cross peak signals, respectively. Figure 24f–k demonstrates, how the accompanied structural changes in the molecules can be tracked with transient 2D IR T-jump spectroscopy. At short T-jump delays (100 ns), the T-jump difference spectrum shows predominately temperature-induced disruption of H-bonds (red ellipse), which cause a blue-shift of the sheet vibrations. Following that initial response, the difference spectra gradually evolve up to delays as large as 7 ms, (g)–(k), where almost a quantitative agreement with the stationary difference spectrum, (e), is obtained. Through a kinetic analysis of a full series of such T-jump 2D IR spectra, as well as a comparison with MD simulations, it was possible to obtain a detailed understanding of the progressive unfolding mechanism of ubiquitin. It was revealed that ubiquitin unfolds in three distinguishable phases, i.e. a first stage of thermal excitation of the solvent, a second stage of microsecond downhill unfolding, and a third stage of ms-unfolding over a barrier (Fig. 24b). Based on the close matching of experimental and MD simulation results it was furthermore proposed that the T-jump 2D IR spectroscopy can be used to discriminate between different theoretically proposed unfolding pathways.

T-jump 2D IR spectroscopy has also been used to study structural changes of other biological sample systems. Also, Tokmakoff et al. have studied the folding mechanism of the N-terminal domain of ribosomal proteins and compared the experimental results to Markov-state based MD-simulations [262]. Native and denaturized structures could be correctly predicted and time-constants for the temperature-induced interconversion were successfully modeled. Other applications by the same group considered protein-protein binding and T-jump induced dissociation dynamics of insulin homodimers [271]. A detailed comparison of the observable kinetics at different starting temperatures indicated that the dissociation is characterized by two processes, the influence of which varies with temperature. Future applications of T-jump 2D IR spectroscopy are moreover likely to involve combinations of T-jump 2D IR spectroscopy with other established methods such isotope labelling and site-specific mutations. Finally, additional future experiments may involve addressing distinguishable conformational states in the folded and unfolded states of various proteins, as well timescales of exchange between them.

In total transient 2D IR spectroscopy in its many different variants has opened up a completely new way of measuring chemical structure and its evolution in sample systems as complex as proteins from picoseconds to milliseconds. These studies are expected to have a profound impact on the understanding of both biological systems, as well as, e.g., functional devices such as solar cell materials. The total potential of transient 2D IR spectroscopy has up to now not been evaluated in all aspects. It is, therefore, assumed that several additional experimental extensions will evolve in future works, some of which are already under way. These additional aspects will be evaluated further in Sect. 5.4.

4.5 Mixed Vibrational-Electronic 2D Spectroscopies

The preceding sections have covered experimental developments that were realized by adding new pulses to a standard 2D IR pulse sequence. Very recently, other variants of multi-dimensional vibrational spectroscopy have been presented that make use of a substitution of selected pulses in the 2D pulse sequence by pulses from other frequency ranges, in particular from the UV/VIS region (> 10000 cm^{-1}). This way, so-called mixed 2D vibrational-electronic (VE) and electronic-vibrational (EV) signals have been obtained, depending on which of the pulses is substituted. Such methods are able to cover large parts of exclusively the cross peak regions in Fig. 1 and reveal information that is very different from standard 2D IR spectroscopy. In this section, as well as the succeding Sect. 4.5.2 these developments are presented, their potential is discussed and selected applications are commented.

4.5.1 2D Vibrational-Electronic (VE) Spectroscopy

To start with, consider the case where the probe pulse in a standard 2D IR sequence is replaced by a near-infrared (NIR) or VIS pulse. This results in the pulse sequence for VE spectroscopy (Fig. 25a). This method has initially been designed to investigate dynamics of intramolecular couplings between vibrational and electronic degrees of freedom and vibronic couplings in metal complexes [272], but the principle is more general and can be applied to other systems as well. Intramolecular properties of that sort are expected to play key-roles in photo-biology and functional materials where molecular struture is often architectured to exhibit a particular function. 2D VE spectroscopy also reports on mode-specific vibrational-electronic frequency-frequency correlations via the observable lineshapes of the signals as well as on intra and intermolecular vibrational relaxation.

The general principle of 2D VE is similar to that of standard 2D IR spectrocopy and involves two phase-stable IR pump pulses that initially excite a coherence and subsequently a population in the sample along with a delayed probe pulse, which is now spectrally located in the NIR/VIS range and a local oscillator for heterodyne-detection. To generate a 2D VE signal the coherence delay between the pump pulses is scanned for a fixed population delay between the second pump and the probe pulse. The temporal evolution along the population time axis then reports on the ultrafast dynamcis of the sample. In addition, the dynamics of the signal may also

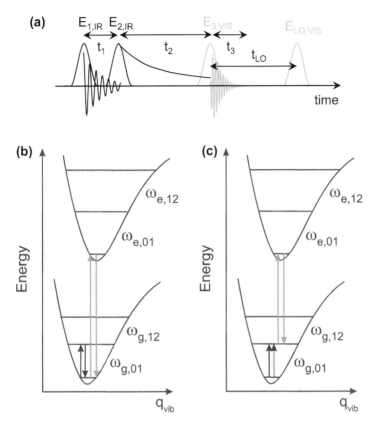

Fig. 25 2D vibrational electronic (VE) spectroscopy. **a** Pulse sequence for the signal generation. Green pulses are NIR/VIS frequencies whereas black pulses represent IR frequencies. Time runs from left to right. **b** and **c** Representative energy level diagrams for 2D VE spectroscopy for ground state bleach (GSB) and excited state absorption (ESA), respectively. Vibrational frequencies in $|g\rangle$ and $|e\rangle$ can be different. Additionally, depending on the spectral width of the probe pulse, multiple vibrational levels can be reached in the excited electronic state. Only non-rephasing diagrams are shown

report on coherent and incoherent energy transfer among the coupled vibrational modes.

Experimentally, 2D VE is realized in the 2D pump-probe geometry to facilitate phase-matching and avoid to scan separately for rephasing and non-rephasing signals to obtain fully absorptive spectra. The signal field emission is intrinsically phase-matched with the probe pulse and thus "self-heterodyned". Mechanistically, signal contributions are different as compared to the response pathways for 2D IR spectroscopy (Fig. 3) and involve resonant vibrational as well as electronic transitions (Fig. 25b and c). As an example, the initial IR excitation can prepare a population in the $v = 1$ state of the electronic ground state ($|g,1\rangle$), which is afterwards probed electronically via the NIR/VIS pulse and the level $|e,1\rangle$, similar to a conventional excited state absorption pathway (Fig. 25c). In a similar manner, also ground state bleach contributions show up in the 2D VE signals (Fig. 25b). These two pathways carry opposite signs and can thus partially cancel in a 2D VE signal. Contrary to 2D IR spectroscopy, no stimulated emission signals occur in the response due to the large frequency difference in the excitation and probing pulses for 2D VE spectroscopy. Note that only exemplary non-rephasing signals are shown in and complementary rephasing signals contribute as well. A more extensive theoretical treatment of the signals involving a full set of response pathways has been given elsewhere [273]. The signal strengths depend on transition dipole moments of both IR, as well as electronic transitions. It is, therefore, important to note that vibrational modes dominate the signal that are coupled to the electronic transitions. However, also non-resonant electronic interactions can contribute to due to the modulation of the polarizability of the sample by IR pre-excitation. That has important consequences since it implies signals from the sample as well as from solvent molecules. The latter response can be separately analyzed or even subtracted. What ultimately counts in the 2D VE signal is the frequency difference between vibrational modes in ground and excited electronic states (Fig. 25b and c). Because the frequency in the excited electronic state can be higher or lower than in the ground state, the magnitude and the sign of the frequency shift both affect the position of the excited state absorption signal with respect to the ground state bleach signal. That is an important difference to 2D IR spectroscopy, where anharmonic shifts in electronic potentials generally shift the excited state absorption bands to lower frequencies, although exceptions exist in special cases [274].

To highlight the applicability of 2D VE spectroscopy, Fig. 26c shows exemplary 2D VE spectra at two indicated population delays of a model compound (Fig. 26d) dissolved in bulk formamide, i.e. $[(CN)_5Fe_{II}CNRu^{III}(NH_3)_5]^-$ (FeRu, upper two panels), along with complementary 2D VE measurements performed on the neat solvent (lower two panels) [272]. The sample exhibits four vibrational modes in the considered spectral range, which are attributable to CN stretching modes (colored arrows in Fig. 26d and top panels for FT IR spectra in c). The 2D VE signals of the FeRu sample exhibit a comparatively narrow width in the ω_1 domain but considerably larger widths in the ω_3 domain due to the involvement of the spectrally broad electronic transition, which is a metal-to-metal charge transfer transition in this case. The signals are dominated by excited state absorption contributions (blue) that are persistent at both delays and change shape as well as intensity with increasing population time. Interestingly, not all modes contribute to the 2D VE signals with an intensity that would reflect their IR absorption spectrum. That observation reflects the different couplings of the modes to the electronic transition. The lineshapes in 2D VE spectra appear strongly different from the ones that are obtained from 2D IR signals. This is a result from the sensitivity of the 2D VE signals on the fluctuations of the coupled vibrational and electronic frequencies. A detailed analysis similar to the CLS method for 2D IR spectra [100] revealed that the CLS exhibit (i) nonlinear components and (ii) positive or negative CLS values for different transitions. Such behavior has been assigned to the different interaction of the modes with their environment, i.e. bridge modes modulates the Fe-Ru separation, whereas the trans modes couple to the solvent [273]. Overall, the loss of correlation was determined as very fast (< 1 ps), which reflects similar timescales as determined for diagonal peaks from 2D IR measurements [273].

(a)

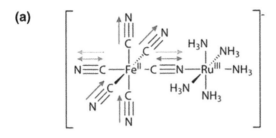

(b)

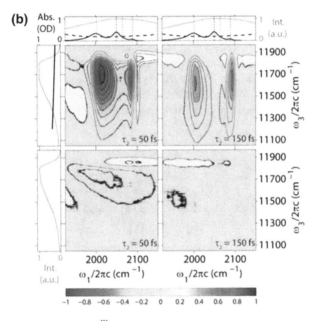

Fig. 26 **a** Sketch of $[(CN)_5Fe_{II}CNRu^{III}(NH_3)_5]^-$ (FeRu) with indications of the vibrational modes that are considered (arrows). **b** 2D VE spectra at different population delays of FeRu dissolved in formamide. Blue signals are excited state absorption and orange/red signals are ground state bleach signals. Top and left panels are FT IR and electronic absorption spectra, respectively. The spectral position of the color-coded vibrations in (**a**) are indicated in the FT IR spectra. Adapted with permission from Ref. [272]. Copyright American Chemical Society (2015)

2D VE spectroscopy appeared only very recently and its full potential has still to be evaluated in detail. Similar as 2D IR spectroscopy, the temporal window of observation is limited by vibrational relaxation and thus information about longer-lived electronic dynamics cannot easily be obtained. The preliminary experiments reported so far utilize comparatively narrow bandwidths of the IR pump and NIR/VIS probe pulses. It can be expected that a significant broadening of the both the ω_1 and ω_3 spectral axes will allow for much more detailed insight into the different signal contributions [275–277]. That way a larger range of Franck-Condon transitions can be detected and the role of anharmonicity on the signals might be addressable. The spectral broadening of the pump and probe pulses is expected to also help clarifying the role of ground state vibrational modes in electronic excitation dynamics during photochemical reactions of large biological molecules such as rhodopsins [278] or light-harvesting complexes [146].

4.5.2 2D Electronic-Vibrational (EV) Spectroscopy

Aiming at combining the strengths of electronic [18, 58, 279] and vibrational 2D spectroscopy [10, 15], Fleming et al. have invented a hybrid technique and pioneered the field of 2D electronic-vibrational (EV) spectroscopy [280–285]. Similar as for 2D VE spectroscopy discussed above, the pulse sequence in 2D EV involves a combination of VIS and IR pulses. The arrangement is such that two phase-coherent VIS pulses predece an IR probe pulse, which is eventually responsible for signal generation (Fig. 27a). The involved VIS excitation allows the investigation of photo-induced chemical reactions or energy transfer in complex molecular systems, while the IR detection guarantees high chemical and structural sensitivity. Experimentally, the coherence delay (t_1) between the two pump pulses is scanned for a fixed population delay to generate a 2D EV spectrum. That generates

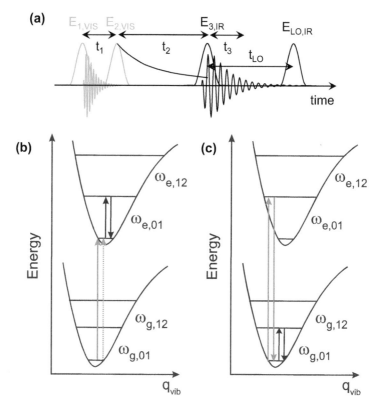

Fig. 27 **a** Pulse sequence for 2D electronic vibrational (EV) spectroscopy. VIS pulses are sketched in green whereas IR pulses and the local oscillator pulse in sketched in black. Time runs from left to right. **b** and **c** Exemplary energy level diagrams for 2D EV spectroscopy for excited state absorption (ESA) and ground state bleach (GSB), respectively. Only non-rephasing diagrams are shown. Vibrational frequencies ω in $|g\rangle$ and $|e\rangle$ can be different

the ω_1 axis, which is in this case the degree for electronic excitation. The second spectral axis (ω_3) is, as almost always, derived by spectrally dispersing the probe pulse. Also in this variant, the signal is heterodyne-detected using a local oscillator, which in most cases is the probe pulse itself, as the reported studies have been performed in the pump-probe geometry. To map out the full molecular dynamics of the sample, the population delay (t_2) is finally scanned to derive information about electronic and vibrational relaxation and the correlations between electronic vibrational frequencies, dynamics of solvent-solute interactions, or the dynamics of electronic and vibrational correlations. Note that 2D EV spectroscopy depends on the electronic relaxation times and may, therefore, provide significantly longer observation windows as compared to 2D VE spectroscopy, what might constitute and important advantage of the method.

The processes that underlie the signal generation can again be visualized using energy level diagrams (Fig. 27b and c). Depending on the average photon energy, the spectral width of the VIS excitation pulse and the vibrational frequency of the modes of the molecule under study, different vibrational levels in the excited electronic state ($|e\rangle$) can contribute to the signal generation processes. Therefore, multiple diagrams for excited state absorption, (a), have to be considered. In contrast, generally only a single ground state bleach diagram is active (Fig. 27c), which is justified as long as the spectral width of the VIS pump pulses does not allow the excitation multiple vibrational levels in the electronic ground state, i.e. the generation of vibrational coherences. Moreover, stimulated emission diagrams between vibrational levels in the excited electronic state may also be considered, but such contributions have not been observed experimentally so far. Note that only exemplary pathways for non-rephasing diagrams are shown and complementary rephasing pathways contribute as well, as demonstrated in more detailed theoretical descriptions of 2D EV spectroscopy [282, 283].

Also, for 2D EV spectroscopy, the signals depend on the frequency shift of vibrational transitions in ground and excited electronic states, similar to the 2D VE methods. However, even in the case of negligible frequency shifts in the two electronic states, a signal could still be measured if the electronic excitation promotes the sample to higher-lying ($v \neq 0$) vibrational states in $|e\rangle$, or if the IR transition dipole moment is different in the two electronic states. Depending on the sign of the frequency-shift between ground and excited electronic states, excited state absorption transitions can show up on the high or the low-frequency side of the ground state bleach signal. Regarding the shape of the signals in 2D EV spectroscopy, the 2D spectra report on the correlation between electronic and vibrational degrees of freedom. This correlation has been investigated in detail with response function approaches for modelling 2D EV signals of simple model systems [282, 285]. In case of mixed electronic and vibrational spectroscopies, these types of correlations can be positive, zero, as well as negative. This is due to the details of the shape of the potential energy surfaces in ground and excited electronic states [285]. In general, correlations are determined by spectral elongation of the signal along the pump and the probe axis, similar to the determination of spectral diffusion in 2D IR methods (Sect. 3.1.3). The dynamical changes in shapes of 2D EV spectra then report on the loss of that correlation with increasing population delay.

Theoretical and experimental investigations have been used to elucidate how the spectral correlations can be related to the solvation correlation function for certain vibrations [282].

Figure 28b–d shows exemplary experimental 2D EV spectra of the laser dye DTTCI (Fig. 28a) dissolved in chloroform at a series of population delays. The single band at about 1400 cm^{-1} is associated with the ground state bleach signal of a backbone C=C stretch mode. No excited state absorption signal is observed due to the very low absorption coefficients of these modes in the excited state [282]. The dynamics of spectral correlations were measured with CLS values with respect to ω_1 (solid line, k_g in (e)) and ω_3 (dashed lines, k'_g in (f)), respectively. The decay of the correlation functions with about 1.8 ps perfectly matched the vibrational dephasing rate of that mode in the ground state, without any long-lived contribution ($(k(')_g > 0$ for $t_2 > 10$ ps). The close matching between experimental data and theoretical predictions, therefore, indicated complete homogeneous broadening of that band in $|g\rangle$. In a detailed theoretical treatment of the signals the authors also showed that the 2D EV spectra are sensitive to the coupling strength of the sample molecule to its

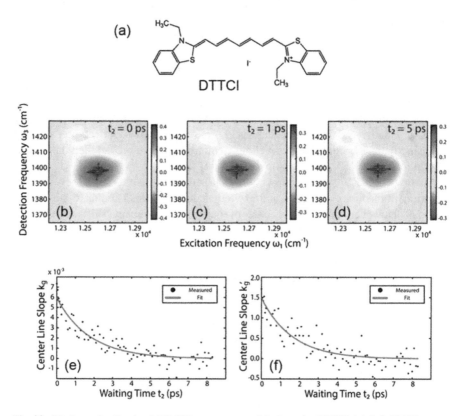

Fig. 28 2D electronic vibrational (2D EV) spectroscopy of the laser dye DTTCI, (**a**). **b–b** 2D EV spectra at indicated population delays. Blue solid and dashed lines measure the centerline slopes with respect to ω_1 and ω_3, respectively. **e** and **f** Spectral diffusion from the 2D EV signals with respect to ω_1 and ω_3, respectively. Solid dots represent experimental data and red lines represent exponential fits. Adapted with permission from Ref. [282]. Copyright American Institute of Physics (2015)

bath in both $|g\rangle$ and $|e\rangle$. The sensitivity of the dynamics also to excited state correlation dynamics originates from the involvement of the initial electronic excitation in the 2D EV spectra. Assuming absent inhomogeneous broadening of the sample a close analysis of the 2D EV spectra revealed that the coupling of the C=C stretch mode is about 50% stronger in the excited state than in the ground state. Such information is not easily available by other methods such as 2D IR or 2D ES and 2D EV may, therefore, provide an advantageous and straight forward way of measuring solvent-solute interactions in excited electronic states, as well as in ground electronic states of more general systems as well.

Also, 2D EV spectroscopy should be viewed as a very young method, the potential of which still needs to be evaluated. However, applications have already been realized, which aim at addressing physical and chemical questions in large bio-organic molecules. Fleming et al. have investigated 2D EV signals of chlorophyll a and b to distinguish different solvation states of the sample [281]. They found differently coordinated magnesium ions that manifest themselves as distinct spectral features in the 2D EV spectra. Additionally, the electronic excitation transfer through light-harvesting complex II (LHC II) could recently be measured with 2D EV spectroscopy for the first time precisely [284]. Using the distinct vibrational bands of chlorophyll a and b, the energy transfer could be observed with high chemical selectivity. A detailed analysis of the signals even made it possible to reveal the underlying relaxation pathways. These results may in future help to evaluate the quality of theoretical predictions for energy transfer pathways for such complex light-harvesting systems. Also, carbonyl carotenoids have been investigated with the help of 2D EV spectroscopy [286]. Lineshape analyses were used to make a precise assignment of the observable excited state vibrational signatures and excluded excited state isomerization reactions in these molecules. Furthermore, the authors were able to identify correlated electronic and vibrational lineshapes on timescales longer than the lifetime of a particular excited state in the sample. The observations were used to argue that the relaxation between different excited states is impulsive and involves a conical interaction. Similar results have been obtained by help of other 2D vibrational methods on closely related samples [287]. The so far presented experiments have thus set the stage for more extensive applications of 2D EV on larger samples as large as pigment-protein complexes.

5 Prospects for 2D IR Spectroscopy

As discussed above, there exists a large range of possible applications for 2D IR spectroscopy to study ultrafast dynamics, chemical reactions, or intermolecular interactions from samples in different environments. However, the general applicability of IR spectroscopy as an analytical tool in chemistry, biology and life sciences is much broacher and spans examples from industrial processes to ultra-sensitive analytics. In the upcoming section a few examples are given, for which 2D IR spectroscopy can be expected to yield a profound impact in the course of the coming years of experimental developments and applications.

5.1 Technical Applications and Advanced Data Acquisition

Most of the reported 2D IR measurements for samples in bulk solution environments are performed with sample flow cells, where a stationary spot of a cuvette is continuously irradiated, the sample is refreshed from shot to shot, and the composition of the sample solution is taken to be isotropic during the entire measurement and averaging time. There exist a couple of experimental situations in which such conditions are not or cannot be met, but where 2D IR spectroscopy would still be very helpful for elucidating dynamics or chemical reactions and interactions between sample constituents. A good example for such a situation is the flow of chemical mixtures in a microfluidic channel. Very recently, this type of application for 2D IR spectroscopy has been demonstrated by Krummel et al. in a study where the authors designed a microfluidic sample cell in conjunction with a high-repetition rate laser system (100 kHz) for measuring 2D IR signals from samples of non-uniform chemical constitution [288]. In that study the ultrafast vibrational dynamics of the OCN⁻ anion in different chemical environments has been investigated as a good model system since the employed solvent-ratio has a profound impact on its vibrational properties. A specific microfluidic channel was designed to create lateral concentration gradients of dimethylformamide/methanol (DMF/MeOH) mixtures in the cell device (Fig. 29a). 2D IR spectra were recorded at different lateral positions of the sample cell to address the different solvent compositions. Figure 29b–d shows excerpts of a full series of 2D IR spectra recorded at the indicated positions and (e) shows the obtained gradient of the two solvents as determined by integrated FT IR signals of some solvent modes [288]. From the spectra it is clear that the spectral position, as well as the shape of the

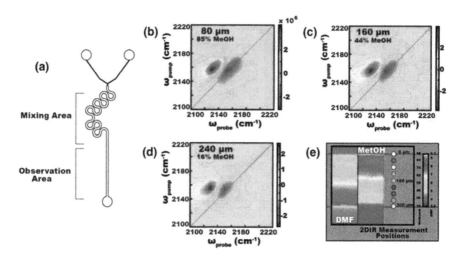

Fig. 29 Combination of 2D IR spectroscopy at high repetition rates (100 kHz) with a microfluidic sample cell shown in (**a**). **b–d** 2D IR spectra of the salt KOCN in a solvent mixture of MeOH/DMF. **e** Chemical map for the distribution of the solvent components across the microfluidic channel. Colored dots indicate the spatial position of the measurements. Adapted with permission from Ref. [288]. Copyright American Chemical Society (2016)

signals drastically depend on the DMF/MeOH ratio. Moreover, a comparison with 2D IR spectra obtained in the pure solvents, as well as spectral diffusion dynamics thereof indicated a complex solvation environment in the mixture, which cannot be modeled by weighted sums of the pure solvent spectra.

The achieved experimental demonstration of the combination of 2D IR spectroscopy with microfluidic conditions is an important step forward in the direction of broadening the applicability of the method as an analytical tool. Starting from this point, a series of additional investigations can be envisioned. As one most obvious point, microfluidics open a versatile route to new variants of transient 2D IR spectroscopy by the application of rapid mixing experiments [289–291]. Such methods are extremely valuable in chemical biology regarding studies of ligand-binding [292] or protein folding [293, 294]. The experiments allow a microsecond temporal resolution of reaction kinetics and minimal sample volumes, which are important for "Lab on a chip" applications [295, 296]. Importantly, the achievable temporal resolution in rapid mixing experiments depends on the spot size of the laser beam thus potentially allowing an interesting combination of 2D IR microscopy and high time-resolution rapid mixing laser spectroscopy in future studies. Other future implementations of 2D IR and microfluidics may involve attacking intermolecular interactions and ultrafast dynamics of reaction intermediates as problems from a chemical engineering point of view [297–299]. Finally, microfluidics is also a powerful method in combinatorial chemistry [289, 300]. As such, 2D IR might be applicable for the identification of new drugs, as well as the clarification of reaction intermediates and altered yields between small and large-scale chemical synthesis.

The fact that a 100 kHz system has been used for the described studies is not solely a side remark. In particular, such conditions allow for very high statistics and thus a comparatively high signal-to-noise ratio. Such laser systems therefore increase the sensitivity of the method and allow high throughput applications with acquisition times for single spectra as low as 5 s [288]. These considerations are particularly important for establishing 2D IR spectroscopy as an analytical tool in chemistry, biology and chemical engineering. This is similarly important for advanced methods such as 2D IR microscopy [301–303], which in combination with microfluidics may be even more powerful than the currently available methods.

Finally, all envisioned applications of 2D IR spectroscopy as an analytical method are likely to involve automated data acquisition [91, 304]. This is particularly important since that approach reduces the work force needed to perform laboratory experiments. As an important development in this direction, Rubtsov et al. have recently devised a versatile and automated multi-color 2D IR spectrometer that can operate between 800–4000 cm^{-1} and thus address any diagonal and cross peak region needed for chemical analytics [305]. Such developments therefore represent significant technological steps forward to broaden the applicability of 2D IR spectroscopy.

5.2 Ultrasensitive Spectroscopy with Plasmonic, Surface-Enhanced 2D IR

2D IR spectroscopy can nowadays be applied to routinely measure vibrational dynamics of samples in bulk solution environments with low absorption coefficients (< 500 M^{-1} cm^{-1}) and with low mM concentrations or even less [306]. Recently, surface-sensitive variants of 2D IR spectroscopy have achieved to resolve signals from only monolayer thin samples at solid-liquid/gas interfaces [15, 173]. It must be noted, however, that in case of a fully covered surface, the number of molecules contributing to the signal is still a factor of about 5 lower compared to a bulk solution sample with, say, 1 mM concentration and a 50 µm path-length [15]. To be able to record vibrational dynamics over multiple integers of the IR-label's vibrational relaxation time constant from monolayer thin samples with even the smallest absorption coefficients (< 100 M^{-1} cm^{-1}), it is, therefore, desirable to obtain a maximum sensitivity of the methods. One way to achieve that is to combine femtosecond 2D IR spectroscopy with established methods from ultra-sensitive IR spectroscopy that exploit the properties of plasmonic substrates [7, 307–310]. That approach has recently been demonstrated by different groups based on controlled nanoantennas [311], or heterogeneous plasmonic substrates [179, 243]. The plasmonic substrates allow the generation of enhanced optical near-fields around polarizable nanostructures, thereby concentrating the interaction volume of the light and the sample. That signal enhancement factors of up to about 500 have been characterized that are achievable under typical conditions for 2D IR spectroscopy [243], and even stronger enhancement has been estimated [311]. These values are generally obtained by comparison of the obtained signals to signals from analogous samples that do not benefit from surface enhancement effects [179, 243, 312]. The strong near-fields therefore open up a couple of exciting applications for ultrafast spectroscopy that have partially been demonstrated very recently, as discussed in the following.

A first and obvious possibility is the exploitation of the increased near-fields for higher-order processes in samples that are either surface-bound, or within the interaction length of the optical near-fields from the nanostructures ($\ll 10$ nm). This effect is generally referred to as "vibrational ladder-climbing" [242, 243, 313–329] and is schematically depicted in Fig. 30a. Taken a given anharmonicity in an electronic ground state potential of an immobilized molecule on a plasmonic substrate, a broadband laser pulse (bandwidth ~ 200 cm^{-1}) can successively excite populations in higher-lying vibrational states by interacting multiple times with the sample. In the first demonstration of this effect in surface-enhanced 2D IR spectroscopy, Hamm et al. employed few nanometer thick, sputter-coated, heterogeneous Gold (Au) layers as plasmonic substrates (Fig. 30b) for 2D ATR IR spectroscopy on p-mercaptobenzonitrile (p-PhCN) monolayers. Such metal layers near the percolation threshold exhibit rather continuous areas of metal patches (light regions), separated by small gaps (< 10 nm in width, dark regions). Polarization-controlled excitation of molecules inside the gaps [179, 243, 330] allowed a dramatic signal enhancement and the clear observation of vibrational ladder-climbing in the 2D ATR IR signals from the CN-stretch vibration of the monolayer up to the 5–6 transition (Fig. 30c and d). In that way, more than

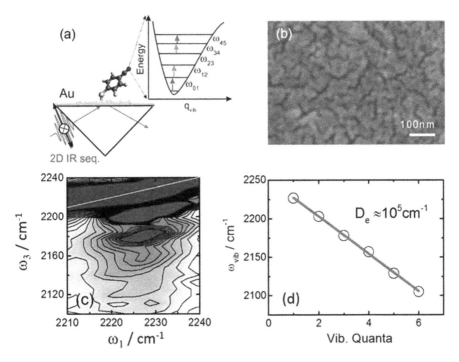

Fig. 30 Vibrational ladder climbing in surface-enhanced 2D ATR IR spectroscopy. **a** Schematic depiction of ladder climbing in an adsorbed sample of para-mercaptobenzonitrile (p-PhCN) on a plasmonic Au surface. **b** Typical plasmonic Au layer with a thickness of 3.5 nm used for surface-enhanced 2D ATR IR spectroscopy. **c** Experimental demonstration of ladder climbing in a 2D ATR IR spectrum of a monolayer of p-PhCN on Au. **d** Determination of the dissociation energy (D_e) of the CN functional group in p-PhCN based on the decreasing energetic positions of excited state absorption signals. Adapted with permission from Ref. [243]. Copyright Royal Society of Chemistry (2016)

10,000 cm^{-1} of vibrational energy were deposited in a single chemical bond. This interesting observation is useful from many different perspectives. First, it is possible to estimate the dissociation energy of the respective chemical bond from such measurements. The energy separation between higher-lying excited vibrational states becomes lower with increasing quantum number of excitation due to the characteristic anharmonicity of the potential energy surface. A linear extrapolation of the vibrational progression up to the continuum states ($\omega = 0$ cm^{-1}) and the area under curve then allows the determination of the dissociation energy D_e [331]. In case of the CN-stretch vibration that value is about 10^5 cm^{-1} for p-PhCN, which is in good agreement with standard dissociation energies ($> 80{,}000$ cm^{-1}) for nitrile bonds in organic molecules [332]. The CN bond is therefore a chemically fairly inert functional group with a very low anharmonicity, but similar experiments can be envisioned on samples with a much larger anharmonicity such as OH-stretching groups in water molecules [239], hydrogen atoms on surfaces [333], or C–H stretch vibrations in organic compounds.

Starting from the presented experiments, more sophisticated variants can be envisioned, which will allow the controlled manipulation of chemical bonds in the electronic ground state of a sample (coherent control) [334–339]. The large amount of vibrational energy that can be deposited in the molecules via vibrational ladder-climbing can be used to overcome energy barriers for isomerization reactions of chemical bonds, or controlled molecular fragmentation [314, 340]. Combinations of plasmonic substrates with pulse-shaping techniques in the mid-IR range [341–345] may further allow the concentration of populations in certain vibrational states [42, 323]. Based on the drastically enhanced near-fields coherent control may therefore be applicable to samples other than strongly absorbing metal carbonyl compounds, thus drastically expanding the applicability of the method.

In the existing initial reports of surface-enhanced 2D IR spectroscopy, fairly simple samples have been applied such as model molecular monolayers [179, 181, 243] or nanometer thin polymer films [311]. The currently obtainable enhancement factors may possibly already now allow ultrafast investigations of large biological molecules such as proteins [307, 346]. Possible applications for the increased sensitivity in 2D IR spectroscopy might involve its combinations with bio-sensing and diagnostics [347–350], or molecular recognition [351–355]. This would be an important development since 2D IR spectroscopy is intrinsically sensitive to intermolecular interactions, which form the basis of the mentioned applications. The developed methods might therefore allow the characterization of the dynamics of recognition, a characterization of binding strengths, or the identification of impurities.

In addition to aspects concerning the range of possible samples to be addressed by ultra-sensitive surface 2D IR spectroscopy, there exists currently a large interest and technical development in surface-enhanced laser spectroscopy [348, 356–361]. Major developments in this field focus on the substrate properties such as type of materials, shapes and physical/chemical aspects. The expectation is that possibly even higher enhancement factors and increased sensitivity of the methods will be achievable in future studies with purposely engineered plasmonic materials [358, 362, 363], metamaterials [357, 364], or even graphene as a substrate [356]. The combination of ultrafast spectroscopy methods and optical near-fields is thus likely to allow unprecedented chemical information from only minimal amounts of samples.

5.3 Ultrahigh Resolution Spectroscopy: 2D IR Nanoscopy

Conventional optics restrict the achievable spatial resolution for the purpose of microscopy to at least several micrometers in the mid-IR spectral range [365, 366]. These dimensions are very large compared to the spatial extensions of molecules and even many nanostructures that exist in blended mixtures of polymers, mesoporous samples, self-assembled monolayers, heterogeneous catalyst assemblies or other functional materials based on organic electronics. With the currently available focus diameters, conventional 2D IR spectroscopy or microscopy will, therefore, yield information from a spatially averaged region in a possibly nanostructured sample, thus blurring important details on length scales that are much smaller than the wavelength of IR light. It is widely known that the spatial resolution can be well extended to dimensions much smaller than the diffraction limit by help of optical near-fields that are generated around single plasmonic nanostructures and afterwards scattered to a detector [367–371]. There exists considerable interest in combining ultrafast vibrational spectroscopy with so-called scattering scanning near-field optical microscopy (s-SNOM) [372].

Combinations of 2D IR and s-SNOM methods have up to now not been realized. However, the combination of metal tips and ultrafast IR spectroscopy has been demonstrated recently as feasible and yields chemically-specific vibrational dynamics from nanometer length scales (nanoscopy) [31, 373, 374]. The spatial resolution of this method can be as good as 20 nm [374]. In proposed ultrafast s-SNOM (Fig. 31a), a femtosecond IR beam is focused on the apex of an AFM tip, where it generates a near-field (shaded orange region) that interacts with the sample for molecular excitation. The near-field generates a polarization in the sample, which is in turn again scattered ($E_{scat}(t)$) off the tip and detected by help of a local oscillator field (E_{LO}, dark red). Instead of multiple excitation beams as conventionally used for 2D IR methods, the extension of femtosecond s-SNOM to multi-dimensional variants will likely involve single-beam combinations of excitation and probing pulses as well as sophisticated phase-cycling schemes ($\varphi_{1,2}$) to isolate unwanted linear responses and other background and scattering signals [59, 60]. s-SNOM 2D IR nanoscopy is expected to allow deconstructing the sample's entire vibrational response into different subsets of contributions, which ideally resemble homogeneously broadened systems. This way, it may reveal the impact of nanometer-sized environments on ultrafast vibrational dynamics of molecules. Such information is of particular importance in the understanding of how crystallinity influences molecular dynamics, how surface morphologies influence molecular orientation, orientational dynamics and intermolecular interactions, as well as charge-transfer. Other points to be addressed are how the nanometer-scale morphology influences vibrational couplings, how biological molecules aggregate at interfaces, or simply which molecular mechanisms are determining for mixing of different sample phases (Fig. 31b), only to mention a few applications. It is also important to note, that the concept is not limited to the IR spectral range, but can be

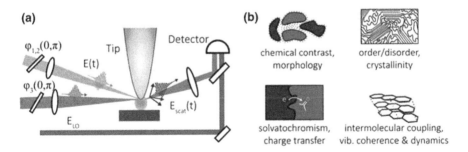

Fig. 31 **a** Experimental configuration for the combination of 2D IR spectroscopy and s-SNOM methods for ultrafast vibrational nanoscopy proposed by Raschke et al.. $\varphi_{1,2}$ denotes the phase of the electric fields, which can be scanned by phase-cycling to isolate signal contributions. $E(t)$, E_{LO} and $E_{scat}(t)$ denote the excitation, local oscillator and scattered signal field, respectively. **b** Some examples of the obtainable information from s-SNOM 2D IR. Adapted with permission from Ref. [31]. Copyright American Chemical Society (2015)

extended towards UV/VIS or THz regions as well to address electronic and very low-frequency, intermolecular modes, respectively.

5.4 Extensions of Transient 2D IR Spectroscopy

Other future developments of 2D vibrational spectroscopy concern the broadening of the range of applications of transient 2D IR methods. In principle, one would like to combine 2D IR spectroscopy with any type of possible perturbation of the sample systems to be able to address a maximum range of scientific questions in chemistry, physics, biology and material science. Depending on the involved mechanism of perturbation and the response of samples, however, different timescales to be studied will become relevant, which may range from picoseconds all the way up to milliseconds, seconds or even minutes and longer. In that way, this approach will involve a likely transition between transient 2D IR spectroscopy to what may be referred to as "real-time 2D IR spectroscopy" [86].

Among the most important additional external stimuli to be implemented is the triggered change of proton concentrations in liquid phases to yield so-called pH-jumps [375–378]. Such triggered variations in the pH are particularly relevant in biological systems such as folding and denaturation of proteins, but also acid-catalyzed chemical reactions and material chemistry. A general way to achieve the pH-jump is to employ photo-acids, which, after the interaction with an ultrashort laser pulse, release protons due to a change of the pK_a value in the excited electronic state [379–381]. Accessible timescales for the proton release are in the nanosecond regime, but depend on the actual mechanism behind the dissociation [382–385]. pH-jumps as large as two or three units have been demonstrated experimentally [384]. In that way pH-jump induced transient 2D IR spectroscopy is expected to enable measuring a large range of structural changes in molecules.

Further methods to perturb the sample are the so-called pressure-jump techniques (p-jump) [386]. Also this variant has previously been developed to study mainly biological samples such as proteins [387] and investigate folding dynamics. But also other applications can be envisioned such as dynamics of adsorbed molecules under two- and three-dimensionally confined conditions [388, 389] or dynamics in polymer samples [390]. Contrary to T- and pH-jumps, the p-jump is not induced by an ultrashort laser pulse, but rather with the help of either mechanical stimuli, or electrical burst diaphragms. This limits the achievable temporal resolution, but still the sub-microsecond range has been achieved experimentally with pressure differences as large as 2500 MPa [386]. With these developments, there exist currently a powerful toolbox of fast sample perturbations, which cover most of the important thermodynamic and kinetic variables.

Making use of the already existing combination of 2D IR spectroscopy and electrochemistry (Sect. 4.1), other variants of transient 2D IR spectroscopy to be developed might involve the laser-induced change of electrode potentials. In that way, molecular dynamics near electrodes might be investigated that are based on a so-called potential-jump (V-jump) of a pre-charged electrode [391, 392]. Also, here, a laser pulse is used initially to act on the electrode material to induce the V-jump. Effectively, it is assumed that the laser pulse heats up the electrode, thereby first evoking a T-jump. This T-jump it thought to subsequently randomize the orientation of the electrolyte molecules near the electrode, thereby establishing a non-equilibrium potential difference to which molecules at the electrode-electrolyte interfaces can respond by re-orientation, or re-structuration of their coordination-sphere and solvation shell [391]. V-jump spectroscopy has already been applied to study frequency shifts of adsorbate molecules at Pt electrodes [391, 392]. It was demonstrated that the time-resolution can be in the range of tens to hundreds of picoseconds. However, the magnitude of the demonstrated jumps is rather low (< 200 mV) [391, 392] and optimizations might be needed to study charge-transfer phenomena. V-jump 2D IR spectroscopy would nevertheless still be extremely useful for studying different orientational dynamics of molecules near electrodes. Such dynamics are particularly relevant in the fields of heterogeneous catalysis, but applications can also be envisioned regarding liquid-crystal displays, batteries and organic electronics.

6 Concluding Remarks

2D IR spectroscopy in all its variants has seen a considerable broadening of possible applications over the last approximately 20 years. In this overview, it was attempted to present some of the highlights from recently presented examples that bear the potential to bring 2D IR spectroscopy to the next level of applications. Currently, a vast range of experiments is possible to be conducted, covering aspects from equilibrated, as well as non-equilibrated ground state sample systems to rapidly evolving systems such as molecules in excited electronic states and samples subjected to other external perturbations such as temperature jumps. Starting from the "classical" applications of 2D IR spectroscopy on bulk samples, for which the elucidation of spectral diffusion, energy transfer, chemical exchange and vibrational couplings in molecules can be considered as almost routine, many additional ways have been found in recent years to obtain that type of information from molecules in a range of different other environments. This includes molecules under three- and two-dimensional confinement, partially even reaching a demonstrated sensitivity that allows measuring sub-monolayer surfaces coverages from molecules with only weakly absorbing IR-labels. Moreover, recent developments also achieved the combination of ultrafast vibrational spectroscopy with microscopy and electro-chemistry, the controlled realization of higher-order pulse sequences for 3D IR spectroscopy, the application of ultra-broadband IR light sources to cover a maximum spectral range of IR-active vibrations, or the application of plasmonic substrates for surface-enhanced 2D IR spectroscopy. All these methods will allow in future applications the resolution of molecular dynamics and interactions under a variety of chemical and physical conditions with unprecedented accuracy and sensitivity.

In addition to the already established developments, different extensions of 2D IR spectroscopy have been devised or at least proposed, all of which will make the application of the method even broader. The combination of 2D IR spectroscopy with microfluidics has recently been realized. This method will allow tackling a series of challenging questions in the field of chemical reactor design, molecular analytics and other industrial processes. Further applications can be expected, e.g. the combination with rapid-mixing experiments as an alternative variant for an externally triggered reaction. Possible other realizations will involve 2D IR spectroscopy in conjunction with combinatorial chemistry. In a similar manner, the combination of 2D IR spectroscopy with microscopy is just taking off and will possibly add a completely new dimension to microscopic sample characterizations in biological systems, polymer samples or functional materials. Currently, it is extremely challenging for other methods to achieve the chemical sensitivity of IR spectroscopy and allow for a similarly high degree in experimental flexibility in characterizing ultrafast dynamics and intermolecular interactions with high spatial resolution. In the same context, one important application of 2D IR spectroscopy has still to be tackled, that is, scattering scanning near-field optical microscopy (s-SNOM) 2D IR as the combination of sub-100 fs temporal resolution and sub-50 nm spatial resolution in a single experiment. Although recent reports aiming in this direction appear very promising [373, 374, 393–395], the realization of 2D IR s-SNOM will still require most likely quite some time. However, this variant will allow the ultimate way to characterize intermolecular interactions and dynamics in space and time and its impact can hardly be overrated.

Despite the impressive achievements and the tremendous current efforts regarding further development of different forms of 2D IR spectroscopy, the method is generally still considered as a specialized technique by many researcher outside the field. In other words, 2D IR has not yet been fully established as a standard analytical tool for characterizing molecular structure, dynamics and intermolecular interactions. An important reason for this is the often comparatively short vibrational lifetimes (pico- to nanoseconds), which intrinsically limit the temporal observation window. Additionally, 2D IR spectra from a large range of vibrational modes contain often only weak or even absent cross peaks, which are due to vibrational coupling or energy transfer. These missing cross peaks are particularly of importance in case of through-space, intermolecular interactions. In general, such interactions are very short-ranged, often only reaching to distances less than a nanometer. Although this is typically also the distance on which 2D NMR experiments reveal structural information about the sample, the strongly nonlinear dependence of dipolar interactions on the (often weak) transition dipole moments complicates the exploitation of intermolecular coupling in IR studies. Finally, performing day-to-day experiments as well as detailed data interpretation often still requires skilled spectroscopists and costly laboratory equipment. In this regard, important developments currently focus on facilitated and automated data acquisition for 2D IR. Progress in this direction will allow a more widespread application of the method and possibly the day-to-day use for a variety of disciplines in life sciences.

Acknowledgements I would like to thank Peter Hamm for many valuable discussions, the productive research atmosphere as well as his continuous and generous support.

References

1. Fayer MD (2013) Ultrafast infrared vibrational spectroscopy. Ultrafast Infrared Vib Spectrosc. doi:10.1201/b13972-2
2. Nibbering ETJ, Fidder H, Pines E (2005) ULTRAFAST CHEMISTRY: using time-resolved vibrational spectroscopy for interrogation of structural dynamics. Annu Rev Phys Chem 56:337–367. doi:10.1146/annurev.physchem.56.092503.141314
3. Butler JM, George MW, Schoonover JR et al (2007) Application of transient infrared and near infrared spectroscopy to transition metal complex excited states and intermediates. Coord Chem Rev 251:492–514. doi:10.1016/j.ccr.2006.12.002
4. Tolstoij VP, Chernyshova IV, Skryshevsky VA (2003) Handbook of infrared spectroscopy of ultrathin films. Wiley, Hoboken
5. Stuart B (2013) Infrared Spectrosc Fundam Appl. doi:10.1017/CBO9781107415324.004
6. Radu I, Schleeger M, Bolwien C, Heberle J (2009) Time-resolved methods in biophysics. 10. Time-resolved FT-IR difference spectroscopy and the application to membrane proteins. Photochem Photobiol Sci 8:1517–1528. doi:10.1039/b9pp00050j
7. Ataka K, Kottke T, Heberle J (2010) Thinner, smaller, faster: IR techniques to probe the functionality of biological and biomimetic systems. Angew Chemie Int Ed 49:5416–5424. doi:10.1002/anie.200907114
8. Nibbering ETJ, Elsaesser T (2004) Ultrafast vibrational dynamics of hydrogen bonds in the condensed phase. Chem Rev 104:1887–1914. doi:10.1021/cr020694p
9. Zaera F (2014) New advances in the use of infrared absorption spectroscopy for the characterization of heterogeneous catalytic reactions. Chem Soc Rev 43:7624–7663. doi:10.1039/c3cs60374a
10. Hamm P, Zanni MT (2011) Concepts and methods of 2D infrared spectroscopy. Cambridge University Press, New York
11. Hamm P (2009) For structural biology, try infrared instead. Structure 17:149–150. doi:10.1016/j.str.2009.01.002
12. Hill JR, Dlott DD, Rella CW et al (1996) Ultrafast infrared spectroscopy in biomolecules: active site dynamics of heme proteins. Biospectroscopy 2:277–299. doi:10.1002/(SICI)1520-6343(1996)2:53.3.CO;2-4
13. Fayer MD (2001) Ultrafast infrared and Raman spectroscopy. CRC Press, Boca Raton
14. Cho M (2002) Ultrafast vibrational spectroscopy in condensed phases. PhysChemComm 5:40. doi:10.1039/b110898k
15. Kraack JP, Hamm P (2016) Surface-sensitive and surface-specific ultrafast two-dimensional vibrational spectroscopy. Chem Rev. doi:10.1021/acs.chemrev.6b00437
16. Hamm P, Lim M, Hochstrasser RM (1998) Structure of the amide I band of peptides measured by femtosecond nonlinear-infrared spectroscopy. J Phys Chem B 102:6123–6138. doi:10.1021/jp9813286
17. Abramavicius D, Palmieri B, Voronine DV et al (2009) Coherent multidimensional optical spectroscopy of excitons in molecular aggregates; quasiparticle versus supermolecule perspectives. Chem Rev 109:2350–2408. doi:10.1021/cr800268n
18. Cho M (2008) Coherent two-dimensional optical spectroscopy. Chem Rev 108:1331–1418. doi:10.1021/cr078377b
19. Simpson J (2008) Organic Structure Determination Using 2-D NMR Spectroscopy. Academic Press, New York, Oxford
20. Jeener J, Meier BH, Bachmann P, Ernst RR (1979) Investigation of exchange processes by two-dimensional NMR spectroscopy. J Chem Phys 71:4546–4553. doi:10.1063/1.438208
21. Brey WS (2012) Pulse methods in 1D & 2D liquid-phase NMR. Elsevier, New York
22. Friebolin H, Becconsall JK (1993) Basic one-and two-dimensional NMR spectroscopy. VCH, Weinheim
23. Macura S, Ernst RR (1980) Elucidation of cross relaxation in liquids by two-dimensional N.M.R. spectroscopy. Mol Phys 41:95–117. doi:10.1080/00268978000102601
24. Reppert M, Tokmakoff A (2016) Computational amide I 2D IR spectroscopy as a probe of protein structure and dynamics. Annu Rev Phys Chem 67:359–386. doi:10.1146/annurev-physchem-040215-112055

25. Hamm P, Lim M, DeGrado WF, Hochstrasser RM (1999) The two-dimensional IR nonlinear spectroscopy of a cyclic penta-peptide in relation to its three-dimensional structure. Proc Natl Acad Sci USA 96:2036–2041. doi:10.1073/Pnas.96.5.2036

26. Kleckner IR, Foster MP (2011) An introduction to NMR-based approaches for measuring protein dynamics. Biochim Biophys Acta Proteins Proteom 1814:942–968. doi:10.1016/j.bbapap.2010.10.012

27. Haller JD, Schanda P (2013) Amplitudes and time scales of picosecond-to-microsecond motion in proteins studied by solid-state NMR: a critical evaluation of experimental approaches and application to crystalline ubiquitin. J Biomol NMR 57:263–280. doi:10.1007/s10858-013-9787-x

28. Sapienza P, Lee A (2010) Using NMR to study fast dynamics in proteins: methods and applications. Curr Opin Pharmacol 10:723–730. doi:10.1016/j.coph.2010.09.006.Using

29. Johnson PJM, Koziol KL, Hamm P (2017) Quantifying biomolecular recognition with site-specific 2D infrared probes. J Phys Chem Lett 8:2280–2284. doi:10.1021/acs.jpclett.7b00742

30. Ghosh A, Ostrander JS, Zanni MT (2017) Watching proteins wiggle: mapping structures with two-dimensional infrared spectroscopy. Chem Rev. doi:10.1021/acs.chemrev.6b00582

31. Muller EA, Pollard B, Raschke MB (2015) Infrared chemical nano-imaging: accessing structure, coupling, and dynamics on molecular length scales. J Phys Chem Lett 6:1275–1284. doi:10.1021/acs.jpclett.5b00108

32. Fayer MD (2009) Dynamics of liquids, molecules, and proteins measured with ultrafast 2D IR vibrational echo chemical exchange spectroscopy. Annu Rev Phys Chem 60:21–38. doi:10.1146/annurev-physchem-073108-112712

33. Cho M (2008) Coherent two-dimensional optical spectroscopy. Chem Rev 108:1331–1418. doi:10.1021/cr078377b

34. Remorino A, Hochstrasser RM (2012) Three-dimensional structures by two- dimensional vibrational spectroscopy. Acc Chem Res 45:1896–1905. doi:10.1021/ar3000025

35. Zanni MT, Hochstrasser RM (2001) Two-dimensional infrared spectroscopy: a promising new method for the time resolution of structures. Curr Opin Struct Biol 11:516–522

36. Kim YS, Hochstrasser RM (2009) Applications of 2D IR spectroscopy to peptides, proteins, and hydrogen-bond dynamics. J Phys Chem B 113:8231–8251. doi:10.1021/jp8113978

37. Woutersen S, Hamm P (2002) Nonlinear two-dimensional vibrational spectroscopy of peptides. J Phys Condens Matter 1035:R1035–R1062

38. Hamm P, Helbing J, Bredenbeck J (2008) Two-dimensional infrared spectroscopy of photoswitchable peptides. Annu Rev Phys Chem 59:291–317. doi:10.1146/annurev.physchem.59.032607.093757

39. Bredenbeck J, Helbing J, Kolano C, Hamm P (2007) Ultrafast 2D–IR spectroscopy of transient species. ChemPhysChem 8:1747–1756. doi:10.1002/cphc.200700148

40. Simpson N, Hunt NT (2015) Ultrafast 2D-IR spectroscopy of haemoproteins. Int Rev Phys Chem 34:361–383. doi:10.1080/0144235X.2015.1061793

41. Hunt NT (2009) 2D-IR spectroscopy: ultrafast insights into biomolecule structure and function. Chem Soc Rev 38:1837–1848. doi:10.1039/b819181f

42. Strasfeld DB, Shim S-H, Zanni MT (2009) New Advances in mid-IR pulse shaping and its application to 2D IR spectroscopy and ground-state coherent control. Adv Chem Phys 141:1–28

43. Chen H, Bian H, Li J et al (2012) Ultrafast multiple-mode multiple-dimensional vibrational spectroscopy. Int Rev Phys Chem 31:469–565. doi:10.1080/0144235X.2012.733116

44. Le Sueur AL, Horness RE, Thielges M (2015) Applications of two-dimensional infrared spectroscopy. Analyst 140:4336–4349. doi:10.1039/C5AN00558B

45. Finkelstein IJ, Zheng J, Ishikawa H et al (2007) Probing dynamics of complex molecular systems with ultrafast 2D IR vibrational echo spectroscopy. Phys Chem Chem Phys 9:1533–1549. doi:10.1039/b618158a

46. Ge N-H, Hochstrasser RM (2002) Femtosecond two-dimensional infrared spectroscopy: IR-COSY and THIRSTY. PhysChemComm 5:17. doi:10.1039/b109935c

47. Ganim Z, Hoi SC, Smith AW et al (2008) Amide I two-dimensional infrared spectroscopy of proteins. Acc Chem Res 41:432–441. doi:10.1021/ar700188n

48. Mukamel S (2000) Multidimensional femtosecond correlation spectroscopies of electronic and vibrationnal excitations. Annu Rev Phys Chem 51:691–729

49. Shim S-H, Zanni MT (2009) How to turn your pump-probe instrument into a multidimensional spectrometer: 2D IR and Vis spectroscopies via pulse shaping. Phys Chem Chem Phys 11:748–761. doi:10.1039/b813817f

50. Zheng J, Kwak K, Fayer MD (2007) Ultrafast 2D IR vibrational echo spectroscopy. Acc Chem Res 40:75–83. doi:10.1021/ar068010d
51. Van Wilderen LJGW, Bredenbeck J (2015) From ultrafast structure determination to steering reactions: mixed IR/Non-IR multidimensional vibrational spectroscopies. Angew Chemie Int Ed 54:11624–11640. doi:10.1002/anie.201503155
52. Cho M (2008) Coherent two-dimensional optical spectroscopy. Bull Korean Chem Soc 108:1331–1418. doi:10.1021/cr078377b
53. Wang J (2017) Ultrafast two-dimensional infrared spectroscopy for molecular structures and dynamics with expanding wavelength range and increasing sensitivities: from experimental and computational perspectives. Int Rev Phys Chem 36:377–431. doi:10.1080/0144235X.2017.1321856
54. Jonas DM (2003) Two-dimensional femtosecond spectroscopy. Annu Rev Phys Chem 54:425–463. doi:10.1146/annurev.physchem.54.011002.103907
55. Cho M (2009) Two-dimensional optical spectroscopy, 1st edn. CRC Press/Taylor & Francis Group, Boca Raton/London
56. Goodno GD, Dadusc G, Miller RJ (1998) Ultrafast heterodyne-detected transient-grating spectroscopy using diffractive optics. JOSA B 15:1791–1794
57. Maznev AA, Nelson KA, Rogers JA (1998) Optical heterodyne detection of laser-induced gratings. Opt Lett 23:1319–1321. doi:10.1364/OL.23.001319
58. Fuller FD, Ogilvie JP (2013) Experimental implementations of two-dimensional fourier transform electronic spectroscopy. Annu Rev Phys Chem 66:667–690. doi:10.1146/annurev-physchem-040513-103623
59. Baiz CR, Schach D, Tokmakoff A (2014) Ultrafast 2D IR microscopy. Opt Express 22:18724–18735. doi:10.1364/OE.22.018724
60. Ostrander JS, Serrano AL, Ghosh A, Zanni MT (2016) Spatially resolved two-dimensional infrared spectroscopy via wide-field microscopy. ACS Photonics 3:1315–1323. doi:10.1021/acsphotonics.6b00297
61. Helbing J, Hamm P (2010) Compact implementation of Fourier transform two-dimensional IR spectroscopy without phase ambiguity. J Opt Soc Am B 28:171–178. doi:10.1364/JOSAB.28.000171
62. Mukamel S (1995) Principles of nonlinear optical spectroscopy. Oxford University Press, New York
63. Boyd RW (2008) Nonlinear optics, 3rd edn. Academic Press, San Diego
64. Shen YR (2002) The principles of nonlinear optics. Wiley-VCH Verlag, New York
65. Grimberg BI, Lozovoy VV, Dantus M, Mukamel S (2002) Ultrafast Nonlinear spectroscopic techniques in the gas phase and their density matrix representation. J Phys Chem A 106:697–718
66. Khalil M, Demirdöven N, Tokmakoff A (2003) Coherent 2D IR spectroscopy: molecular structure and dynamics in solution. J Phys Chem A 107:5258–5279. doi:10.1021/jp0219247
67. Khalil M, Demirdöven N, Tokmakoff A (2003) Obtaining absorptive line shapes in two-dimensional infrared vibrational correlation spectra. Phys Rev Lett 90(047401):1–4. doi:10.1103/PhysRevLett.90.047401
68. Kim YS, Hochstrasser RM (2006) Comparison of linear and 2D IR spectra in the presence of fast exchange. J Phys Chem B 110:8531–8534. doi:10.1021/jp060935n
69. Roberts ST, Loparo JJ, Tokmakoff A (2006) Characterization of spectral diffusion from two-dimensional line shapes. J Chem Phys 125(084502):1–8. doi:10.1063/1.2232271
70. Guo Q, Pagano P, Li Y-L et al (2015) Line shape analysis of two-dimensional infrared spectra. J Chem Phys 142:212427. doi:10.1063/1.4918350
71. Kwak K, Park S, Finkelstein IJ, Fayer MD (2007) Frequency-frequency correlation functions and apodization in two-dimensional infrared vibrational echo spectroscopy: a new approach. J Chem Phys 127:124503. doi:10.1063/1.2772269
72. Rosenfeld DE, Fayer MD (2012) Excitation transfer induced spectral diffusion and the influence of structural spectral diffusion. J Chem Phys 137(064109):1–18. doi:10.1063/1.4742762
73. Donaldson PM, Guo R, Fournier F et al (2007) Direct identification and decongestion of Fermi resonances by control of pulse time ordering in two-dimensional IR spectroscopy. J Chem Phys doi 10(1063/1):2771176
74. Kurochkin DV, Naraharisetty SRG, Rubtsov IV (2007) A relaxation-assisted 2D IR spectroscopy method. Proc Natl Acad Sci USA 104:14209–14214. doi:10.1073/pnas.0700560104

191   Springer

75. Rubtsova NI, Rubtsov IV (2015) Vibrational energy transport in molecules studied by relaxation-assisted two-dimensional infrared spectroscopy. Annu Rev Phys Chem 66:717–738. doi:10.1146/annurev-physchem-040214-121337

76. Golonzka O, Khalil M, Demirdöven N, Tokmakoff A (2001) Coupling and orientation between anharmonic vibrations characterized with two-dimensional infrared vibration echo spectroscopy. J Chem Phys 115:10814–10828. doi:10.1063/1.1417504

77. Woutersen S, Mu Y, Stock G, Hamm P (2001) Subpicosecond conformational dynamics of small peptides probed by two-dimensional vibrational spectroscopy. Proc Natl Acad Sci USA 98:11254–11258. doi:10.1073/pnas.201169498

78. Zanni MT, Ge NH, Kim YS, Hochstrasser RM (2001) Two-dimensional IR spectroscopy can be designed to eliminate the diagonal peaks and expose only the crosspeaks needed for structure determination. Proc Natl Acad Sci USA 98:11265–11270. doi:10.1073/pnas.201412998

79. Krummel AT, Mukherjee P, Zanni MT (2003) Inter and intrastrand vibrational coupling in DNA studied with heterodyned 2D-IR spectroscopy. J Phys Chem B 107:9165–9169. doi:10.1021/jp035473h

80. Buchanan LE, Dunkelberger EB, Zanni MT (2012) Examining amyloid structure and kinetics with 1D and 2D infrared spectroscopy and isotope labeling BT - protein folding and misfolding: shining light by infrared spectroscopy. In: Fabian H, Naumann D (eds) Protein fold. Misfolding. Springer, Berlin, pp 217–237

81. Buchanan LE, Carr JK, Fluitt AM et al (2014) Structural motif of polyglutamine amyloid fibrils discerned with mixed-isotope infrared spectroscopy. Proc Natl Acad Sci 111:5796–5801. doi:10.1073/pnas.1401587111

82. Moran A, Mukamel S (2004) The origin of vibrational mode couplings in various secondary structural motifs of polypeptides. Proc Natl Acad Sci USA 101:506–510. doi:10.1073/pnas.2533089100

83. Bereau T, Meuwly M (2014) Computational Two-Dimensional Infrared Spectroscopy without Maps: N-Methylacetamide in Water. J Phys Chem B 118:8135–8147. doi: 10.1021/jp5011692

84. Woutersen S, Hamm P (2000) Structure determination of trialanine in water using polarization sensitive two-dimensional vibrational spectroscopy. J Phys Chem B 104:11316–11320. doi:10.1021/jp001546a

85. Krummel AT, Zanni MT (2006) DNA Vibrational Coupling Revealed with Two-Dimensional Infrared Spectroscopy : Insight into Why Vibrational Spectroscopy Is Sensitive to DNA Structure. J Phys Chem B 110:13991–14000. doi: 10.1021/jp062597w

86. Baiz CR, Reppert M, Tokmakoff A (2013) Introduction to protein 2D IR spectroscopy. In: Fayer MD (ed) Ultrafast infrared Vib. Spectrosc. CRC Press/Taylor & Francis Group, Boca Raton/New York, pp 361–405

87. Baiz CR, Peng CS, Reppert ME et al (2012) Coherent two-dimensional infrared spectroscopy: quantitative analysis of protein secondary structure in solution. Analyst 137:1793–1799. doi:10.1039/c2an16031e

88. Waegele MM, Culik RM, Gai F (2011) Site-specific spectroscopic reporters of the local electric field, hydration, structure, and dynamics of biomolecules. J Phys Chem Lett 2:2598–2609. doi:10.1021/jz201161b

89. Kim H, Cho M (2013) Infrared probes for studying the structure and dynamics of biomolecules. Chem Rev 113:5817–5847. doi:10.1021/cr3005185

90. Buchanan LE, Dunkelberger EB, Tran HQ et al (2013) Mechanism of IAPP amyloid fibril formation involves an intermediate with a transient beta-sheet. Proc Natl Acad Sci USA 110:19285–19290. doi:10.1073/pnas.1314481110

91. Strasfeld DB, Ling YL, Shim S-H, Zanni MT (2008) Tracking fiber formation in human islet amyloid polypeptide with automated 2D-IR spectroscopy. J Am Chem Soc 130:6698–6699. doi:10.1021/ja801483n

92. Shim S-H, Gupta R, Ling YL et al (2009) Two-dimensional IR spectroscopy and isotope labeling defines the pathway of amyloid formation with residue-specific resolution. Proc Natl Acad Sci 106:6614–6619. doi:10.1073/pnas.0805957106

93. Middleton CT, Marek P, Cao P et al (2012) Two-dimensional infrared spectroscopy reveals the complex behaviour of an amyloid fibril inhibitor. Nat Chem 4:355–360. doi:10.1038/nchem.1293

94. Moran SD, Woys AM, Buchanan LE et al (2012) Two-dimensional IR spectroscopy and segmental 13C labeling reveals the domain structure of human D-crystallin amyloid fibrils. Proc Natl Acad Sci 109:3329–3334. doi:10.1073/pnas.1117704109

95. Kratochvil HT, Carr JK, Matulef K et al (2016) Instantaneous ion configurations in the K+ ion channel selectivity filter revealed by 2D IR spectroscopy. Science 353:1040–1044. doi:10.1126/science.aag1447

96. Thielges MC, Axup JY, Wong D et al (2011) Two-dimensional IR spectroscopy of protein dynamics using two vibrational labels: a site-specific genetically encoded unnatural amino acid and an active site ligand. J Phys Chem B 115:11294–11304. doi:10.1021/jp206986v

97. Ma J, Pazos IM, Zhang W et al (2015) Site-specific infrared probes of proteins. Annu Rev Phys Chem 66:357–377. doi:10.1146/annurev-physchem-040214-121802

98. Krummel AT, Zanni MT (2008) Evidence for coupling between nitrile groups using DNA templates: a promising new method for monitoring structures with infrared spectroscopy. J Phys Chem B 112:1336–1338. doi:10.1021/jp711558a

99. Ganim Z, Jones K, Tokmakoff A (2010) Biomolecular structures: from isolated molecules to the cell crowded medium. Phys Chem Chem Phys 12:3579–3588. doi:10.1039/c004156b

100. Kwak K, Rosenfeld DE, Fayer MD (2008) Taking apart the two-dimensional infrared vibrational echo spectra: more information and elimination of distortions. J Chem Phys 128(204505):1–10. doi:10.1063/1.2927906

101. Roy S, Pshenichnikov MS, Jansen TLC (2011) Analysis of 2D CS spectra for systems with non-gaussian dynamics. J Phys Chem B 115:5434–5440. doi:10.1021/jp109742p

102. Woutersen S, Pfister R, Hamm P et al (2002) Peptide conformational heterogeneity revealed from nonlinear vibrational spectroscopy and molecular-dynamics simulations. J Chem Phys 117:6833–6840. doi:10.1063/1.1506151

103. Woutersen S, Hamm P (2002) Nonlinear two-dimensional vibrational spectroscopy of peptides. J Phys Condens Matter 14:R1035–R1062. doi:10.1088/0953-8984/14/39/202

104. Woutersen S, Hamm P (2001) Isotope-edited two-dimensional vibrational spectroscopy of trialanine in aqueous solution. J Chem Phys 114:2727–2737. doi:10.1063/1.1336807

105. King JT, Baiz CR, Kubarych KJ (2010) Solvent-dependent spectral diffusion in a hydrogen bonded "vibrational aggregate". J Phys Chem A 114:10590–10604. doi:10.1021/jp106142u

106. Kiefer LM, King JT, Kubarych KJ (2015) Dynamics of rhenium photocatalysts revealed through ultrafast multidimensional spectroscopy. Acc Chem Res 48:1123–1130. doi:10.1021/ar500402r

107. Perakis F, Hamm P (2011) Two-dimensional infrared spectroscopy of supercooled water. J Phys Chem B 115:5289–5293

108. Perakis F, De Marco L, Shalit A et al (2016) Vibrational spectroscopy and dynamics of water. Chem Rev 116:7590–7607. doi:10.1021/acs.chemrev.5b00640

109. Ishikawa H, Finkelstein IJ, Kim S et al (2007) Neuroglobin dynamics observed with ultrafast 2D-IR vibrational echo spectroscopy. Proc Natl Acad Sci USA 104:16116–16121. doi:10.1073/pnas.0707718104

110. Chung JK, Thielges MC, Fayer MD (2011) Dynamics of the folded and unfolded villin headpiece (HP35) measured with ultrafast 2D IR vibrational echo spectroscopy. Proc Natl Acad Sci USA 108:3578–3583. doi:10.1073/pnas.1100587108

111. Sokolowsky KP, Bailey HE, Fayer MD (2014) New Divergent dynamics in the isotropic to nematic phase transition of liquid crystals measured with 2D IR vibrational echo spectroscopy. J Chem Phys 194502:1–38. doi:10.1063/1.4901081

112. Sokolowsky KP, Fayer MD (2013) Dynamics in the isotropic phase of nematogens using 2D IR vibrational echo measurements on natural-abundance 13CN and extended lifetime probes. J Phys Chem B 117:15060–15071. doi:10.1021/jp4071955

113. King JT, Ross MR, Kubarych KJ (2012) Ultrafast α-like relaxation of a fragile glass-forming liquid measured using two-dimensional infrared spectroscopy. Phys Rev Lett 108:1–5. doi:10.1103/PhysRevLett.108.157401

114. Eaves JD, Loparo JJ, Fecko CJ et al (2005) Hydrogen bonds in liquid water are broken only fleetingly. Proc Natl Acad Sci USA 102:13019–13022. doi:10.1073/pnas.0505125102

115. Ren Z, Ivanova AS, Couchot-Vore D, Garrett-Roe S (2014) Ultrafast Structure and Dynamics in Ionic Liquids: 2D-IR Spectroscopy Probes the Molecular Origin of Viscosity. J Phys Chem Lett 5:1541–1546. doi:10.1021/jz500372f

116. Yamada SA, Bailey HE, Tamimi A, et al (2017) Dynamics in a room-temperature ionic liquid from the cation perspective: 2D IR vibrational echo spectroscopy. J Am Chem Soc. doi:10.1021/jacs.6b12011

117. Tamimi A, Fayer MD (2016) Ionic liquid dynamics measured with 2D IR and IR pump-probe experiments on a linear anion and the influence of potassium cations. J Phys Chem B 120:5842–5854. doi:10.1021/acs.jpcb.6b00409

118. Rosker MJ, Dantus M, Zewail AH (1988) Femtosecond real-time probing of reactions. I. The technique. J Chem Phys 89:6113

119. Zewail AH (1988) Laser femtochemistry. Science (80-) 242:1645–1653

120. Motzkus M, Pedersen S, Zewail AH (1996) Femtosecond real-time probing of reactions. 19. nonlinear (DFWM) techniques for probing transition states of uni- and bimolecular reactions. J Phys Chem 100:5620–5633

121. Herbst J, Heyne K, Diller R (2002) Femtosecond infrared spectroscopy of bacteriorhodopsin chromophore isomerization. Science (80-) 297:822–825

122. Kukura P, McCamant DW, Yoon S et al (2005) Structural observation of the primary isomerization in vision with femtosecond-stimulated Raman. Science (80-) 310:1006–1009

123. Vos MH, Lambry JC, Robles SJ et al (1991) Direct observation of vibrational coherence in bacterial reaction centers using femtosecond absorption-spectroscopy. Proc Natl Acad Sci USA 88:8885–8889

124. Rose TS, Rosker MJ, Zewail AH (1988) Femtosecond real-time observation of wave packet oscillations (resonance) in dissociation reactions. J Chem Phys 88:6672–6673

125. Dantus M, Bowman RM, Gruebele M, Zewail AH (1989) Femtosecond real-time probing of reactions. V. The reaction of IHgI. J Chem Phys 91:7437–7450

126. Kraack JP, Buckup T, Motzkus M (2013) Coherent high-frequency vibrational dynamics in the excited electronic state of all-trans retinal derivatives. J Phys Chem Lett 383–387. doi:10.1021/jz302001m

127. Kraack JP, Buckup T, Hampp N, Motzkus M (2011) Ground- and excited-state vibrational coherence dynamics in bacteriorhodopsin probed with degenerate four-wave-mixing experiments. ChemPhysChem 12:1851–1859. doi:10.1002/cphc.201100032

128. Kraack JP, Wand A, Buckup T et al (2013) Mapping multidimensional excited state dynamics using pump-impulsive- vibrational-spectroscopy and pump-degenerate-four-wave-mixing. Phys Chem Chem Phys 15:14487–14501. doi:10.1039/c3cp50871d

129. Gallagher Faeder SM, Jonas DM, Faeder SMG (1999) Two-dimensional electronic correlation and relaxation spectra: theory and model calculations. J Phys Chem A 103:10489–10505. doi:10.1021/jp9925738

130. Aue W, Bartholdi E, Ernst R (1976) Two-dimensional spectroscopy. Application to nuclear magnetic resonance. J Chem Phys 64:2229–2246. doi:10.1063/1.432450

131. Zheng J, Kwak K, Asbury J et al (2005) Ultrafast dynamics of solute- solvent complexation observed at thermal equilibrium in real time. Science 309:1338–1343

132. Zheng J, Fayer MD (2007) Hydrogen bond lifetimes and energetics for solute/solvent complexes studied with 2D-IR vibrational echo spectroscopy. J Am Chem Soc 94305:4328–4335

133. Kwak K, Zheng J, Cang H, Fayer MD (2006) Ultrafast two-dimensional infrared vibrational echo chemical exchange experiments and theory. J Phys Chem B 110:19998–20013. doi:10.1021/jp0624808

134. Kwak K, Park S, Fayer MD (2007) Dynamics around solutes and solute-solvent complexes in mixed solvents. Proc Natl Acad Sci USA 104:14221–14226. doi:10.1073/pnas.0701710104

135. Rosenfeld DE, Kwak K, Gengeliczki Z, Fayer MD (2010) Hydrogen bond migration between molecular sites observed with ultrafast 2D IR chemical exchange spectroscopy. J Phys Chem B 114:2383–2389. doi:10.1021/jp911452z

136. Kim YS, Hochstrasser RM (2005) Chemical exchange 2D IR of hydrogen-bond making and breaking. Proc Natl Acad Sci USA 102:11185–11190. doi:10.1073/pnas.0504865102

137. Woutersen S, Mu Y, Stock G, Hamm P (2001) Hydrogen-bond lifetime measured by time-resolved 2D-IR spectroscopy: N-methylacetamide in methanol. Chem Phys 266:137–147. doi:10.1016/S0301-0104(01)00224-5

138. Chuntonov L, Pazos IM, Ma J, Gai F (2015) Kinetics of exchange between zero-, one-, and two-hydrogen-bonded states of methyl and ethyl acetate in methanol. J Phys Chem B 150313152915006. doi:10.1021/acs.jpcb.5b00745

139. Park S, Odelius M, Gaffney KJ (2009) Ultrafast dynamics of hydrogen bond exchange in aqueous ionic solutions. J Phys Chem B 113:7825–7835. doi:10.1021/jp9016739

140. Olschewski M, Lindner J, Vöhringer P (2013) A hydrogen-bond flip-flop through a bjerrum-type defect. Angew Chemie Int Ed 52:2602–2605. doi:10.1002/anie.201208625

141. Zheng J, Kwak K, Xie J, Fayer MD (2006) Ultrafast carbon-carbon single-bond rotational isomerization in room-temperature solution. Science 313:1951–1955. doi:10.1126/science.1132178
142. Ishikawa H, Kwak K, Chung JK et al (2008) Direct observation of fast protein conformational switching. Proc Natl Acad Sci USA 105:8619–8624. doi:10.1073/pnas.0803764105
143. Park S, Ji M, Gaffney KJ (2010) Ligand exchange dynamics in aqueous solution studied with 2DIR spectroscopy. J Phys Chem B 114:6693–6702. doi:10.1021/jp100833t
144. Sun Z, Zhang W, Ji M et al (2013) Contact ion pair formation between hard acids and soft bases in aqueous solutions observed with 2DIR spectroscopy. J Phys Chem B 117:15306–15312. doi:10.1021/jp4033854
145. Woutersen S, Bakker HJ (1999) Resonant intermolecular transfer of vibrational energy in liquid water. Nature 402:507–509. doi:10.1038/990058
146. Mirkovic T, Ostroumov EE, Anna JM, et al (2016) Light absorption and energy transfer in the antenna complexes of photosynthetic organisms. Chem Rev. doi:10.1021/acs.chemrev.6b00002
147. Chen H, Wen X, Li J, Zheng J (2014) Molecular distances determined with resonant vibrational energy transfers. J Phys Chem A 118:2463–2469. doi:10.1021/jp500586h
148. Chen H, Bian H, Li J et al (2015) Vibrational energy transfer: an angstrom molecular ruler in studies of ion pairing and clustering in aqueous solutions. J Phys Chem B 119:4333–4349. doi:10.1021/jp512320a
149. Li J, Chen H, Miranda A et al (2016) Non-resonant vibrational energy transfer on metal nanoparticle/liquid interface. J Phys Chem C 120:25173–25179. doi:10.1021/acs.jpcc.6b03777
150. Lin Z, Rubtsov IV (2012) Constant-speed vibrational signaling along polyethyleneglycol chain up to 60-Å distance. Proc Natl Acad Sci 109:1413–1418. doi:10.1073/pnas.1116289109
151. Bian H, Wen X, Li J et al (2011) Ion clustering in aqueous solutions probed with vibrational energy transfer. Proc Natl Acad Sci USA 108:4737–4742. doi:10.1073/pnas.1019565108
152. Chuntonov L (2016) 2D-IR spectroscopy of hydrogen-bond-mediated vibrational excitation transfer. Phys Chem Chem Phys 18:13852–13860. doi:10.1039/C6CP01640E
153. De Marco L, Thämer M, Reppert M, Tokmakoff A (2014) Direct observation of intermolecular interactions mediated by hydrogen bonding. J Chem Phys 141(034502):1–10. doi:10.1063/1.4885145
154. Elsaesser T (2009) Two-dimensional infrared spectroscopy of intermolecular hydrogen bonds in the condensed phase. Acc Chem Res 42:1220–1228. doi:10.1021/ar900006u
155. Laaser JE, Christianson R, Oudenhoven TA et al (2014) Dye self-association identified by intermolecular couplings between vibrational modes as revealed by infrared spectroscopy, and implications for electron injection. J Phys Chem C 118:5854–5861
156. Oudenhoven TA, Joo Y, Laaser JE et al (2015) Dye aggregation identified by vibrational coupling using 2D IR spectroscopy. J Chem Phys 142(212449):1–12. doi:10.1063/1.4921649
157. Ostrander JS, Knepper R, Tappan AS et al (2017) Energy transfer between coherently delocalized states in thin films of the explosive pentaerythritol tetranitrate (PETN) revealed by two-dimensional infrared spectroscopy. J Phys Chem 121:1352–1361. doi:10.1021/acs.jpcb.6b09879
158. Kraack JP, Frei A, Alberto R, Hamm P (2017) Ultrafast vibrational energy-transfer in catalytic monolayers at solid-liquid interfaces. J Phys Chem Lett 8:2489–2495. doi:10.1021/acs.jpclett.7b01034
159. Hagfeldt A, Boschloo G, Sun L et al (2010) Dye-sensitized solar cells. Chem Rev 110:6595–6663. doi:10.1021/cr900356p
160. Hagfeldt A, Graetzel M (1995) Light-induced redox reactions in nanocrystalline systems. Chem Rev 95:49–68. doi:10.1021/cr00033a003
161. Ashford DL, Gish MK, Vannucci AK et al (2015) Molecular chromophore-catalyst assemblies for solar fuel applications. Chem Rev 115:13006–13049. doi:10.1021/acs.chemrev.5b00229
162. Esswein AJ, Nocera DG (2007) Hydrogen production by molecular photocatalysis. Chem Rev 107:4022–4047. doi:10.1021/cr050193e
163. White JL, Baruch MF, Pander JE et al (2015) Light-driven heterogeneous reduction of carbon dioxide: photocatalysts and photoelectrodes. Chem Rev 115:12888–12935. doi:10.1021/acs.chemrev.5b00370
164. Anthony JE (2006) Functionalized acenes and heteroacenes for organic electronics. Chem Rev 106:5028–5048. doi:10.1021/cr050966z
165. Bendikov M, Wudl F, Perepichka DF (2004) Tetrathiafulvalenes, oligoacenenes, and their buckminsterfullerene derivatives: the brick and mortar of organic electronics. Chem Rev 104:4891–4945. doi:10.1021/cr030666m

166. Ostroverkhova O (2016) Organic optoelectronic materials: mechanisms and applications. Chem Rev 116:13279–13412. doi:10.1021/acs.chemrev.6b00127

167. Kiefer LM, Kubarych KJ (2016) NOESY-Like 2D-IR spectroscopy reveals non-gaussian dynamics. J Phys Chem Lett 7:3819–3824. doi:10.1021/acs.jpclett.6b01803

168. Yue Y, Qasim LN, Kurnosov AA et al (2015) Band-selective ballistic energy transport in alkane oligomers: toward controlling the transport speed. J Phys Chem B 119:6448–6456. doi:10.1021/acs.jpcb.5b03658

169. Kurochkin DV, Naraharisetty SRG, Rubtsov IV (2005) Dual-frequency 2D IR on interaction of weak and strong IR modes. J Phys Chem A 109:10799–10802. doi:10.1021/jp055811+

170. Lin Z, Rubtsova NI, Kireev VV, Rubtsov IV (2015) Ballistic energy transport in PEG oligomers. Acc Chem Res 48:2547–2555. doi:10.1051/epjconf/20134105039

171. Gray DE (1957) American institute of physics handbook. McGraw-Hill, Boca Raton

172. Bredenbeck J, Ghosh A, Smits M, Bonn M (2008) Ultrafast two dimensional-infrared spectroscopy of a molecular monolayer. J Am Chem Soc 130:2152–2153. doi:10.1021/ja710099c

173. Rosenfeld DE, Gengeliczki Z, Smith BJ et al (2011) Structural dynamics of a catalytic monolayer probed by ultrafast 2D IR vibrational echoes. Science 334:634–639. doi:10.1126/science.1211350

174. Rosenfeld DE, Nishida J, Yan C et al (2012) Dynamics of functionalized surface molecular monolayers studied with ultrafast infrared vibrational spectroscopy. J Phys Chem C 116:23428–23440. doi:10.1021/jp307677b

175. Rosenfeld DE, Nishida J, Yan C et al (2013) Structural dynamics at monolayer-liquid interfaces probed by 2D IR spectroscopy. J Phys Chem C 117:1409–1420. doi:10.1021/jp311144b

176. Nishida J, Yan C, Fayer MD (2014) Dynamics of molecular monolayers with different chain lengths in air and solvents probed by ultrafast 2D IR spectroscopy. J Phys Chem C 118:523–532. doi:10.1021/jp410683h

177. Yan C, Yuan R, Nishida J, Fayer MD (2015) Structural influences on the fast dynamics of alkyl-siloxane monolayers on SiO 2 surfaces measured with 2D IR spectroscopy. J Phys Chem C 119:16811–16823. doi:10.1021/acs.jpcc.5b05641

178. Yan C, Yuan R, Pfalzgraff WC et al (2016) Unraveling the dynamics and structure of functionalized self-assembled monolayers on gold using 2D IR spectroscopy and MD simulations. Proc Natl Acad Sci 113:4929–4934. doi:10.1073/pnas.1603080113

179. Kraack JP, Kaech A, Hamm P (2016) Surface-enhancement in ultrafast 2D ATR IR spectroscopy at the metal-liquid interface. J Phys Chem C 120:3350–3359. doi:10.1021/acs.jpcc.5b11051

180. Kraack JP, Lotti D, Hamm P (2015) 2D attenuated total reflectance infrared spectroscopy reveals ultrafast vibrational dynamics of organic monolayers at metal-liquid interfaces. J Chem Phys 142:212413. doi:10.1063/1.4916915

181. Kraack JP, Lotti D, Hamm P (2015) Surface-enhanced, multi-dimensional attenuated total reflectance spectroscopy. In: Hayes SC, Bittner ER (eds) Proc SPIE, Phys Chem Interfaces Nanomater, vol XIV, pp 95490S

182. Zhang Z, Piatkowski L, Bakker HJ, Bonn M (2011) Ultrafast vibrational energy transfer at the water/air interface revealed by two-dimensional surface vibrational spectroscopy. Nat Chem 3:888–893. doi:10.1038/nchem.1158

183. Piatkowski L, Eisenthal KB, Bakker HJ (2009) Ultrafast intermolecular energy transfer in heavy water. Phys Chem Chem Phys 11:9033–9038. doi:10.1039/b908975f

184. Wang J, Clark ML, Li Y et al (2015) Short-range catalyst-surface interactions revealed by heterodyne two-dimensional sum frequency generation spectroscopy. J Phys Chem Lett 6:4204–4209. doi:10.1021/acs.jpclett.5b02158

185. Li Y, Wang J, Clark ML et al (2016) Characterizing interstate vibrational coherent dynamics of surface adsorbed catalysts by fourth-order 3D SFG spectroscopy. Chem Phys Lett 650:1–6. doi:10.1016/j.cplett.2016.02.031

186. Li Z, Wang J, Li Y, Xiong W (2016) Solving the "Magic Angle" challenge in determining molecular orientation at interfaces. J Phys Chem C 120:20239–20246. doi:10.1021/acs.jpcc.6b08093

187. Laaser JE, Zanni MT (2013) Extracting structural information from the polarization dependence of one- and two-dimensional sum frequency generation spectra. J Phys Chem A 117:5875–5890. doi:10.1021/jp307721y

188. Nihonyanagi S, Kusaka R, Inoue K et al (2015) Accurate determination of complex χ(2) spectrum of the air/water interface. J Chem Phys 143(124707):1–4. doi:10.1063/1.4931485

189. Nihonyanagi S, Mondal JA, Yamaguchi S, Tahara T (2013) Structure and dynamics of interfacial water studied by heterodyne-detected vibrational sum-frequency generation. Annu Rev Phys Chem 64:579–603. doi:10.1146/annurev-physchem-040412-110138

190. Shen YR, Ostroverkhov V (2006) Sum-frequency vibrational spectroscopy on water interfaces: polar orientation of water molecules at interfaces. Chem Rev 106:1140–1154. doi:10.1021/cr040377d

191. Ishiyama T, Imamura T, Morita A (2014) Theoretical studies of structures and vibrational sum frequency generation spectra at aqueous interfaces. Chem Rev 114:8447–8470

192. Laaser JE, Zanni MT (2013) Extracting structural information from the polarization dependence of one- and two-dimensional sum frequency generation spectra. J Phys Chem A 117:5875–5890

193. Ho J-J, Skoff DR, Ghosh A, Zanni MT (2015) Structural characterization of single-stranded DNA monolayers using two-dimensional sum frequency generation spectroscopy. J Phys Chem B 119:10586–10596. doi:10.1021/acs.jpcb.5b07078

194. Laaser JE, Skoff DR, Ho J et al (2014) Two-dimensional sum-frequency generation reveals structure and dynamics of a surface-bound peptide. J Am Chem Soc 136:956–962

195. Zhang Z, Piatkowski L, Bakker HJ, Bonn M (2011) Communication: interfacial water structure revealed by ultrafast two-dimensional surface vibrational spectroscopy. J Chem Phys 135:18–21. doi:10.1063/1.3605657

196. Hsieh C-S, Okuno M, Hunger J et al (2014) Aqueous heterogeneity at the air/water interface revealed by 2D-HD-SFG spectroscopy. Angew Chem Int Ed Engl 53:8146–8149. doi:10.1002/anie.201402566

197. Nagata Y, Mukamel S (2011) Spectral diffusion at the water/lipid interface revealed by two-dimensional fourth-order optical spectroscopy: a classical simulation study. J Am Chem Soc 133:3276–3279. doi:10.1021/ja110748s

198. Singh PC, Inoue KI, Nihonyanagi S et al (2016) Femtosecond hydrogen bond dynamics of bulk-like and bound water at positively and negatively charged lipid interfaces revealed by 2D HD-VSFG spectroscopy. Angew Chemie Int Ed 55:10621–10625. doi:10.1002/anie.201603676

199. Livingstone RA, Nagata Y, Bonn M, Backus EHG (2015) Two types of water at the water-surfactant interface revealed by time-resolved vibrational spectroscopy. J Am Chem Soc 137:14912–14919. doi:10.1021/jacs.5b07845

200. Singh PC, Nihonyanagi S, Yamaguchi S, Tahara T (2012) Ultrafast vibrational dynamics of water at a charged interface revealed by two-dimensional heterodyne-detected vibrational sum frequency generation. J Chem Phys 137(094706):1–6. doi:10.1063/1.4747828

201. Inoue KI, Nihonyanagi S, Singh PC et al (2015) 2D heterodyne-detected sum frequency generation study on the ultrafast vibrational dynamics of H2O and HOD water at charged interfaces. J Chem Phys 142(212431):1–12. doi:10.1063/1.4918644

202. Piatkowski L, Zhang Z, Backus EHG et al (2014) Extreme surface propensity of halide ions in water. Nat Commun 5:4083. doi:10.1038/ncomms5083

203. Fecko CJ, Eaves JD, Loparo JJ et al (2003) Ultrafast hydrogen-bond dynamics in the infrared. Science 301:1698–1702

204. Kraack JP, Lotti D, Hamm P (2014) Ultrafast, multidimensional attenuated total reflectance spectroscopy of adsorbates at metal surfaces. J Phys Chem Lett 5:2325–2329

205. Kraack JP, Kaech A, Hamm P (2017) Molecule-specific interactions of diatomic adsorbates at metal-liquid interfaces. Struct Dyn 4(044009):1–14. doi:10.1063/1.4978894

206. Zamadar M, Asaoka S, Grills DC, Miller JR (2013) Giant infrared absorption bands of electrons and holes in conjugated molecules. Nat Commun 4:2818. doi:10.1038/ncomms3818

207. Bertie JE, Lan Z (1996) Infrared intensities of liquids XX: the intensity of the OH stretching band of liquid water revisited, and the best current values of the optical constants of H2O(l) at 25 C between 15,000 and 1 cm^{-1}. Appl Spectrosc 50:1047–1057. doi:10.1366/0003702963905385

208. Yan C, Nishida J, Yuan R, Fayer MD (2016) Water of hydration dynamics in minerals gypsum and bassanite: ultrafast 2D IR spectroscopy of rocks. J Am Chem Soc 138:9694–9703. doi:10.1021/jacs.6b05589

209. Bakulin AA, Cringus D, Pieniazek PA et al (2013) Dynamics of water confined in reversed micelles: multidimensional vibrational spectroscopy study. J Phys Chem B 117:15545–15558. doi:10.1021/jp405853j

210. Bagchi B (2005) Water dynamics in the hydration layer around proteins and micelles. Chem Rev 105:3197–3219. doi:10.1021/cr020661+

211. Volkov VV, Palmer DJ, Righini R (2007) Distinct water species confined at the interface of a phospholipid membrane. Phys Rev Lett 99:1–4. doi:10.1103/PhysRevLett.99.078302
212. Bakulin AA, Selig O, Bakker HJ, et al (2015) Real-time observation of organic cation reorientation in methylammonium lead iodide perovskites. J Phys Chem Lett 3663–3669. doi:10.1021/acs.jpclett.5b01555
213. Mizuno N, Misono M (1998) Heterogeneous catalysis. Chem Bond Surf Interfaces 98:199–217. doi:10.1016/B978-044452837-7.50005-8
214. Medders GR, Paesani F (2014) Water dynamics in metal – Organic frameworks: effects of heterogeneous confinement predicted by computational spectroscopy. J Phys Chem Lett 5:2897–2902
215. Nishida J, Tamimi A, Fei H et al (2014) Structural dynamics inside a functionalized metal-organic framework probed by ultrafast 2D IR spectroscopy. Proc Nat Acad Sci USA 111:18442–18447. doi:10.1073/pnas.1422194112
216. Yoon M, Srirambalaji R, Kim K (2012) Homochiral metal-organic frameworks for asymmetric heterogeneous catalysis. Chem Rev 112:1196–1231. doi:10.1021/cr2003147
217. Kreno LE, Leong K, Farha OK et al (2012) Metal-organic framework materials as chemical sensors. Chem Rev 112:1105–1125. doi:10.1021/cr200324t **(Washington, DC, United States)**
218. Lotti D, Hamm P, Kraack JP (2016) Surface-sensitive spectro-electrochemistry using ultrafast 2D ATR IR spectroscopy. J Phys Chem C 120:2883–2892. doi:10.1021/acs.jpcc.6b00395
219. Lambert DK (1988) Vibrational stark effect of CO on Ni(100), and CO in the aqueous double layer: experiment, theory, and models. J Chem Phys 89:3847. doi:10.1063/1.454860
220. Hush NS, Reimers JR (1995) Vibrational stark spectroscopy. 1. basic theory and application to the CO stretch. J Phys Chem 99:15798–15805. doi:10.1021/j100043a018
221. Chattopadhyay A, Boxer SG (1995) Vibrational stark effect spectroscopy. J Am Chem Soc 117:1449–1450
222. Lambert DK (1996) Vibrational stark effect of adsorbates at electrochemical interfaces. Electrochim Acta 41:623–630. doi:10.1016/0013-4686(95)00349-5
223. Fried SD, Boxer SG (2015) Measuring electric fields and noncovalent interactions using the vibrational stark effect. Acc Chem Res 48:998–1006. doi:10.1021/ar500464j
224. Mojet BL, Ebbesen SD, Lefferts L (2010) Light at the interface: the potential of attenuated total reflection infrared spectroscopy for understanding heterogeneous catalysis in water. Chem Soc Rev 39:4643–4655. doi:10.1039/c0cs00014k
225. El Khoury Y, van Wilderen LJGW, Vogt T et al (2015) A spectroelectrochemical cell for ultrafast two-dimensional infrared spectroscopy. Rev Sci Instrum 86(083102):1–5. doi:10.1063/1.4927533
226. El Khoury Y, van Wilderen LJGW, Bredenbeck J (2015) Ultrafast 2D-IR spectroelectrochemistry of flavin mononucleotide. J Chem Phys 142:212416. doi:10.1063/1.4916916
227. Nishida J, Yan C, Fayer MD (2017) Enhanced nonlinear spectroscopy for monolayers and thin films in near-Brewster ' s angle reflection pump-probe geometry. J Chem Phys 146:94201. doi:10.1063/1.4977508
228. Liu W-T, Shen YR (2014) In situ sum-frequency vibrational spectroscopy of electrochemical interfaces with surface plasmon resonance. Proc Natl Acad Sci 111:1293–1297. doi:10.1073/pnas.1317290111
229. Peremans A, Tadjeddine A, Zheng W-Q et al (1996) Vibrational dynamics of CO at single-crystal platinum electrodes in aqueous and non-aqueous electrolytes. Surf Sci 368:384–388. doi:10.1016/S0039-6028(96)01080-1
230. Peremans A, Tadjeddine A (1994) Vibrational spectroscopy of electrochemically deposited hydrogen on platinum. Phys Rev Lett 73:3010–3013
231. Roberts ST, Loparo JJ, Ramasesha K, Tokmakoff A (2011) A fast-scanning Fourier transform 2D IR interferometer. Opt Commun 284:1062–1066. doi:10.1016/j.optcom.2010.10.049
232. Garrett-Roe S, Hamm P (2009) What can we learn from three-dimensional infrared spectroscopy? Acc Chem Res 42:1412–1422. doi:10.1021/ar900028k
233. Mukamel S, Piryatinski A, Chernyak V (1999) Two-dimensional Raman echoes: femtosecond view of molecular structure and vibrational coherence. Acc Chem Res 32:145–154. doi:10.1021/ar960206y
234. Tokmakoff A, Lang M, Larsen D et al (1997) Two-dimensional Raman spectroscopy of vibrational interactions in liquids. Phys Rev Lett 79:2702–2705. doi:10.1103/PhysRevLett.79.2702
235. Tokmakoff A, Fleming GR (1997) Two-dimensional Raman spectroscopy of the intermolecular modes of liquid CS2. J Chem Phys 106:2569–2582. doi:10.1063/1.473361

236. Frostig H, Bayer T, Dudovich N et al (2015) Single-beam spectrally controlled two-dimensional Raman spectroscopy. Nat Photonics 9:339–343. doi:10.1038/nphoton.2015.64

237. Garrett-Roe S, Perakis F, Rao F, Hamm P (2011) Three-dimensional infrared spectroscopy of isotope-substituted liquid water reveals heterogeneous dynamics. J Phys Chem B 115:6976–6984. doi:10.1021/jp201989s

238. Garrett-Roe S, Hamm P (2009) Purely absorptive three-dimensional infrared spectroscopy. J Chem Phys 130:164510. doi:10.1063/1.3122982

239. Perakis F, Borek JA, Hamm P (2013) Three-dimensional infrared spectroscopy of isotope-diluted ice Ih. J Chem Phys. doi:10.1063/1.4812216

240. Ding F, Zanni MT (2007) Heterodyned 3D IR spectroscopy. Chem Phys 341:95–105. doi:10.1016/j.chemphys.2007.06.010

241. Fulmer EC, Ding F, Zanni MT (2005) Heterodyned fifth-order 2D-IR spectroscopy of the azide ion in an ionic glass. J Chem Phys 122:34302. doi:10.1063/1.1810513

242. Kemlin V, Bonvalet A, Daniault L, Joffre M (2016) Transient two-dimensional infrared spectroscopy in a vibrational ladder. J Phys Chem Lett 3377–3382. doi:10.1021/acs.jpclett.6b01535

243. Kraack JP, Hamm P (2016) Vibrational ladder-climbing in surface-enhanced, ultrafast infrared spectroscopy. Phys Chem Chem Phys 18:16088–16093. doi:10.1039/C6CP02589G

244. Kraack, J. P. (2013). Multi-dimensional Ultrafast Spectroscopy of Vibrational Coherence Dynamics in Excited Electronic States of Polyenes, Ruprecht-Karls-University of Heidelberg (Doctoral dissertation).

245. Borek JA, Perakis F, Hamm P (2014) Testing for memory-free spectroscopic coordinates by 3D IR exchange spectroscopy. Proc Natl Acad Sci 111:10462–10467. doi:10.1073/pnas.1406967111

246. Kwac K, Lee C, Jung Y et al (2006) Phenol-benzene complexation dynamics: quantum chemistry calculation, molecular dynamics simulations, and two dimensional IR spectroscopy. J Chem Phys. doi:10.1063/1.2403132

247. Lane TJ, Shukla D, Beauchamp KA, Pande VS (2013) To milliseconds and beyond: challenges in the simulation of protein folding. Curr Opin Struct Biol 23:58–65. doi:10.1016/j.sbi.2012.11.002

248. Shukla D, Hernandez CX, Weber JK, Pande VS (2015) Markov state models provide insights into dynamic modulation of protein function. Acc Chem Res 48:414–422. doi:10.1021/ar5002999

249. Lane TJ, Bowman GR, Beauchamp KA et al (2011) Markov state model reveals folding and functional dynamics in ultra-long MD trajectories. J Am Chem Soc 133:18413–18419. doi:10.1021/ja207470h.Markov

250. Xiong W, Laaser JE, Paoprasert P et al (2009) Transient 2D IR spectroscopy of charge injection in dye-sensitized nanocrystalline thin films. J Am Chem Soc 131:18040–18041. doi:10.1021/ja908479r

251. Bredenbeck J, Helbing J, Behrendt R et al (2003) Transient 2D-IR spectroscopy : snapshots of the nonequilibrium ensemble during the picosecond conformational transition of a small peptide. J Phys Chem B 107:8654–8660. doi:10.1021/jp034552q

252. Bredenbeck J, Helbing J, Hamm P (2004) Labeling vibrations by light: ultrafast transient 2D-IR spectroscopy tracks vibrational modes during photoinduced charge transfer. J Am Chem Soc 126:990–991. doi:10.1021/ja0380190

253. Hunt NT (2014) Transient 2D-IR spectroscopy of inorganic excited states. Dalt Trans 43:17578–17589. doi:10.1039/C4DT01410C

254. Kiefer LM, King JT, Kubarych KJ (2014) Equilibrium excited state dynamics of a photoactivated catalyst measured with ultrafast transient 2DIR. J Phys Chem A 118:9853–9860. doi:10.1021/jp508974w

255. Di Donato M, Ragnoni E, Lapini A et al (2015) Femtosecond transient infrared and stimulated Raman spectroscopy shed light on the relaxation mechanisms of photo-excited peridinin. J Chem Phys 142:212409. doi:10.1063/1.4915072

256. Di Donato M, Centellas MS, Lapini A et al (2014) Combination of transient 2D-IR experiments and Ab initio computations sheds light on the formation of the charge-transfer state in photoexcited carbonyl carotenoids. J Phys Chem B 118:9613–9630

257. Delor M, Sazanovich IV, Towrie M, Weinstein JA (2015) Probing and exploiting the interplay between nuclear and electronic motion in charge transfer processes. Acc Chem Res 150319121125006. doi:10.1021/ar500420c

258. Kolano C, Helbing J, Kozinski M et al (2006) Watching hydrogen-bond dynamics in a beta-turn by transient two-dimensional infrared spectroscopy. Nature 444:469–472. doi:10.1038/nature05352

259. Cervetto V, Pfister R, Bredenbeck J et al (2008) Transient IR and 2D-IR spectroscopy of thiopeptide isomerization. J Phys Chem B 112:8395–8405. doi:10.1016/B978-044452821-6/50058-7

260. Chung HS, Ganim Z, Jones KC, Tokmakoff A (2007) Transient 2D IR spectroscopy of ubiquitin unfolding dynamics. Proc Natl Acad Sci USA 104:14237–14242. doi:10.1073/pnas.0700959104

261. Jones KC, Peng CS, Tokmakoff A (2013) Folding of a heterogeneous β-hairpin peptide from temperature-jump 2D IR spectroscopy. Proc Natl Acad Sci USA 110:2828–2833. doi:10.1073/pnas.1211968110

262. Baiz CR, Lin YS, Peng CS et al (2014) A molecular interpretation of 2D IR protein folding experiments with markov state models. Biophys J 106:1359–1370. doi:10.1016/j.bpj.2014.02.008

263. Chung HS, Khalil M, Smith AW, Tokmakoff A (2007) Transient two-dimensional IR spectrometer for probing nanosecond temperature-jump kinetics. Rev Sci Instrum. doi:10.1063/1.2743168

264. Peng CS, Baiz CR, Tokmakoff A (2013) Direct observation of ground-state lactam-lactim tautomerization using temperature-jump transient 2D IR spectroscopy. Proc Natl Acad Sci USA 110:9243–9248. doi:10.1073/pnas.1303235110

265. Bredenbeck J, Helbing J, Hamm P (2004) Transient two-dimensional infrared spectroscopy: exploring the polarization dependence. J Chem Phys 121:5943–5957. doi:10.1063/1.1779575

266. Andresen ER, Hamm P (2009) Site-specific difference 2D-IR spectroscopy of bacteriorhodopsin. J Phys Chem B 113:6520–6527. doi:10.1021/jp810397u

267. Bredenbeck J, Helbing J, Hamm P (2005) Solvation beyond the linear response regime. Phys Rev Lett 95:1–4. doi:10.1103/PhysRevLett.95.083201

268. Baiz CR, Nee MJ, McCanne R, Kubarych KJ (2008) Ultrafast nonequilibrium Fourier-transform two-dimensional infrared spectroscopy. Opt Lett 33:2533–2535. doi:10.1364/OL.33.002533

269. Bredenbeck J, Helbing J, Nienhaus K et al (2007) Protein ligand migration mapped by nonequilibrium 2D-IR exchange spectroscopy. Proc Natl Acad Sci USA 104:14243–14248. doi:10.1073/pnas.0607758104

270. Van Wilderen LJGW, Messmer AT, Bredenbeck J (2014) Mixed IR/Vis two-dimensional spectroscopy: chemical exchange beyond the vibrational lifetime and sub-ensemble selective photochemistry. Angew Chemie-Int Ed 53:2667–2672. doi:10.1002/anie.201305950

271. Zhang XX, Jones KC, Fitzpatrick A et al (2016) Studying protein–protein binding through T-jump induced dissociation: transient 2D IR spectroscopy of insulin dimer. J Phys Chem B 120:5134–5145. doi:10.1021/acs.jpcb.6b03246

272. Courtney TL, Fox ZW, Estergreen L, Khalil M (2015) Measuring coherently coupled intramolecular vibrational and charge transfer dynamics with two-dimensional vibrational-electronic spectroscopy. J Phys Chem Lett 6:1286–1292. doi:10.1021/acs.jpclett.5b00356

273. Courtney TL, Fox ZW, Slenkamp KM, Khalil M (2015) Two-dimensional vibrational-electronic spectroscopy. J Chem Phys 143:154201. doi:10.1063/1.4932983

274. Dahms F, Fingerhut BP, Nibbering ETJ et al (2017) Large-amplitude transfer motion of hydrated excess protons mapped by ultrafast 2D IR spectroscopy. Science (80-) 5144:1–9. doi:10.1126/science.aan5144

275. Balasubramanian M, Courtney TL, Gaynor JD, Khalil M (2016) Compression of tunable broadband mid-IR pulses with a deformable mirror pulse shaper. J Opt Soc Am B 33:2033–2037. doi:10.1364/JOSAB.33.002033

276. Gaynor JD, Courtney TL, Balasubramanian M, Khalil M (2016) Coherent Fourier transform two-dimensional electronic-vibrational spectroscopy using an octave-spanning mid-IR probe. Opt Lett 41:2895–2898. doi:10.1364/UP.2016.UTu4A.7

277. Maiti KS (2015) Broadband two dimensional infrared spectroscopy of cyclic amide 2-Pyrrolidinone. Phys Chem Chem Phys 17:24998–25003. doi:10.1039/C5CP04272K

278. Wand A, Gdor I, Zhu J, et al (2013) Shedding new light on retinal protein photochemistry. Annu Rev Phys Chem 64:null. doi:10.1146/annurev-physchem-040412-110148

279. Brixner T, Mančal T, Stiopkin IV, Fleming GR (2004) Phase-stabilized two-dimensional electronic spectroscopy. J Chem Phys 121:4221. doi:10.1063/1.1776112

280. Oliver TAA, Lewis NHC, Graham R (2014) Correlating the motion of electrons and nuclei with two-dimensional electronic–vibrational spectroscopy. Proc Nat Acad Sci USA. doi:10.5452/maax7dd

281. Lewis NHC, Fleming GR (2016) Two-dimensional electronic-vibrational spectroscopy of chlorophyll a and b. J Phys Chem Lett 831–837. doi:10.1021/acs.jpclett.6b00037

282. Lewis NHC, Dong H, Oliver TAA, Fleming GR (2015) Measuring correlated electronic and vibrational spectral dynamics using line shapes in two-dimensional electronic–vibrational spectroscopy. J Chem Phys. doi:10.1063/1.4919686

283. Lewis NHC, Dong H, Oliver TAA, Fleming GR (2015) A method for the direct measurement of electronic site populations in a molecular aggregate using two-dimensional electronic-vibrational spectroscopy. J Chem Phys. doi:10.1063/1.4931634

284. Lewis NHC, Gruenke NL, Oliver TAA, et al (2016) Observation of electronic excitation transfer through light harvesting complex II using two-dimensional electronic–vibrational spectroscopy. J Phys Chem Lett 4197–4206. doi:10.1021/acs.jpclett.6b02280

285. Dong H, Lewis NHC, Oliver TAA, Fleming GR (2015) Determining the static electronic and vibrational energy correlations via two-dimensional electronic-vibrational spectroscopy. J Chem Phys. doi:10.1063/1.4919684

286. Oliver TAA, Fleming GR (2015) Following coupled electronic-nuclear motion through conical intersections in the ultrafast relaxation of β-Apo-8'-carotenal. J Phys Chem B 119:11428–11441. doi:10.1021/acs.jpcb.5b04893

287. Buckup T, Kraack JP, Marek MS, Motzkus M (2013) Vibronic coupling in excited electronic states investigated with resonant 2D Raman spectroscopy. EPJ Web Conf 41:5018

288. Tracy KM, Barich MV, Carver CL et al (2016) High-throughput two-dimensional infrared (2D IR) spectroscopy achieved by interfacing microfluidic technology with a high repetition rate 2D IR spectrometer. J Phys Chem Lett 7:4865–4870. doi:10.1021/acs.jpclett.6b01941

289. Dittrich PS, Manz A (2006) Lab-on-a-chip: microfluidics in drug discovery. Nat Rev Drug Discov 5:210–218. doi:10.1038/nrd1985

290. Johnson TJ, Ross D, Locascio LE (2002) Rapid microfluidic mixing. Anal Chem 74:45–51. doi:10.1021/ac010895d

291. Wong SH, Ward MCL, Wharton CW (2004) Micro T-mixer as a rapid mixing micromixer. Sens Actuators B Chem 100:359–379. doi:10.1016/j.snb.2004.02.008

292. Olson JS (1981) [38] Stopped-flow, rapid mixing measurements of ligand binding to hemoglobin and red cells. Methods Enzymol 76:631–651. doi:10.1016/0076-6879(81)76148-2

293. Roder H, Wüthrich K (1986) Protein folding kinetics by combined use of rapid mixing techniques and NMR observation of individual amide protons. Proteins Struct Funct Genet 1:34–42. doi:10.1002/prot.340010107

294. Roder H, Maki K, Latypov RF et al (2006) Early events in protein folding explored by rapid mixing methods. Chem Rev 106:1836–1861. doi:10.1002/9783527619498.ch15

295. Chow AW (2002) Lab-on-a-chip: opportunities for chemical engineering. AIChE J 48:1590–1595. doi:10.1002/aic.690480802

296. Hansen C, Quake SR (2003) Microfluidics in structural biology: smaller, faster… better. Curr Opin Struct Biol 13:538–544. doi:10.1016/j.sbi.2003.09.010

297. Stone HA, Stroock AD, Ajdari A (2004) Engineering flows in small devices/microfluidics toward a lab-on-a-chip. Annu Rev Fluid Mech 36:381–411. doi:10.1146/annurev.fluid.36.050802.122124

298. Jensen K (2001) Microreaction engineering—is small better? Chem Eng Sci 56:293–303. doi:10.1016/S0009-2509(00)00230-X

299. Stone HA, Kim S (2001) Microfluidics: basic issues, applications, and challenges. AIChE J 47:1250–1254. doi:10.1002/aic.690470602

300. Watts P, Haswell SJ (2003) Microfluidic combinatorial chemistry. Curr Opin Chem Biol 7:380–387. doi:10.1016/S1367-5931(03)00050-4

301. Luther BM, Tracy KM, Gerrity M et al (2016) 2D IR spectroscopy at 100 kHz utilizing a Mid-IR OPCPA laser source. Opt Express 24:4117–4127. doi:10.1364/OE.24.004117

302. Greetham GM, Donaldson PM, Nation C et al (2016) A 100 kHz time-resolved multiple-probe femtosecond to second infrared absorption spectrometer. Appl Spectrosc 70:645–653

303. Chalus O, Bates PK, Smolarski M, Biegert J (2009) Mid-IR short-pulse OPCPA with micro-joule energy at 100 kHz. Opt Express 17:3587–3594. doi:10.1364/OE.17.003587

304. Shim S-H, Strasfeld DB, Ling YL, Zanni MT (2007) Automated 2D IR spectroscopy using a mid-IR pulse shaper and application of this technology to the human islet amyloid polypeptide. Proc Natl Acad Sci 104:14197–14202. doi:10.1073/pnas.0700804104

305. Leger JD, Nyby CM, Varner C et al (2014) Fully automated dual-frequency three-pulse-echo 2DIR spectrometer accessing spectral range from 800 to 4000 wavenumbers. Rev Sci Instrum. doi:10.1063/1.4892480

306. Kuroda DG, Bauman JD, Challa JR et al (2013) Snapshot of the equilibrium dynamics of a drug bound to HIV-1 reverse transcriptase. Nat Chem 5:174–181. doi:10.1038/nchem.1559
307. Ataka K, Heberle J (2007) Biochemical applications of surface-enhanced infrared absorption spectroscopy. Anal Bioanal Chem 388:47–54. doi:10.1007/s00216-006-1071-4
308. Ataka K, Stripp ST, Heberle J (2013) Surface-enhanced infrared absorption spectroscopy (SEIRAS) to probe monolayers of membrane proteins. Biochim Biophys Acta-Biomembr 1828:2283–2293. doi:10.1016/j.bbamem.2013.04.026
309. Neubrech F, Pucci A, Cornelius TW et al (2008) Resonant plasmonic and vibrational coupling in a tailored nanoantenna for infrared detection. Phys Rev Lett 101(157403):1–4. doi:10.1103/PhysRevLett.101.157403
310. Neubrech F, Pucci A (2013) Plasmonic enhancement of vibrational excitations in the infrared. IEEE J Sel Top Quantum Electron 19:4600809. doi:10.1109/JSTQE.2012.2227302
311. Selig O, Siffels R, Rezus YLA (2015) Ultrasensitive ultrafast vibrational spectroscopy employing the near field of gold nanoantennas. Phys Rev Lett 114(233004):1–5. doi:10.1103/PhysRevLett.114.233004
312. Donaldson PM, Hamm P (2013) Gold nanoparticle capping layers: structure, dynamics, and surface enhancement measured using 2D-IR spectroscopy. Angew Chem Int Ed 52:634–638. doi:10.1002/anie.201204973
313. Arrivo SM, Dougherty TP, Grubbs WT, Heilweil EJ (1995) Ultrafast infrared spectroscopy of vibrational CO-stretch up-pumping and relaxation dynamics of W(CO)6. Chem Phys Lett 235:247–254
314. Witte T, Hornung T, Windhorn L et al (2003) Controlling molecular ground-state dissociation by optimizing vibrational ladder climbing. J Chem Phys 118:2021–2024. doi:10.1063/1.1540101
315. Windhorn L, Witte T, Yeston JS et al (2002) Molecular dissociation by mid-IR femtosecond pulses. Chem Phys Lett 357:85–90. doi:10.1016/S0009-2614(02)00444-X
316. Falvo C, Debnath A, Meier C (2013) Vibrational ladder climbing in carboxy-hemoglobin: effects of the protein environment. J Chem Phys. doi:10.1063/1.4799271
317. Debnath A, Falvo C, Meier C (2013) State-selective excitation of the CO stretch in carboxyhemoglobin by mid-IR laser pulse shaping: a theoretical investigation. J Phys Chem A 117:12884–12888. doi:10.1021/jp410473u
318. Witte T, Yeston JS, Motzkus M et al (2004) Femtosecond infrared coherent excitation of liquid phase vibrational population distributions (v > 5). Chem Phys Lett 392:156–161. doi:10.1016/j.cplett.2004.05.052
319. Maas DJ, Duncan DI, Vrijen RB et al (1998) Vibrational ladder climbing in NO by (sub)picosecond frequency-chirped infrared laser pulses. Chem Phys Lett 290:75–80. doi:10.1016/S0009-2614(98)00531-4
320. Kleiman VD, Arrivo SM, Melinger JS, Heilweil EJ (1998) Controlling condensed-phase vibrational excitation with tailored infrared pulses. Chem Phys 233:207–216
321. Nuernberger P, Vieille T, Ventalon C, Joffre M (2011) Impact of pulse polarization on coherent vibrational ladder climbing signals. J Phys Chem B 115:5554–5563. doi:10.1021/jp1113762
322. Ventalon C, Fraser JM, Vos MH et al (2004) Coherent vibrational climbing in carboxyhemoglobin. Proc Natl Acad Sci USA 101:13216–13220. doi:10.1073/pnas.0401844101
323. Strasfeld DB, Shim SH, Zanni MT (2007) Controlling vibrational excitation with shaped Mid-IR pulses. Phys Rev Lett 99:1–4. doi:10.1103/PhysRevLett.99.038102
324. Wodtke AM, Matsiev D, Auerbach D (2008) Energy transfer and chemical dynamics at solid surfaces: the special role of charge transfer. Prog Surf Sci 83:167–214. doi:10.1016/j.progsurf.2008.02.001
325. Golibrzuch K, Bartels N, Auerbach DJ, Wodtke AM (2015) The dynamics of molecular interactions and chemical reactions at metal surfaces: testing the foundations of theory. Annu Rev Phys Chem 66:399–425. doi:10.1146/annurev-physchem-040214-121958
326. Krüger BC, Meyer S, Kandratsenka A et al (2016) Vibrational inelasticity of highly vibrationally excited NO on Ag(111). J Phys Chem Lett 7:441–446. doi:10.1021/acs.jpclett.5b02448
327. Silva M, Jongma R, Field RW, Wodtke AM (2001) The dynamics of "stretched molecules": experimental studies of highly vibrationally excited molecules with stimulated emission pumping. Annu Rev Phys Chem 52:811–852. doi:10.1146/annurev.physchem.52.1.811
328. Kneba M, Wolfrum J (1980) Bimolecular reactions of vibrationally excited molecules. Annu Rev Phys Chem 31:47–79

329. Moore C, Smith I (1979) Vibrational–rotational excitation. Chemical reactions of vibrationally excited molecules. Faraday Discuss Chem Soc 67:146–161. doi:10.1039/DC9796700146

330. Enders D, Nagao T, Pucci A et al (2011) Surface-enhanced ATR-IR spectroscopy with interface-grown plasmonic gold-island films near the percolation threshold. Phys Chem Chem Phys 13:4935–4941. doi:10.1039/c0cp01450h

331. Lessinger L (1994) Morse oscillators, Birge–Sponer extrapolation, and the electronic absorption spectrum of I2. J Chem Educ 71:388. doi:10.1021/ed071p388

332. Luo Y-R (2007) Comprehensive handbook of chemical bond energies. CRC Press, Boca Raton

333. Huber CJ, Egger SM, Spector IC et al (2015) 2D-IR spectroscopy of porous silica nanoparticles: measuring the distance sensitivity of spectral diffusion. J Phys Chem C 119:25135–25144. doi:10.1021/acs.jpcc.5b05637

334. Wohlleben W, Buckup T, Herek JL, Motzkus M (2005) Coherent control for spectroscopy and manipulation of biological dynamics. ChemPhysChem 6:850–857

335. Herek JL, Wohlleben W, Cogdell RJ et al (2002) Quantum control of energy flow in light harvesting. Nature 417:533–535

336. Prokhorenko VI, Halpin A, Miller RJD (2011) Coherently-controlled two-dimensional spectroscopy: evidence for phase induced long-lived memory effects. Faraday Discuss 153:27–39. doi:10.1039/c1fd00095k

337. Buckup T, Hauer J, Mohring J, Motzkus M (2009) Multidimensional spectroscopy of beta-carotene: vibrational cooling in the excited state. Arch Biochem Biophys 483:219–223

338. Prokhorenko VI, Nagy AM, Waschuk SA et al (2006) Coherent control of retinal isomerization in bacteriorhodopsin. Science (80-) 313:1257–1261. doi:10.1126/science.1130747

339. Kraack JP, Motzkus M, Buckup T (2011) Selective nonlinear response preparation using femtosecond spectrally resolved four-wave-mixing. J Chem Phys 135:224505

340. Windhorn L, Yeston JS, Witte T et al (2003) Getting ahead of IVR: a demonstration of mid-infrared induced molecular dissociation on a sub-statistical time scale. J Chem Phys 119:641–645. doi:10.1063/1.1587696

341. Shim S-H, Strasfeld DB, Fulmer EC, Zanni MT (2006) Femtosecond pulse shaping directly in the mid-IR using acousto-optic modulation. Opt Lett 31:838–840. doi:10.1364/OL.31.000838

342. Shim S-H, Strasfeld DB, Zanni MT (2006) Generation and characterization of phase and amplitude shaped femtosecond mid-IR pulses. Opt Express 14:13120–13130. doi:10.1364/UP.2010.TuE28

343. Witte T, Kompa KL, Motzkus M (2003) Femtosecond pulse shaping in the mid infrared by difference-frequency mixing. Appl Phys B Lasers Opt 76:467–471. doi:10.1007/s00340-003-1118-6

344. Middleton CT, Strasfeld DB, Zanni MT (2009) Polarization shaping in the mid-IR and polarization-based balanced heterodyne detection with application to 2D IR spectroscopy. Opt Express 17:14526–14533. doi:10.1364/OE.17.014526

345. Strasfeld DB, Middleton CT, Zanni MT (2009) Mode selectivity with polarization shaping in the mid-IR. New J Phys. doi:10.1088/1367-2630/11/10/105046

346. Jiang X, Zaitseva E, Schmidt M et al (2008) Resolving voltage-dependent structural changes of a membrane photoreceptor by surface-enhanced IR difference spectroscopy. Proc Natl Acad Sci USA 105:12113–12117. doi:10.1073/pnas.0802289105

347. Dennis AM, Delehanty JB, Medintz IL (2016) Emerging physicochemical phenomena along with new opportunities at the biomolecular-nanoparticle interface. J Phys Chem Lett 7:2139–2150. doi:10.1021/acs.jpclett.6b00570

348. Howes PD, Rana S, Stevens MM (2014) Plasmonic nanomaterials for biodiagnostics. Chem Soc Rev 43:3835–3853. doi:10.1039/c3cs60346f

349. Samanta D, Sarkar A (2011) Immobilization of bio-macromolecules on self-assembled monolayers: methods and sensor applications. Chem Soc Rev 40:2567–2592. doi:10.1039/c0cs00056f

350. Anker JN, Hall WP, Lyandres O et al (2008) Biosensing with plasmonic nanosensors. Nat Mater 7:442–453. doi:10.1038/nmat2162

351. Leblanc RM (2006) Molecular recognition at Langmuir monolayers. Curr Opin Chem Biol 10:529–536. doi:10.1016/j.cbpa.2006.09.010

352. Spinke J, Liley M, Angermaierj HGL, Knoll W (1993) Molecular recognition at self -assembled monolayers : the construction of multicomponent multilayers. Langmuir 9:1821–1825. doi:10.1021/la00031a033

353. Zhang X, Yadavalli VK (2012) Functional self-assembled DNA nanostructures for molecular recognition. Nanoscale 4:2439–2446. doi:10.1039/c2nr11711h

354. Sampson NS, Mrksich M, Bertozzi CR (2000) Surface molecular recognition. Proc Nat Acad Sci USA 98:2000–2001
355. Aprile A, Ciuchi F, Pinalli R et al (2016) Probing molecular recognition at the solid-gas interface by sum-frequency vibrational spectroscopy. J Phys Chem Lett 7:3022–3026. doi:10.1021/acs.jpclett. 6b01300
356. Rodrigo D, Limaj O, Janner D et al (2016) Mid-infrared plasmonic biosensing with graphene. Science 349:165–168. doi:10.1017/CBO9781107415324.004
357. Wu C, Khanikaev AB, Adato R et al (2012) Fano-resonant asymmetric metamaterials for ultra-sensitive spectroscopy and identification of molecular monolayers. Nat Mater 11:69–75. doi:10. 1038/nmat3161
358. Adato R, Aksu S, Altug H (2015) Engineering mid-infrared nanoantennas for surface enhanced infrared absorption spectroscopy. Mater Today 18:436–446. doi:10.1016/j.mattod.2015.03.001
359. Huck C, Vogt J, Sendner M et al (2015) Plasmonic enhancement of infrared vibrational signals: nanoslits versus nanorods. ACS Photonics 2:1489–1497. doi:10.1021/acsphotonics.5b00390
360. Li M, Cushing SK, Wu N (2015) Plasmon-enhanced optical sensors: a review. Analyst 140:386–406. doi:10.1039/c4an01079e
361. Huck C, Neubrech F, Vogt J et al (2014) Surface-enhanced infrared spectroscopy using nanometer-sized gaps. ACS Nano 8:4908–4914
362. Adato R, Altug H (2013) In-situ ultra-sensitive infrared absorption spectroscopy of biomolecule interactions in real time with plasmonic nanoantennas. Nat Commun 4:2154. doi:10.1038/ncomms3154
363. Howes PD, Chandrawati R, Stevens MM (2014) Colloidal nanoparticles as advanced biological sensors. Science (80-) 346:1247390-1–1247390-10. doi:10.1016/0250-6874(86)80002-6
364. Cheng F, Yang X, Gao J (2015) Ultrasensitive detection and characterization of molecules with infrared plasmonic metamaterials. Sci Rep 5:14327. doi:10.1038/srep14327
365. Betzig E, Trautman JK (1992) Near-field optics: microscopy, spectroscopy, and surface modification beyond the diffraction limit. Science 257:189–195. doi:10.1126/science.257.5067.189
366. Hell SW, Wichmann J (1994) Breaking the diffraction resolution limit by stimulated-emission - stimulated-emission-depletion fluorescence microscopy. Opt Lett 19:780–782. doi:10.1364/OL.19. 000780
367. Courjon D, Bainier C (1999) Near field microscopy and near field optics. Reports Prog Phys 57:989–1028. doi:10.1088/0034-4885/57/10/002
368. Schuller JA, Barnard ES, Cai W et al (2010) Plasmonics for extreme light concentration and manipulation. Nat Mater 9:193–204. doi:10.1038/nmat2630
369. Kravtsov V, Ulbricht R, Atkin JM, Raschke MB (2016) Plasmonic nanofocused four-wave mixing for femtosecond near-field imaging. Nat Nanotechnol 11:1–7. doi:10.1038/nnano.2015.336
370. Centrone A (2015) Infrared imaging and spectroscopy beyond the diffraction limit. Annu Rev Anal Chem 8:101–126. doi:10.1146/annurev-anchem-071114-040435
371. Centrone A, Lahiri B, Holland G (2013) Chemical imaging beyond the diffraction limit using photothermal induced resonance microscopy. 27:6–9
372. Keilmann F, Hillenbrand R (2004) Near-field microscopy by elastic light scattering from a tip. Philos Trans A Math Phys Eng Sci 362:787–805. doi:10.1098/rsta.2003.1347
373. Atkin JM, Sass PM, Teichen PE, et al (2015) Nanoscale probing of dynamics in local molecular environments. J Phys Chem Lett. acs.jpclett.5b02093. doi:10.1021/acs.jpclett.5b02093
374. Xu XG, Raschke MB (2013) Near-field infrared vibrational dynamics and tip-enhanced decoherence. Nano Lett 13:1588–1595. doi:10.1021/nl304804p
375. Kim SK, Wang J-K, Zewail AH (1994) Femtosecond pH jump: dynamics of acid—base reactions in solvent cages. Chem Phys Lett 228:369–378. doi:10.1016/0009-2614(94)00951-1
376. Donten ML, Hassan S, Popp A et al (2015) pH-Jump induced leucine zipper folding beyond the diffusion limit. J Phys Chem B 119:1425–1432. doi:10.1021/jp511539c
377. Kohse S, Neubauer A, Pazidis A et al (2013) Photoswitching of enzyme activity by laser-induced pH-jump. J Am Chem Soc 135:9407–9411. doi:10.1021/ja400700x
378. Nunes RMD, Pineiro M, Arnaut LG (2009) Photoacid for extremely long-lived and reversible pH-jumps. J Am Chem Soc 131:9456–9462. doi:10.1021/ja901930c
379. Donten ML, Hamm P (2011) PH-Jump overshooting. J Phys Chem Lett 2:1607–1611. doi:10.1021/jz200610n
380. Genosar L, Cohen B, Huppert D (2000) Ultrafast direct photoacid-base reaction. J Phys Chem A 104:6689–6698. doi:10.1021/jp0003171

381. Pines E, Huppert D (1983) Ph jump—a relaxational approach. J Phys Chem 87:4471–4478. doi:10.1021/j100245a029

382. Donten ML, Hamm P, VandeVondele J (2011) A consistent picture of the proton release mechanism of oNBA in water by ultrafast spectroscopy and ab initio molecular dynamics. J Phys Chem B 115:1075–1083. doi:10.1021/jp109053r

383. Donten ML, Hamm P (2013) PH-jump induced alpha-helix folding of poly-L-glutamic acid. Chem Phys 422:124–130. doi:10.1016/j.chemphys.2012.11.023

384. Kohse S, Neubauer A, Lochbrunner S, Kragl U (2015) Improving the time resolution for remote control of enzyme activity by a nanosecond laser-induced pH jump. ChemCatChem 6:3511–3517. doi:10.1002/cctc.201402442

385. Abbruzzetti S, Sottini S, Viappiani C, Corrie JET (2006) Acid-induced unfolding of myoglobin triggered by a laser pH jump method. Photochem Photobiol Sci 5:621–628. doi:10.1039/b516533d

386. Dumont C, Emilsson T, Gruebele M (2009) Reaching the protein folding speed limit with large, sub-microsecond pressure jumps. Nat Methods 6:515–519. doi:10.1038/Nmeth.1336

387. Smeller L, Heremans K (1999) 2D FT-IR spectroscopy analysis of the pressure-induced changes in proteins. Vib Spectrosc 375–378. doi:10.1016/S0924-2031(98)00075-7

388. Chenevarin S, Thibault-Starzyk F (2004) Two-dimensional IR pressure-jump spectroscopy of adsorbed species for zeolites. Angew Chemie Int Ed 43:1155–1158. doi:10.1002/anie.200352754

389. Rivallan M, Seguin E, Thomas S et al (2010) Platinum sintering on H-ZSM-5 followed by chemometrics of CO adsorption and 2D pressure-jump IR spectroscopy of adsorbed species. Angew Chemie Int Ed 49:785–789. doi:10.1002/anie.200905181

390. Noda I, Story GM, Marcott C (1999) Pressure-induced transitions of polyethylene studied by two-dimensional infrared correlation spectroscopy. Vib Spectrosc 19:461–465. doi:10.1016/S0924-2031(98)00080-0

391. Yamakata A, Uchida T, Kubota J, Osawa M (2006) Laser-induced potential jump at the electrochemical interface probed by picosecond time-resolved surface-enhanced infrared absorption spectroscopy. J Phys Chem B 110:6423–6427. doi:10.1021/jp060387d

392. Yamakata A, Osawa M (2008) Dynamics of double-layer restructuring on a platinum electrode covered by CO: laser-induced potential transient measurement. J Phys Chem C 112:11427–11432. doi:10.1021/jp8018149

393. Muller EA, Pollard B, Bechtel HA et al (2016) Infrared vibrational nano-crystallography and -imaging. Sci Adv 2:e1601006. doi:10.1126/sciadv.1601006

394. Pollard B, Maia FCB, Raschke MB, Freitas RO (2016) Infrared vibrational nanospectroscopy by self-referenced interferometry. Nano Lett 16:55–61. doi:10.1021/acs.nanolett.5b02730

395. Pollard B, Muller EA, Hinrichs K, Raschke MB (2014) Vibrational nano-spectroscopic imaging correlating structure with intermolecular coupling and dynamics. Nat Commun 5:3587. doi:10.1038/ncomms4587

Topics in Current Chemistry (2018) 376:35
https://doi.org/10.1007/s41061-018-0213-4

REVIEW

Multidimensional Vibrational Coherence Spectroscopy

Tiago Buckup[1] · Jérémie Léonard[2]

Received: 12 April 2018 / Accepted: 31 July 2018
© Springer Nature Switzerland AG 2018

Abstract

Multidimensional vibrational coherence spectroscopy has been part of laser spectroscopy since the 1990s and its role in several areas of science has continuously been increasing. In this contribution, after introducing the principals of vibrational coherence spectroscopy (VCS), we review the three most widespread experimental methods for multidimensional VCS (multi-VCS), namely femtosecond stimulated Raman spectroscopy, pump-impulsive vibrational spectroscopy, and pump-degenerate four wave-mixing. Focus is given to the generation and typical analysis of the respective signals in the time and spectral domains. Critical aspects of all multidimensional techniques are the challenges in the data interpretation due to the existence of several possible contributions to the observed signals or to optical interferences and how to overcome the corresponding difficulties by exploiting experimental parameters including higher-order nonlinear effects. We overview how multidimensional vibrational coherence spectroscopy can assist a chemist in understanding how molecular structural changes and eventually photochemical reactions take place. In order to illustrate the application of the techniques described in this chapter, two molecular systems are discussed in more detail in regard to the vibrational dynamics in the electronic excited states: (1) carotenoids as a non-reactive system and (2) stilbene derivatives as a reactive system.

Keywords Ultrafast laser spectroscopy · Multidimensional spectroscopy · Raman · Vibrational spectroscopy · Coherence spectroscopy · Excited states · Vibronic coupling · Photoisomerization

---

Chapter 5 was originally published as Buckup, T. & Léonard, J. Topics in Current Chemistry (2018) 376: 35. https://doi.org/10.1007/s41061-018-0213-4.

✉ Tiago Buckup
  tiago.buckup@pci.uni-heidelberg.de

  Jérémie Léonard
  jeremie.leonard@ipcms.unistra.fr

[1] Physikalisch-Chemisches Institut, Universität Heidelberg, Im Neuenheimer Feld 229, 69120 Heidelberg, Germany

[2] Université de Strasbourg, CNRS, Institut de Physique et Chimie des Matériaux de Strasbourg, UMR 7504, and Labex NIE, 67034 Strasbourg, France

Published online: 24 August 2018

1 Introduction

The dynamics and function of complex molecular systems can be understood as resulting from the interaction between three different interacting sub-units composed of the electronic, vibrational, and environmental degrees of freedom. Focusing particularly on photoreactions—i.e., excited state reactions—they may be described as (non radiative) transitions between distinct electronic states of one or several molecules. Depending on the nature of the coupling responsible for such a transition in a given case, the photoreaction may correspond to a charge or energy transfer, or an internal conversion, an intersystem crossing, etc. Within the Born–Oppenheimer (BO) approximation, electronic states are characterized by multidimensional electronic potential energy surfaces (PESs) representing the energy of the electronic subsystem as a function of all internal nuclear coordinates (3 N-6 for N nuclei) treated as fixed, external parameters. The vibrational degrees of freedom, which drive the molecular system along the photoreactive path from the Franck–Condon state (i.e., initially produced by the photon absorption) to the photoproduct, contribute to the reaction coordinate. All other molecular (and solvent, in condensed phase) vibrational degrees of freedom constitute a large thermal bath—the environment—responsible for very fast energy relaxation and dissipation. There is not only fundamental interest in understanding how these three sub-units interact to perform a given photoreaction and govern its quantum yield, but also in understanding how such interactions shape the molecular functionality.

The mechanism of photoreactions may be elucidated by identifying the conformations and vibrational dynamics of transient electronic states successively populated, from the Franck–Condon state to the vibrationally and thermally relaxed photoproduct. Vibrational spectroscopies (i.e., Raman and IR spectroscopies) have long been exploited to reveal vibrational activity and conformations of stationary or transient molecular states. In this contribution, we will review recent experimental developments, which implement in various ways stimulated Raman scattering to monitor vibrational dynamics along the course of a photoreaction, with a time resolution typically below 100 fs, therefore allowing to resolve the vibrational activity accompanying ultrafast photoreactions. In condensed phases, the vibrational energy relaxation and dissipation to the environment occurs on the 0.1–10 ps time scale, which is in many cases faster than the photoreaction itself. We shall consider as "ultrafast" the photoreactions occurring on a similar time scale or faster. One difficulty inherent to their investigations comes from the fact that there is no time scale separation between the various relevant processes.

To investigate the dynamics and functions of molecular systems on ultrashort time scales, time-resolved UV–VIS spectroscopy has been used since the advent of femtosecond laser light sources. Time-resolved transient absorption—or so-called pump-probe spectroscopy—exploits the non-linear response of the complex system described above upon interaction with coherent laser light pulses. More generally, in the regime of weak-field light–matter interaction, the use of short, coherent laser pulses enables the preparation of controlled, coherent superpositions of molecular (i.e., electronic and vibrational) quantum states. The quantum evolution of these initial coherent states may be followed spectroscopically, i.e., interrogated by the interaction with another laser pulse, until decoherence (interaction with the bath) occurs on typical time scales of a few femtoseconds (electronic decoherence) up to a few picoseconds (vibrational decoherence) [1–4]. After electronic decoherence has occurred, the time evolution of the laser-induced populations can be further observed spectroscopically.

Over the last decade, the investigation of coherence in molecular processes occurring in the condensed phase has become a frontier research topic in molecular quantum physics [5–11]. In this chapter, we will describe an application of femtosecond coherent multidimensional spectroscopy which engineers vibrational coherence in molecular systems and uses it as a spectroscopic tool. More precisely, the goal is to follow in time the vibrational coherence imprinted in the electronic excited states by the non-linear interaction with coherent laser light and exploit the peculiar spectroscopic signatures of such vibrationally coherent molecular states. This type of spectroscopy can be performed under several distinct experimental implementations, which have been named differently. Here we will summarize all these implementations under the name multidimensional vibrational coherence spectroscopy, or multi-VCS. This contribution does not deal, however, with other kinds of vibrational spectroscopies based on so-called rephasing mechanisms like 2D Raman or 2D infrared spectroscopies. These topics are reviewed in different contributions in this collection.

We will first shortly describe how VCS results from the third-order interaction (and higher-order for multi-VCS) of the molecular system with femtosecond light pulses. This initial description will assist the reader in understanding how the higher-dimensional versions of VCS are able to report on structural dynamics in excited states. The main experimental implementations in the time-domain (pump-Impulsive Vibrational Spectroscopy-pump-IVS, pump-Degenerate Four Wave Mixing-pump-DFWM, and Population-controlled IVS) and in the frequency-domain (Femtosecond Stimulated Raman Scattering-FSRS) will be discussed and compared. Finally, we will illustrate the success of multi-VCS at revealing a mechanistic understanding of ultrafast photoreactions in a selection of molecular systems: carotenoids and stilbene derivatives.

2 Introduction to Vibrational Coherence Spectroscopy (VCS)

The principle of vibrational coherence spectroscopy was demonstrated in molecules as soon as picosecond and then femtosecond laser pulses became available. In particular, laser pulses which are shorter than the period of nuclear motions in a molecule have a spectrum larger than the corresponding vibrational level spacing. With such a laser pulse, coherent superposition of vibrational levels, also referred to as vibrational wavepackets, may be produced impulsively in essentially any molecule. The spectroscopic signature of such vibrational wavepackets allows tracking molecular structural dynamics accompanying ultrafast photoreactions in molecules. Early investigations of vibrational wavepacket signatures in simple molecules in gas phase have pioneered the so-called field of femtochemistry [12].

The physical mechanism of VCS may be introduced by discussing a conceptually simple pump-probe experimental scheme. Let us consider a particular example where a short-enough, resonant pump pulse impulsively excites a molecular system, thus producing a non-stationary population, and the probe pulse is used subsequently to measure the absorbance of this pump-induced population. Considering the interaction with the pump laser as a perturbation of the molecular system, a very general result of the perturbation theory is the following. At the first order of the perturbative expansion, a coherent superposition of the two states coupled by the perturbation is produced. Such a superposition is called a "coherence". Instead, a "population" (e.g., depopulation of the ground S_0 state and population of the excited S_1 state) is produced at the second order only. With the vocabulary of non-linear optics, this is rephrased as: A first interaction with the pump field (denoted by its wavevector k_1) creates an *electronic* coherence while a second interaction (k_2) creates an *electronic* population. Since the pump pulse is spectrally broad, both interactions (i.e., with k_1 and k_2) may occur with distinct spectral components of the laser spectrum, resulting in the population of distinct vibrational levels in the same electronic state, as illustrated in Fig. 1a, b. Under impulsive excitation (i.e., the pump pulse duration is shorter than the vibrational period and dephasing), a coherent superposition of vibrational states is produced, which in this case is labeled *vibrational* coherence. In addition, the second interaction may act on the ground-state component of the coherence and couple it to the excited state (Fig. 1a), but it may also act on the excited state component of the coherence and couple it back to the ground state (Fig. 1b). As

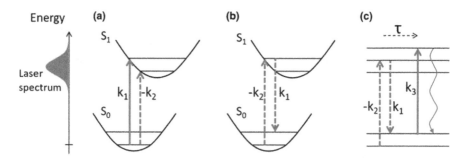

Fig. 1 Generation and probing of vibrational wavepackets in molecules with impulsive pump-probe spectroscopy (see e.g., Ref. [25] for a detailed discussion). Provided the laser spectrum is broader than the vibrational level spacing, a vibrational wavepacket is produced at the second order of the perturbation theory (i.e., two "interactions") either **a** in the excited state (i.e., one interaction with wave k_1 on the "ket" side and the other with wave $-k_2$ on the "bra" side of the density matrix element $|0> <0|$ representing the initial S_0 population, thus generating a population $|1> <1|$ in S_1) or **b** in the ground state (both interactions on the bra side, generating a vibrational wavepacket in S_0). **c** A vibrational wavepacket produced by the first two interactions, for instance in S_0 via an impulsive stimulated Raman process, is probed after a waiting time τ via a third interaction with the "probe" wave k_3. This third interaction generates a third-order coherence, which radiates a fourth wave (*wavy arrow*) in the direction $k_1 - k_2 + k_3$, imposed by the phase matching condition

a result, the pump-induced vibrational coherence is a vibrational wavepacket in the S_1 electronic state in the first case, or in S_0 in the latter case. Since these two cases correspond to two terms of the perturbative expansion at the same order, a short-enough laser pump pulse will in general produce vibrational wavepackets both in the ground and excited states [13–15]. Particular chirp (i.e., time-ordering of the spectral components within the short pump pulse) may be engineered to favor S_1 or S_0 wavepacket formation [16–21]. The impulsive formation of a ground-state wavepacket upon interaction with an ultrashort resonant laser pulse is a Raman process, named "resonant impulsive stimulated Raman scattering" RISRS [22, 23]. It also operates with a non-resonant laser pulse (ISRS) [24].

In the sequential pump-probe scheme discussed here (i.e., the probe pulse does not overlap temporally with the pump pulse), the first two interactions with the pump field are followed, after a "waiting" time τ, by a third interaction with the probe field, as illustrated in Fig. 1c. The latter couples (i.e., creates a "coherence" between) the pump-induced population and another electronic state. The resulting (third-order) polarization of the molecular system radiates a field that builds up constructively in a fourth wave (to be detected), which propagates in a specific direction $k_1 - k_2 + k_3$ imposed by the phase matching condition [25]. In the present case, where the first two interactions occur with the same pump pulse $k_1 = k_2$, the fourth wave propagates collinearly with the probe pulse k_3. Hence, one detects the co-propagating third and fourth waves, which means detecting the fourth wave heterodyned by the incident probe field. Finally, this experiment may be described as the measurement of the linear absorbance of the probe beam (i.e., first order in the probe field), by the non-stationary population induced (at the second order of perturbation) by the pump [26].

The vibrational wavepackets produced by the quantum superposition of vibrational states results in the classical oscillation of the vibrational degrees of freedom (bonds elongations, torsions, etc.). These oscillatory molecular motions induce an oscillatory modulation of all the linear optical properties of the system such as absorbance, dichroism, birefringence, etc. [15, 22]. Hence, the pump-induced change in the spectrum or polarization state of the transmitted probe oscillates accordingly as a function of the waiting time τ. The sequential pump-probe experiment described here and performed with ultrashort laser pulses, so as to generate and probe vibrational wavepackets both in the ground and excited states, is referred to as impulsive vibrational spectroscopy (IVS). Fourier transformation of the oscillatory signals reveals the Raman activity of the system. For a ground-state wavepacket—RISRS or ISRS—the resulting vibrational spectrum coincides with the molecular Raman spectrum as measured directly in the frequency domain by conventional Raman spectroscopy [27–29]. The limited bandwidth of the laser pulses may, however, result in the attenuation of the relative intensity of the highest-frequency vibrational modes or even prevent from detecting them. In practice, 10–12 fs pulses are short enough to trigger and detect vibrational activities up to the 3000 cm^{-1} range. Hence, VCS of ground-state wavepackets is a time-domain equivalent of frequency-domain Raman spectroscopy. For excited-state wavepackets, time–frequency representations of mode-specific coherent oscillations was recently proposed as a spectroscopic tool for detecting CInt's [30].

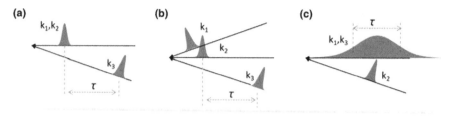

Fig. 2 Multiple implementations of VCS have been demonstrated in the time-domain with two (**a**) or three (**b**) laser pulses, or in the frequency domain (**c**), as used in FSRS. The waiting time τ is the time delay between the second and third interactions in cases **a** and **b**, while τ refers to the pump-pulse duration in case (**c**). In all cases, τ defines the vibrational coherence observation time window

In fact, a broad variety of four-wave mixing spectroscopies have been developed for VCS, as depicted schematically in Fig. 2, and the IVS scheme introduced above to illustrate the physical mechanism at the origin of VCS is only one possible (Fig. 2a) implementation of VCS. Enhanced experimental control on the first two interactions (k_1 and k_2) may be achieved by engineering these two interactions with two distinct, non-collinear laser pulses instead of one (Fig. 2b). In this three-beam experiment, the signal is still generated in the $k_1 - k_2 + k_3$ direction which is, however, no longer parallel to the probe beam k_3. Consequently, the signal is detected on a dark background, in the so-called homodyne detection scheme. Note that homodyne and heterodyne detection schemes have been briefly reviewed in the first contribution of this collection and explained in detail in the literature [31]. This implementation has been named "transient grating", because the signal can be understood as originating from the diffraction of the probe beam by the non-stationary population grating imprinted in the sample by the interaction with the non-collinear, interfering, first two beams. Two types of experimental realizations have been devised and named time-resolved coherent anti-Stokes Raman scattering (CARS) when pulses with different spectra are used, [32] or degenerate four-wave mixing (DFWM) [33] when all three laser pulses are derived from the same initial femtosecond, spectrally broad pulse.

Alternatively, VCS may also be performed directly in the frequency domain according to an experimental scheme implementing stimulated Raman scattering and illustrated in Fig. 2c. The major difference is that the pump pulse is temporally significantly longer (typically a few ps) and spectrally narrower than the probe pulse, and both pump and probe pulses overlap temporally, in contrast to the sequential scheme discussed above. The typical non-linear process at work in this implementation is the following. The vibrational wavepacket is produced by one interaction with the pump pulse (k_1) and one with the probe pulse (k_2). This vibrational wavepacket is subsequently interrogated by another interaction with the long, spectrally narrow pump pulse (k_3). The signal is here again detected in the direction of the probe (i.e., self-heterodyned). The result of this interaction mechanism is that the probe spectrum is amplified at frequencies that correspond to the difference between the pump frequency and the frequencies of the vibrational mode initially populated with the probe pulse. This coherent stimulated Raman technique reveals in the frequency domain the spectroscopic signatures of vibrational wavepackets in both the ground and excited state, [34] thus providing, in principle, the same spectroscopic information as IVS does in the time domain. In this frequency-domain VCS approach, the Raman pump-pulse duration is the observation window which limits the achievable spectral resolution, in the same way as the time window considered for Fourier transformation with respect to the waiting time τ limits the spectral resolution of Raman spectra recorded in the time domain.

A fundamental difference between time-domain and frequency-domain VCS is that in the latter case, the three light–molecule interactions are no longer engineered sequentially as in IVS or DFWM, but the interaction with the probe wave may occur at any moment before, between or after both interactions with the pump pulse. In fact, these various time-ordering options correspond to distinct terms in the third-order perturbative expansion, and all of them contribute to the third-order polarization [35] while in the sequential scheme by experimental design, only those where the probe interaction is the last one contribute. Hence, a drawback of the frequency-domain VCS is the background signal generated by contributions other than the stimulated Raman signal of interest. The features and challenges of each technique will be presented in more detail in the next section.

3 Multidimensional Vibrational Coherence Spectroscopy (Multi-VCS)

All the VCS approaches introduced above engineer a third-order light–molecule interaction which simultaneously reveals the vibrational activity of both electronic states coupled by a resonant pump laser field, i.e., the ground and the Franck–Condon excited state. This may pose a challenge to distinguish between vibrational signatures specific of each electronic state. Above all, this does not generally give access to the vibrational signatures of other possible transient states produced along the course of a photoreaction. In this regard, VCS has been further developed to be sensitive specifically to the excited states and to successive transient states by adding an "actinic" pulse. The role of this actinic pulse is to trigger a photoreaction prior to generating and probing a vibrational wavepacket by a subsequent third-order VCS scheme applied after a given time delay T. The vibrational activity can thus be monitored along the successive structures and electronic states achieved by the molecular system during the course of its photoreaction. This leads to the multidimensional character of the VCS, where usually one axis displays Raman frequencies while the other axis shows the photoreaction time delay T.

Several implementations of multidimensional VCS (multi-VCS) have been proposed, called "transient CARS" [36], "pump-IVS" [37], "pump-DFWM" [38, 39] and, in the spectral domain, femtosecond stimulated Raman spectroscopy (FSRS) [40], ultrafast Raman loss spectroscopy (URLS) [42–43]. Some of these are schematically introduced in Fig. 3. Here, it is important to mention a potential problem in the semantics of the word "pump": The actinic pulse has received different names by different groups and techniques along all years, e.g., "pump", "initial pump", or simply "excitation pulse". The word "pump" in front of the techniques names, e.g., "pump-IVS" or "pump-DFWM" denotes the actinic pulse which triggers the

Fig. 3 Pulse scheme of multidimensional vibrational coherence spectroscopies. In all schemes, the actinic pulse is used to trigger a photoreaction by populating a transient excited state, which is then probed by VCS, after a waiting time T. *From top to the bottom*: FSRS is also often called time-resolved FSRS due to the ability of detecting stimulated Raman spectra in dependence of the delay T between the actinic pulse (*black*) and the Raman pump-probe pulse pair (*red* and *green*, respectively). Pump-IVS is a three-pulse experiment with an actinic pulse (*black*) delayed by T from a "repump" pulse (*red*), followed by a white light probe pulse (*multicolor*). Pump-DFWM uses an actinic pulse (*black*) delayed by T from two spectrally degenerate broadband pulses (called pump and Stokes; both in *red*), followed by an equally spectrally degenerate broadband probe pulse (also in *red*). Population controlled pump-IVS is an extension of pump-IVS, which adds an additional depletion pulse (*blue*) between the repump (*red*) and probe pulses (*multicolor*)

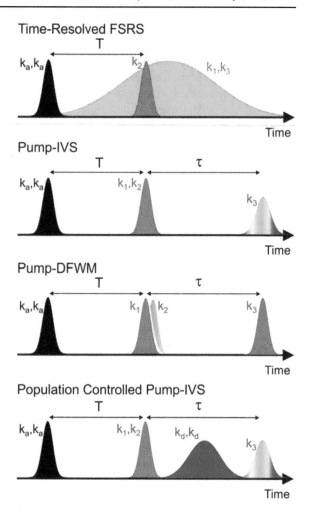

photoreaction, prior to the IVS or DFWM third-order scheme introduced in the previous section. This should not confound the reader with the pump pulse present in the IVS or DFWM techniques themselves, which triggers the vibrational coherence to be probed.

Independent of the technique, the actinic pulse always excites the electronic population via two interactions with its electric field. Since these first two interactions originate from a single actinic beam, they do not affect the phase matching of the subsequent VCS signal: $k_{\text{multi-VCS}} = k_a - k_a + k_1 - k_2 + k_3$. Strictly speaking, multi-VCS engineers a fifth-order light–molecule interaction [44], but again within a sequential scheme, the initial actinic pulse first triggers a photoreaction, i.e., creates a non-stationary excited-state population, which is subsequently probed by a third-order VCS scheme revealing the Raman activity as a function of the photoreaction time T. Importantly, the transition dipole moment, which operates in the T-delayed VCS scheme, does not necessarily involve the ground state anymore. This is a major difference between VCS and multi-VCS. Indeed, with multi-VCS, vibrational activity may be triggered and probed in a transient state by using laser spectra on resonance with one specific electronic transition of this very state, which may not be resonant with the ground state absorption. This allows taking advantage of the so-called resonant enhancement for generating a vibrational wavepacket specifically in this transient state.

For completeness, we note that another implementation of multi-VCS uses what we call here the actinic pump to specifically trigger a *vibrational wavepacket* in the ground state (by ISRS), followed by the subsequent three-pulse VCS scheme to generate and monitor an excited-state vibrational wavepacket from this non-stationary ground state population. This enables correlating the vibrational activity on the excited state to that of the ground state in a so-called two-dimensional resonance Raman (2DRR) spectroscopy. Another chapter in this collection is dedicated to this technique. Therefore here, we limit the discussion to a scheme where the effect of the actinic pulse can be simply understood as initiating a photoreaction by populating an excited state.

3.1 Femtosecond Stimulated Raman Scattering (FSRS)

FSRS is a frequency-domain spectroscopy based on stimulated Raman scattering generated after an actinic pulse (Fig. 3, top panel). The signal is generated in the same direction as the probe beam, leading to a self-heterodyne detection geometry. The delay T between the probe pulse and the actinic pulse is scanned and Raman spectra are detected in dependence of this delay. The first implementation of FSRS in its three-beam configuration used a 10-Hz laser and had a Raman resolution of only 76 cm^{-1} [45]. After almost 25 years of experimental development, the state-of-the-art FSRS setup nowadays offers a Raman resolution of about 10 cm^{-1} while using actinic pulses with durations of less than 100 fs, and covering the whole ultraviolet and visible spectral range [47–50].

The probe pulse is spectrally broad and usually red-shifted with respect to the narrow spectrum of the Raman pump. The advantage of a broadband implementation is that a full Raman spectrum, i.e., typically from 200 to 3000 cm^{-1}, can be directly recorded and the high peak intensity of femtosecond pulses enhances the efficiency of the stimulated Raman process generating vibrational wavepackets [40]. While the actinic pulse is electronic resonant with the ground-state absorption, the Raman pump-probe pair is usually off-resonant in most experiments. The ability of tuning the spectrum of the Raman pump-probe pair allows to probe specific electronic transitions, other than those involving the ground state [51]. Thus, all Raman modes in the excited state or subsequent transient states can be measured, not only those Franck–Condon active. Being a spectral domain technique where only the scan of the actinic pulse T delay is required, the acquisition of several spectra is intrinsically faster than time domain techniques where an additional probe τ delay must be scanned (Fig. 3). While FSRS has been already demonstrated with a single laser shot acquisition [40], typical acquisition times of FSRS transients are only a few minutes [52].

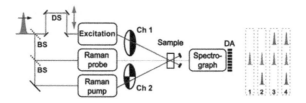

Fig. 4 Scheme of the FSRS experimental setup and pulse-chopping scheme. All three beams are focused and overlapped on the sample, while only the actinic pulse (*Excitation*) is delayed (*T*). The chopping of the actinic beam and Raman pump beam with different frequencies allows automatic subtraction of the signal background. In this scheme, the phase 1 of the chopping detects only the probe background. In phase 2, the stimulated Raman signal of ground state (stationary), including solvent, is detected. In phase 3, a transient absorption signal at delay *T* is detected. Finally, at phase 4 all signals are collected together. The FSRS signal is calculated from signal $4 - (3+2)$. See Fig. 5 for more details. Reprinted from Ref. [53] with permission of Springer

Fig. 5 a Typical baseline correction of FSRS difference spectrum. "TA" stands for "transient absorption". **b** Negative features originating from depleted ground-state Raman signals contribute to the FSRS difference spectrum. This feature is removed by adding back an appropriately scaled background Raman spectrum, optimized for one of the solvent modes. Correction of this feature of the FSRS spectrum yields the FSRS signal of interest Reprinted with permission from Ref. [54]. Copyright 2017 American Chemical Society

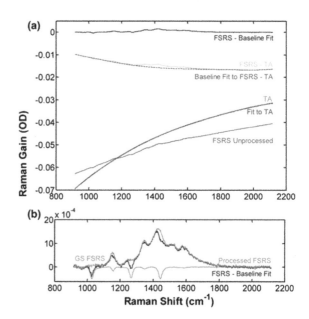

A major challenge in FSRS is the separation of the stimulated Raman spectrum from the non-coherent background spectrally overlapping with the signal. The lack of automatic methods to extract the excited-state Raman spectrum from the ground-state Raman spectrum, probe background, and transient absorption signal has been a major obstacle in the practical development of FSRS as an analytical tool. Several experimental and numerical approaches have been developed in this regard. Since FSRS is usually performed with kHz laser sources, the chopping of the actinic and Raman pulses at different frequencies (Fig. 4) has been shown to separate to some degree the different overlapping signal contributions [53]. This allows for the subtraction of the transient absorption (TA) baseline from the FSRS raw data ("unprocessed", see Fig. 5a), but does not eliminate all baseline distortions or solvent contributions in FSRS (Fig. 5b) [54]. The intensity modulation of the spectrally narrow Raman pulse has been identified as a major artifact in FSRS, often requiring ad hoc scaling of baselines (see e.g., Fig. 5a FSRS-baseline fit). This artifact can be corrected by a factor numerically calculated by including transient absorption changes measured under similar experimental intensities [55]. More recently, the use of carefully crafted Raman spectra in form of watermarks has been used to easily identify Raman resonances in the raw signal [56]. The last word in the correction of the baseline distortions in FSRS has not been spoken yet and it is an active research focus in the field of multidimensional VCS.

3.2 Pump-IVS

Pump-IVS is based on the combination of *impulsive* stimulated Raman scattering (ISRS) with an actinic excitation. As FSRS, it can be easily implemented since the signal is generated in the direction of the probe beam. However, it requires scanning two time delays, namely the photoreaction time T between the actinic pulse and the "repump" pulse (or Raman pump, or even impulsive pump, simply called "pump" hereafter), and the waiting time τ between the pump and probe pulse, during which the wavepacket dynamics evolve. Typical acquisition times for pump-IVS are of several tens of minutes for laser systems with kHz repetition rates. The acquired data must then be post-processed by Fourier transformation along τ to reveal the T-dependent impulsive Raman spectrum. The spectral range of the Raman spectrum depends exclusively on the bandwidth of the pump spectrum and the length of the transients measured along the τ delay. The pump bandwidth limits the highest upper Raman frequency and the length of the τ scanning interval defines the lowest detectable Raman frequency. Raman spectra from as low as few tens of wavenumbers up to 3000 cm^{-1} have been demonstrated with pump-IVS [19, 57, 58]. Detection of very low frequency Raman modes (< 200 cm^{-1}) is another central advantage of time-domain methods in comparison with spectral domain methods like FSRS, which are constrained by optical filters to spectrally cut the Raman pump from the Raman spectrum. Moreover, since the probe spectrum in pump-IVS is normally spectrally resolved via a grating spectrometer, Raman spectra can be obtained at several detection wavelengths, depending only on the bandwidth of the probe spectrum.

Experimentally, pump-IVS setups are based on non-collinear optical parametric amplifiers (nc-OPA) [59, 60] to generate the broadband spectra of the actinic and pump pulses. Pulse durations below 10 fs have been successfully used [19, 57, 58]. The probe pulse is usually a white light supercontinuum with a very broad spectrum spanning from about 300 nm up to 750 nm or more (depending on the crystal used for white light generation). The pump spectrum is usually spectrally resonant with excited-state absorption bands. Typically, the transients along the τ delay are measured for different T delays with respect to actinic pulse with and without the pump pulse by using a chopper (Fig. 6). This assists in the separation of the transients originated in the excited-state manifold from the ground-state or solvent contributions (induced by the actinic pulse alone). The extraction of the excited-state signal induced by the VCS pump pulse can be easily obtained if the pump spectrum is

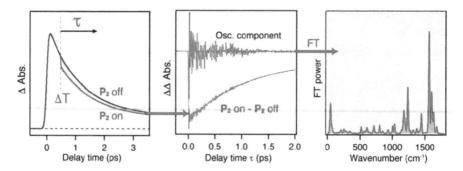

Fig. 6 Scheme of the pump-IVS signal analysis procedure. *From left to right* Two transients are detected, one with pump pulse ("P_2 on") and one with the pump pulse blocked ("P_2 off"). The transients are subtracted from each other ("P_2 on-P_2 off"). The residual of this subtraction is finally Fourier transformed to obtain the Raman spectrum Reprinted from Ref. [57] with the permission of AIP Publishing LLC

off-resonant with any ground-state absorption. Pump spectra resonant (or near-resonant) with the ground-state absorption will lead instead to contamination of the pure "excited-state" signal since the pump pulse will also efficiently generate vibrational wavepackets in the ground electronic state as well as in the electronically excited state. Under these conditions, a way to partially mitigate this effect is by additionally chopping the actinic pulse to subtract the vibrational wavepackets induced without the actinic pump.

3.3 Pump-DFWM

Pump-degenerate four-wave mixing is based on the same concept as pump-IVS, but the two interactions with the second pulse in pump-IVS (repump) is split in two pulses. This four-beam geometry offers several advantages with respect to the three beams used in pump-IVS. One of them is the delay between the two "repump" field interactions (called here "pump" and "Stokes"), which can be used for selective nonlinear response preparation and to separate optical beating artifacts from molecular vibrations [20]. The four-beam geometry generates the signal in a background-free configuration (homodyne detection), i.e., not in the direction of the probe as in pump-IVS, which typically leads to a superior signal-to-noise ratio and shorter acquisition times [58].

Similar to pump-IVS, pump-DFWM has been usually performed with kHz laser sources. A typical setup consists of two non-collinear OPAs to generate the broadband spectra of the actinic and DFWM pulses independently. Like in pump-IVS, time-resolved signals are recorded as a function of the actinic pulse delay (T) and of the probe delay (τ). In pump-DFWM, however, the transients along the τ delay are not subtracted from any reference, e.g., with and without actinic pump or by any other method, due to the homodyne detection (see below). Beyond that, the analysis of each transient at a given T delay containing oscillatory and non-oscillatory contributions is essentially the same as for pump-IVS, and illustrated in Fig. 7.

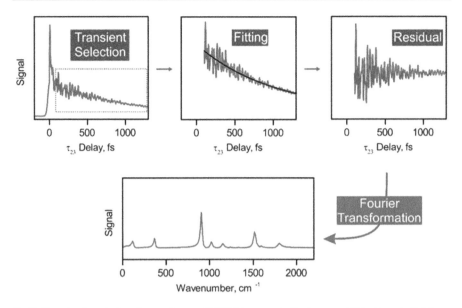

Fig. 7 The signal analysis scheme of pump-DFWM is very similar to pump-IVS. At each T delay a kinetic trace is recorded, which contains oscillatory and non-oscillatory contributions. They can be separated by a polynomial fitting performed for probe delays $\tau > 100$ fs, typically. Vibrational spectra are then obtained by fast Fourier transformation (FFT) of the oscillatory contribution along the τ delay after zero-padding, windowing and apodization Adapted from Ref. [170]

As briefly mentioned above, pump-DFWM differs from pump-IVS and FSRS in the signal detection and the dependence on the molecule concentration in the sample. The pump-DFWM signal is detected in a homodyne configuration. This contrasts to the self-heterodyne detection used in the other two techniques, where the signal created in the sample is generated in the same direction as the (Raman) probe beam. (Self-)heterodyne detection offers amplification of weak optical signals (see the contribution "Introduction to State of the Art Multidimensional Time-resolved Spectroscopy Methods" in the collection), but its signal-to-noise ratio often suffers from local oscillator fluctuations, inserting random noise as well as signal distortion that can only be corrected under e.g., specific balanced detection and averaging schemes [30, 61]. Pump-DFWM is not prone to such signal distortions, instead it may suffer from interferences (see below on the Challenges in Multidimensional VCS) leading to an ambiguous assignment of (low) frequency modes and artificially broadened Raman lines [20]. The second major difference between pump-DFWM and other techniques is the nonlinear concentration dependence: while pump-IVS and FSRS depend linearly on the probed concentration, pump-DFWM (as any other homodyne nonlinear technique) shows a quadratic dependence. This nonlinear dependence leads to a distortion of the evolution of the amplitude of Raman modes during the T delay, which can only be linearized by using a heterodyne detection [58]. A major drawback of such nonlinear concentration dependence is the difficulty of detecting minor components in a multi-component sample, since the contribution of major component(s) will dominate the signal.

4 Challenges in Multidimensional VCS

A fundamental challenge of multi-VCS in the detection of vibrational coherence in the electronically excited manifold is the small amount of excited-state population compared to the ground state or other non-actinic activated molecules (like solvent or buffer molecules). The amount of excited molecules can be changed by the actinic pulse energy, but in general is well below 20–30% to avoid saturation or other nonlinear effects. Usually, it is the interplay of the involved transition dipole moments (i.e., resonant or non-resonant with the ground state or with the small excited-state population) and the spectral overlap with optical signals originated by other states, that will determine whether a VCS signal of a given electronic state is detectable or not. In practice, very seldom a VCS signal will contain the vibrational signature of only one single electronic state. The only exception for that is electronically nonresonant experiments involving only the electronic ground state (no actinic pump). In spectrally resonant experiments, for example, the VCS spectrum is often resonant with distinct transitions during the course of a photoreaction, since the involved transition dipole moments and FC overlap with higher lying states are continuously changing. This leads inevitably to a modulation of the VCS signal due to a modification of the resonance enhancement conditions. The overlap of VCS signals originated by different electronic states is one of several causes for the interpretation challenges in VCS experiments and has been the motivation of new techniques to disentangle them.

One approach to assist the assignment of a Raman mode to the ground- or excited-state manifold is directly related to the properties of the oscillatory signal itself. Ground-state vibrational modes often show a longer-living dynamics than transient excited states, where the dephasing may be faster, due to e.g., a photoinduced reaction or internal conversion to the ground state [63–64]. In time-domain techniques, this leads to vibrational wavepackets with shorter dephasing times (or broader Raman peaks in spectral domain techniques like FSRS) normally associated to excited state modes, while longer dephasing times are taken as coming from the ground state. Of course, longer and shorter are relative quantities, which can only correctly be interpreted when taking into account the duration of other vibrational modes and electronic population times of the involved electronic states. An additional approach often used to identify the nature of vibrational wavepackets is to rely on the phase of the oscillatory signal [13, 66–68]. For example, in spectrally dispersed time-resolved experiments, probing at wavelengths red- and blue-detuned from the center of the absorption spectrum often leads to oscillatory signals with a π-phase difference due to vibrational wavepacket dynamics. This information, when combined with the ground- and excited-state absorption bands, may assist in the assignment of a Raman mode to the respective electronic state (see e.g. [30]).

Separation and assignment of contributions originating from different excited states or electronic transitions can be acutely challenging in multi-VCS, especially when the respective excited-state absorption spectra overlap. However, when they do not overlap, the assignment can be facilitated by exploiting the tuneability of

VCS laser spectrum. Importantly, when the VCS pump pulse(s), i.e., k_1 and k_2 in Figs. 2 or 3, are resonant with different electronic transitions from the same electronic state, different Raman activities may be observed. This has been demonstrated in pump-DFWM applied to all-*trans*-retinal (ATR) and retinal Schiff Base (RSB) in solution to obtain the Raman active modes at different excited-state electronic transitions (Fig. 8) [69]. In this example, the DFWM spectrum was tuned to different electronic transitions in the excited state absorption (ESA) bands of ATR (Fig. 8a) and RSB (Fig. 8b). While ATR displays a single ESA and an excited state stimulated emission (ESE) band (Fig. 8a), RSB shows two ESA bands above 17,000 cm^{-1} and a shallow ESE below 15,000 cm^{-1}. These two bands observed for RSB are due to different transitions from the S_1 state to two different high-lying electronic states, which are allowed only for RSB. The Fourier transformed signal of pump-DFWM at different DFWM detuning shows how different the Raman active modes are in each case (Fig. 8c–e): For example, in ATR, the C=C bond structure at 1510 and 1580 cm^{-1} shows a strong shift and amplitude change when the DFWM spectrum is detuned from resonance with the ESA band at about 19,000 cm^{-1} (Fig. 8c) to the red-shifted spectrum resonant with the ESE band (Fig. 8d). In RSB, the tuning of the DFWM spectra from one ESA band to a red-shifted one (Fig. 8b), shows that the C=N bond observed around 1700 cm^{-1} is active at only one electronic transition (Fig. 8e, f).

Detuning of the VCS laser spectrum can be also exploited in the absence of the actinic pulse to assist in the assignment of ground- and excited-state modes in multi-VCS. In this case, the detuning of the VCS spectrum from non-resonant, over

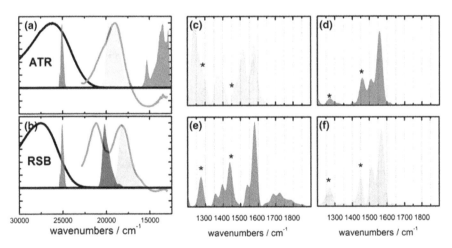

Fig. 8 a Ground-state absorption of ATR in ethanol (*black*), UV excitation (25,000 cm^{-1}, *violet*), DFWM spectra for ATR (*cyan* and *red*), and transient absorption (TA) spectrum (*orange*) of ATR in ethanol at $T = 510$ fs after UV excitation. **b** Ground-state absorption of RSB in ethanol (*black*), UV excitation (*violet*), DFWM spectra (*blue* and *green*), and TA spectrum of RSB (*orange*) in ethanol at $T = 510$ fs after UV excitation. Pump-DFWM data was acquired at $T = 1$ ps for ATR **c** (*cyan* DFWM spectrum) and **d** (*red* DFWM spectrum). Pump-DFWM data were acquired at $T = 2$ ps for RSB **e** (*blue* DFWM spectrum) and **f** (*green* DFWM spectrum). Vibrational bands assigned to solvent dynamics are indicated with *asterisks* Figure adapted with permission from Ref. [69]. Copyright 2013 American Chemical Society

near-resonant, up to completely electronically resonant leads to different degrees of vibrational coherence in the excited state. While electronically non-resonant excitation cannot induce any vibrational coherence in the excited state, near resonant excitation, for example, can induce low-frequency vibrational coherence only. Complete resonant excitation is able to induce high- as well as low-frequency coherence in the excited state. This dependence on the spectral overlap between an absorption spectrum and VCS laser spectrum is different for the groundstate, where low- as well as high-frequency modes will be always induced. By comparing how the amplitude of specific vibrational modes decreases when the excitation becomes non-resonant, it is possible to pinpoint which vibrational modes are present only in the excited state or in both states, and how specific vibrational modes are being activated (via direct laser interaction or via coherent excitation from other vibrational mode) [10, 19, 28, 70]. For example, this has been applied to retinal protonated Schiff base (RPSB) to show that low-frequency modes are not active in the electronic ground state but are coherently activated by vibrational energy redistribution from high-frequency modes directly excited in the electronic excited state [71]. In FSRS, the detuning of the Raman pump wavelength was exploited to record the excited-state Raman spectrum in the absence of the actinic pump (Fig. 9) [34]. By tuning the Raman pump wavelength from red to blue towards the $S_0 \rightarrow S_1$ resonance, the S_1 modes become stronger (at about 200, 300, 650, and 850 cm^{-1}), while the S_0 contributions (peaks with negative amplitude in Fig. 9b) remain effectively unchanged. This method has also been recently applied to record unambiguously for the first time the low frequency Raman modes of the S_2 state of all-*trans*-β-carotene (see Sect. 5.2) [51].

An additional central aspect in the signal interpretation of VCS-detected signals is interferences between signal contributions. Several kinds of interferences can be

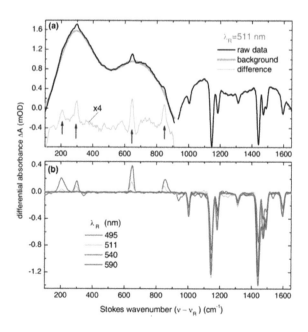

Fig. 9 FSRS difference spectra of *trans*-azobenzene in *n*-hexane measured without actinic pump. **a** Raw data and FSRS difference spectra measured with a Raman pump at 511 nm. **b** FSRS difference spectra measured as function of the Raman pump detuning. Negative bands correspond to the conventional S_0 signal, while positive peaks originate from S_1. The S_1 contributions vanish when the Raman pump is detuned farther from the $S_0 \rightarrow S_1$ resonance, i.e., from blue to red detuning Reproduced from Ref. [34] with permission from AIP Publishing

experimentally observed which may overlap with the vibrational coherences of interest [73–75]. For example, *interference beats* originate from the interference between vibrational modes of two different electronic states, and are an intrinsic beat of electronically resonant techniques, since vibrational coherence will be excited on both electronic states involved. *Polarization beats* originate from interference between nonlinear polarization contributions at the detector and are thus a pure optical effect. The origin of the polarization beating can be e.g., the emission, by different kinds of molecules, of polarizations oscillating at different vibrational frequencies and interfering at the detector. We refer the reader to references [20, 31] for more technical details.

In general, interference between oscillatory signals resulting in additional low- or high-frequency oscillations hampers the correct interpretation of VCS experiments. *Polarization beating*, for example, is of special importance when the low-frequency region (< 800 cm^{-1}) in FFT spectra of transient dynamics is interpreted [20]. In the spectral domain, band positions of molecular normal modes and signal contributions from polarization beating between molecular high-frequency contributions can overlap, hampering an unambiguous interpretation and assignment. Polarization beating is an interference effect present in all types of VCS methods, with homodyne and, to a lesser degree, (self-)heterodyne detection. The discrimination between molecular normal modes and polarization beating is possible in techniques where the vibrational wavepacket generation originates from two different pulses like DFWM and CARS, where the delay τ_{12} between pump and Stokes pulses can be exploited. The separation of different response pathways has been demonstrated for several polyatomic molecules in solution using chirped, spectrally resolved DFWM (Fig. 10) [20]. Two organic dye molecules, rhodamine B and S-9, show strong oscillatory beating centered at $\tau_{12} = 0$ when no chirp is applied (Fig. 10a, c, respectively). When chirp is applied, the maxima of interference contributions are shifted with a given amount depending on the chirp applied, while beating due to vibrational wavepackets are not shifted with the delay τ_{12} between pump and Stokes pulses (Fig. 10b, d).

Finally, *interference beating* can be also addressed by applying higher-order nonlinear techniques. Population-controlled pump-IVS is an extension of pump-IVS to suppress undesired vibrational coherence contributions to the optical signal (Fig. 3, bottom panel) [77–78]. It is based on the same three-beam geometry to excite impulsively Raman vibrations, but adds an additional pulse named "dump" to interact with the sample between the pump and the probe pulses, formally resulting in a seventh-order nonlinear time-resolved method. PC-pump-IVS follows the same approach as pump-depletion-probe experiments in visible [80–84] and infrared [85, 86] to disentangle population relaxation pathways, and applies to the depletion of the vibrational coherence in the excited state manifold. In PC-pump-IVS, the interaction of the narrowband dump pulse has the effect of only changing the population of the excited state and decreasing the respective Raman signal. By measuring the transients along the τ delay with different combination of pulses (Fig. 11a), it is possible to subtract the contribution of the ground-state and solvent vibrational coherence (Fig. 11b). A key aspect of this technique is the bandwidth of the dump pulse: The spectrally narrow dump pulse with a pulse duration of few hundreds of femtoseconds is not able to induce any additional vibrational coherence, leading to a pure

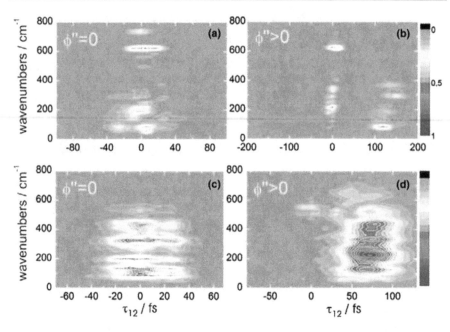

Fig. 10 Disentanglement of real molecular vibrational bands from polarization interferences depicted for **a, b** Rh-B and for **c, d** S-9 in methanol. The disentanglement is possible by comparing the maximum of a given frequency in respect to τ_{12} when the excitation chirp (ϕ'') is varied. By using positively chirped ($\phi'' > 0$) DFWM pulses, oscillatory contributions due to polarization beating will appear for slightly delayed pump and Stokes pulses ($\tau_{12} > 0$), while real molecular vibrational bands will not be shifted in τ_{12} in comparison to the non-chirped excitation ($\phi'' = 0$). While for Rh-B (**a, b**), several frequencies can be assigned to molecular modes, in particular at 205 and 620 cm^{-1}, for S-9 (**c, d**), only two weak contributions at about 505 and 555 cm^{-1} are real molecular vibrational normal modes Reprinted from Ref. [20] with the permission of AIP Publishing

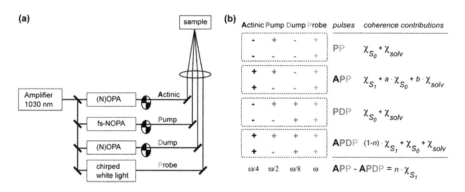

Fig. 11 Scheme of the population-controlled pump-IVS; **a** experimental setup and **b** pulse chopping scheme. By chopping all pulses but the probe pulse, it is possible to extract the pure signal of the excited state S_1 in this scheme Reprinted from Ref. [77] with permission of ACS https://pubs.acs.org/doi/abs/10.1021%2Fjp5075863

depletion of electronic population. PC-pump-IVS has been demonstrated with dump pulses with a duration of 200 fs (i.e. about 75 cm^{-1} FWHM spectral width) [77].

5 Application of Multidimensional VCS

5.1 How can Multidimensional VCS Help a Chemist?

The most natural way of describing how multi-VCS can assist a chemist is by comparing it to Raman spectroscopy. In its essence, multi-VCS delivers a sequence of Raman spectra obtained after an actinic pulse interacts with the sample. In Raman spectroscopy, structural and interaction information comes from the energy, width, and amplitude of Raman bands, which assist in the identification of specific molecular species and conformations. Multi-VCS provides the same information and goes beyond by measuring how Raman frequencies shift and amplitudes evolve in time (for example, Refs. [88–89]). Hence, it reveals time-resolved structural information on transient molecular species along the course of a photoreaction. In this regard, multi-VCS follows the same conceptual approach as transient spontaneous Raman measurements with picosecond pulses [91–93] pioneered by Lauberau et al. more than four decades ago [94]. Multi-VCS differs from transient picosecond Raman spectroscopy, however, in many aspects. The first one is the signal intensity of the spontaneous Raman signal detected in the transient picosecond Raman spectroscopy, which is intrinsically much weaker than the coherent/stimulated signal detected in multi-VCS methods. The second one is related to the Fourier relation between the spectral resolution and pulse duration. Transient picosecond Raman spectroscopy is limited to molecular processes much slower than picoseconds due to the intrinsic time duration of the narrow-band Raman pulse, if a spectral resolution of few wavenumbers is desired. For example, a 1-ps Raman probe pulse leads to an intrinsic band broadening of more than 10 cm^{-1}. It is also interesting to note that the vibrational content in multi-VCS can be very similar to the one detected by transient infrared absorption spectroscopies, in particular for complex systems where symmetry rules are relaxed and vibrations can be probed by Raman as well as infrared interactions.

Generally, following how the frequency of Raman bands change in dependence of the actinic pulse time delay T gives information on how bond strengths are modified during a photochemical reaction [37, 89, 95]. Frequency blue-shifts indicate stiffening of the vibrational motion or relaxation within an anharmonic potential, while red shifts may hint at elongation of a given chemical bond. On the other hand, the lack of any shift or amplitude changes is usually taken as the signature of non-reactive coordinates. The changes in the amplitude of Raman bands in multi-VCS report on the formation and breaking of chemical bonds after interaction with the actinic pulse. When correlated with frequencies changes, they become the central feature in the mapping of structural changes during photochemical reactions. Moreover, the evolution of the amplitude of a band does not need to follow a first-order kinetics and decay or grow exponentially, it can also show very complex dynamics. For example, modulation of the amplitude of a Raman band with the frequency of another Raman band has been identified as the signature of anharmonic vibrational coupling between the two corresponding modes [88]. Information on how vibrational energy redistributes can be retrieved from the width as well as frequency changes of Raman bands. Strong coupling between molecular vibrations leads to strong dephasing and damping of specific bands and can be exploited to record how e.g., intramolecular vibrational relaxation (IVR) and vibrational cooling—e.g., solute–solvent interaction—take place in complex systems [97–98]. Initially, very fast photoreactive systems evolve within few tens or hundreds of femtoseconds away from the Franck–Condon region, leaving very broad multi-VCS bands in the Raman spectrum.

Also, the ability of multi-VCS to extract information on the molecular dynamics is enhanced when the evolution of Raman bands is combined with other techniques [99]. This is particularly interesting when methods sensitive to other molecular degrees-of-freedom are considered, like transient absorption and the detection of the electronic population dynamics. For example, rise or decay times of specific Raman bands can be correlated to respective times observed in the electronic population

Table 1 Selected molecular systems investigated with multidimensional VCS in the last decade

Molecular system	Technique	References
Based on carotenoids	Pump-CARS Pump-DFWM Pump-IVS FSRS PC-pump-IVS	[39, 55, 58, 78, 101–105]
Based on retinal	Pump-DFWM Pump-IVS FSRS PC-pump-IVS	[69, 71, 77, 107–109]
Based on stilbene	Pump-IVS FSRS PC-pump-IVS	[34, 37, 48, 77, 110, 111]
Dye molecules in solution	Pump-IVS FSRS	[21, 112]
TIPS-pentacene	Pump-IVS	[113]
Aromatic amino acid residues	Pump-IVS FSRS	[114, 115]
Green fluorescence protein	Pump-IVS FSRS	[88, 109, 116]
Photoactive yellow protein	Pump-IVS FSRS	[99, 117]
Bis(phenylethynyl)benzene	URLS	[42, 118]
Tetraphenylethylene	URLS	[43]
Photoactive flavoproteins	FSRS	[97, 119]
Myoglobin	FSRS	[96]
Azobenzenes	FSRS	[34, 89]
Cyclohexadiene derivative	FSRS	[95]

evolution to assign Raman spectra to specific electronic states or assist the assignment of electronic states by using known Raman spectra, as will be further illustrated below. Finally, the ability of multi-VCS to reveal detailed structural information on excited electronic states makes it a very powerful experimental approach to assess the accuracy of state-of-the-art computational developments targeting quantum chemical modeling of electronic structures and of molecular photoreactivity in general (Table 1).

A seminal example in multi-VCS, which depicts many of the points mentioned above, is the pioneering FSRS investigation of the photoinduced, sub-ps structural evolution of the protonated Schiff base of retinal (PSBR) in the rhodopsin protein, the visual sensor [120]. Upon light excitation to S_1, the C11=C12 bond of PSBR extends and acquires a single-bond character. This enables a very fast torsional motion to drive the system within ~ 100 fs to a CInt [121], where it decays to S_0, with a predicted C11=C12 bond twist of ~ 90°. By implementing FSRS, the vibrational frequencies associated to the hydrogen out-of-plane motions (in the 800–900 cm^{-1} range) of both hydrogen atoms linked to the H–C11=C12–H isomerizing bond could be observed as a function of time, after decay to the S_0. This experiment revealed a rapid (300-fs time scale) blue shift of the H wagging frequencies by ~ 100 cm^{-1}, which is the signature of a large structural reorganization of PSBR occurring on the ground state PES S_0. More specifically, the blue shift is shown to reveal the stiffening of these vibrations resulting from the planarization of the PSBR backbone, i.e., completion of the C11=C12 isomerization (torsional) motion towards the first vibrationally relaxed photoproduct intermediate, called bathorhodopsin.

5.2 Ultrafast Vibrational Dynamics in the Excited States of Carotenoids

Carotenoids have been an important class of molecules investigated by multi-VCS due to their central role in several biological functions [122]. As chromophores in light-harvesting complexes (LHC), for example, carotenoids are involved in the initial absorption of light and energy transfer to other chromophores (e.g., bacteriochlorophylls). The very strong absorption from the ground state S_0 is due to the π–π* transition to the second electronic state labeled S_2. The transition from the S_0 to the first electronic S_1 state is not one-photon allowed, making the S_1 state a "dark" electronic state. Upon light absorption to S_2, the electronic relaxation to S_1 is very fast (within 100–200 fs for all-trans-β-carotene and lycopene). Further electronic relaxation from S_1 to S_0 takes place in the picosecond time scale, varying with the number N of effective conjugated C=C double bonds. For example, the S_1 state in all-*trans*-β-carotene ($N \sim 10.5$) decays with about 9 ps, while in lycopene ($N = 11$) decays faster with about 4.1 ps [123]. The S_2 to S_1 relaxation mechanism has been intensely debated in the last decades (see for example Ref. [124–128]). One source of discussion has been, for example, the nearly identical lifetime of the S_2 state for several open-chain carotenoids with different numbers N of effective conjugated C=C bonds, which has been interpreted as the result of additional electronic dark states between the S_2 and S_1 states. The experimental identification of these dark states via transient absorption, nevertheless, has been extremely challenging due to the high spectral overlap of the respective absorption bands, the ultrashort time scales involved, but also due to the couplings [130–131] between these states.

Multi-VCS has been applied to carotenoids to assist in the elucidation of the excited state manifold. In general, the excited-state active vibrations of carotenoids are typical of polyenes consisting of strong high-frequency modes like methyl deformation (~ 1000 cm^{-1}), C–C stretching (~ 1140–1200 cm^{-1}), and C=C stretching (~ 1500–1550 cm^{-1}) (Fig. 12). All these vibrational modes are active in all electronic states. However, since carotenoids are polyenes with C$_{2h}$ symmetry, an additional frequency at about 1800 cm^{-1} is observed for the S$_1$ state, due to the adiabatic coupling between the S$_1$ and S$_0$ states [129, 130]. The mapping of the Raman activity, frequency shifts, amplitude rising, and dephasing of these vibrational modes in dependence on the actinic pulse delay has been the main focus of multi-VCS.

One of the first experimental observations of the evolution of the totally symmetric C=C stretching mode at about 1800 cm^{-1} has been pioneered by Hashimoto et al. using an actinic pulse to promote all-*trans*-β-carotene to S$_2$ and followed by a stimulated Raman scattering scheme [132]. With a temporal resolution of 300 fs, the energy flow between the S$_2$ and the S$_1$ states was followed. The initial vibrational relaxation until the $v = 1$ level of the totally symmetric C=C mode of the S$_1$ state was found to be very fast, while further relaxation from $v = 1$ to $v = 0$ was much slower than the internal conversion to the ground state S$_0$. This has been explained by the presence of an additional electronic dark state, which assisted the vibrational relaxation to $v = 1$ of the S$_1$ state. The presence of an electronic dark state being populated within 20 fs in the decay between S$_2$ and S$_1$ states has been also proposed in lycopene based on extensive modeling of the spectroscopic signal observed by pump-DFWM [102].

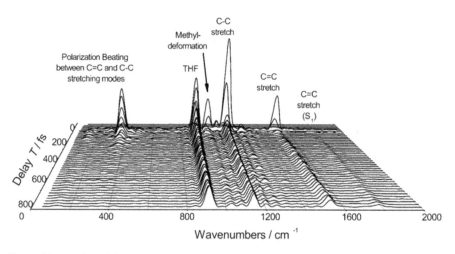

Fig. 12 2D mapping of the vibrational coherence for *all-trans* lycopene, upon photoexcitation in the S$_2$ state. Being a polyene, lycopene shows very strong C=C and C–C stretching modes at 1564 and 1180 cm^{-1}, respectively. The formation of the S$_1$ C=C mode at 1783 cm^{-1} as the S$_1$ state is populated can be clearly seen within the initial 300 fs. *THF*: tetrahydrofuran Reprinted with permission from Ref. [102]. Copyright 2011 American Chemical Society

The vibrational relaxation during the internal conversion between the S_2 and S_1 states for all-*trans*-β-carotene has been also explained by pure vibrational cooling, challenging the presence of any additional electronic dark state assisting this relaxation. The first FSRS measurements of all-*trans*-β-carotene showed that the S_1 C=C stretching mode at 1798 cm^{-1} relaxes via a two-step process identified by two distinct frequency up-shift kinetics [133]. A fast time constant (200 fs) was related to a strong coupling followed by a slower equilibration process (450 fs) to the complete set of vibrational normal modes. Although these first FSRS measurements were not able to resolve the initial 250 fs of the dynamics, more recent comprehensive studies using pump-DFWM and pump-IVS with improved temporal resolution corroborated the ultrafast initial step of the vibrational cooling model of the S_1 C=C. A similar ultrafast vibrational cooling step was also observed for the C=C 1525 cm^{-1} mode for all-*trans*-β-carotene [101] and in other carotenoids [58]. Cooling of the S_1 C=C mode was also observed for carotenoids in a light-harvesting complex by applying FSRS to spirilloxanthin in the native LH1 of *Rhodospirillum rubrum* [134]. After the actinic excitation, the frequency at 1740 cm^{-1} evolved to 1767 cm^{-1} with a time constant of 300 fs.

The vibrational relaxation after deactivation of the S_2 state has been recently further addressed for a series of open-chain carotenoids (like lycopene) with increasing conjugated length N by using pump-DFWM [104]. The simple picture of vibrational cooling, accounted for by a bi-exponential frequency up-shift as discussed in the previous paragraph for all-*trans*-β-carotene, is actually only observed for longer open-chain carotenoids ($N=11$ and 13). Short open-chain carotenoids ($N=9$ and 10) show a down-shift of the C=C stretching mode from about 1580 to about 1510 cm^{-1}, which has been explained as a further indication of coupling between the S_2 and an additional electronic dark state (between the S_1 and S_2 states).

A very intriguing observation in the vibrational dynamics of carotenoids is the additional observation of two vibrational bands in the 1800 cm^{-1} spectral region in several experiments. These bands have been observed with e.g., FSRS at 1770 and 1800 cm^{-1} as well as at 1770 and 1790 cm^{-1} [132, 135], and at 1740 and 1785 cm^{-1} with pump-IVS and pump-DFWM [58] for all-*trans*-β-carotene. In the latter case, they have been explained as the result of vibrationally hot C=C levels of the typical S_1 C=C mode contributing to the signal (agreeing with the above picture). More recently, FSRS has been applied to spirilloxanthin ($N=13$) and two bands at 1743 and 1771 cm^{-1} have been also observed [56]. However, the much longer life time (3 ps) detected for 1771 cm^{-1} in comparison to the S_1 lifetime (1.5 ps) has been interpreted as the signature of an electronic state of another carotenoid conformer present already in the ground state and also excited by the actinic pulse. Whether these two bands are due to inhomogeneous S_0 conformational distributions or to vibrationally hot bands in the S_1 manifold is still unclear.

The first attempt to follow the evolution of the high-frequency Raman spectrum of the S_2 state directly was done by FSRS using a Raman pump and probe spectrally resonant with the S_2 ESA of all-*trans*-β-carotene [136]. A very broad spectrally unresolved Raman band was disentangled with two bands at 1654 and 1739 cm^{-1}. While the 1739 cm^{-1} band was assigned to the C=C mode in S_1, the mode at 1654 cm^{-1} was due to the C=C in the S_2 state. A similar frequency for the

C=C in the S_2 state at 1660 cm^{-1} was also observed for another kind of carotenoid (*trans*-apo-8'-carotenal) with FSRS [137]. It is important to note that, more recently, Kennis et al. showed that the FSRS signal in all-*trans*-β-carotene measurements is distorted by the transient absorption signal of the molecule and careful correction must be performed to extract the correct Raman signal [55]. Non-optimal background subtraction may lead to false vibrational bands. For example, contributions from the conventional transient absorption were taken into account, and the FSRS was corrected, indicating that the spectra observed earlier for S_2 [138] were strongly modulated by the S_2 transient absorption signal. A more careful study of the Raman spectrum of the S_2 state with FSRS and of the S_0 Raman activity by non-resonant stimulated Raman scattering recently showed that all high-frequency Raman bands initially between 800 and 1600 cm^{-1} previously assigned to the S_2 state are due to the dispersive ground-state vibrational bands contaminating the excited state signal (Fig. 13, in particular b) [51, 135].

Finally, low-frequency contributions have also been investigated in carotenoids [20, 51, 58]. Low-frequency bands are notoriously complicated to detect in any multi-VCS, in particular for short-lived electronic states as the S_2 state of carotenoids like all-*trans*-β-carotene and lycopene. Nevertheless, low-frequency modes of the S_2 state were reported by applying resonant stimulated Raman scattering, i.e., without an actinic pump for all-*trans*-β-carotene [51]. The bands at 200, 400, and 600 cm^{-1} in the S_2 have larger amplitude compared to ground state modes, hinting at strong anharmonicities and mixing of low frequencies in the S_2 state (Fig. 13b).

In spite of being investigated by several techniques, the ultrafast vibrational dynamics in the excited states of carotenoids still poses a challenge for multi-VCS

Fig. 13 FSRS on all-*trans*-β-carotene **a** raw FSRS spectrum. **b** Background-corrected FSRS signal showing positive signal from the S_2 state and negative signal from the S_0 ground state Reprinted with permission from Ref. [51]. Copyright 2018 American Chemical Society

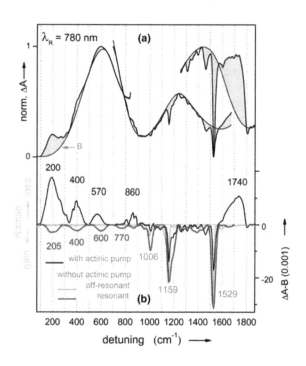

and will require in the future additional experimental and theoretical work to clarify it. Nevertheless, the application to carotenoids illustrates the ability of multi-VCS at revealing the signatures of vibrational dynamics involved in ultrafast, non-reactive, internal conversion. We will describe below another example of application of multi-VCS to the investigation of photoreactive vibrational dynamics.

5.3 Photoisomerization of Stilbene and Derivates

Multidimensional VCS performs vibrational (Raman) spectroscopy of molecular excited states, from the FC state to the photoproduct formation along the photoreaction pathway. Hence, it reports on the time evolution of the molecular structure, which is particularly informative when monitoring photoreactions involving large amplitude motions and structural changes such as C=C double bond photoisomerizations. Here we will illustrate the recent, successful use of multi-VCS for the investigation of the isomerization reaction of stilbene and derivatives (Fig. 14). These photoreactions are not only models for ultrafast C=C photoisomerization but also prototypes of light-to-mechanical energy conversion in well-known synthetic rotary motors [139, 140].

Both the *cis* and *trans* isomers of stilbene, named "c" and "t", respectively, undergo C=C double bond photoisomerization via a common, so-called "phantom" dark state [142–143], which is a perpendicular S_1 transient structure named p*, from which further pyramidalization rapidly drives the system to decay to S_0 via a conical intersection (CInt) [145–146] in a similar way in both cases. Upon photoexcitation of the planar t isomer, the formation of the p* transient state from the t* Franck–Condon state takes ~ 100 ps due to a significant S_1 energy barrier. Photoexcitation of the non-planar c isomer leads either (1) to ultrafast, further C=C bond torsion and sub-ps formation of the same transient p* state, or (2) to planarization enabling a cyclization reaction and dihydrophenanthrene (DHP) formation. Accurate time-resolved structural information along the S_1 reactive paths has long been sought after in order to decipher the photoreaction mechanism in condensed phase [37, 77, 148–149].

The time-resolved, spontaneous, and stimulated Raman spectroscopy of t* and isotopomers has been investigated with outstanding detail and accuracy, offering precious opportunities to benchmark computational methodologies for modeling the excited-state electronic structure and anharmonic PES landscapes [147, 150–154]. Because of its high solubility and strong Raman activity, *trans*-stilbene has also

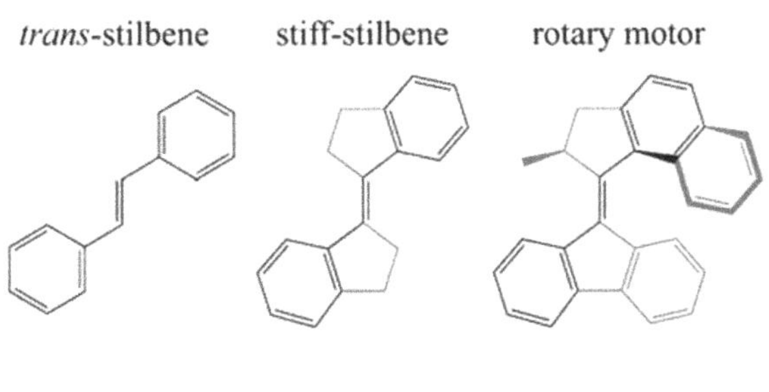

Fig. 14 Chemical structures of stilbene (*left*), stiff-stilbene (*middle*), and a fluorene-based rotary motor (*right*)

trans-stilbene   stiff-stilbene   rotary motor

been used to demonstrate the remarkable sensitivities of state-of-the-art multi-VCS experimental setups producing very high quality spectra [46, 77], as illustrated in Fig. 15. Time-resolved Raman signals observed in the spectral domain [149] or in the time domain [77] were obtained for t* after actinic excitation of t in the UV. Their intensities decay on a time scale corresponding to the t* lifetime (i.e., ~ 80 ps in n-hexane). While the S_0 *trans*-stilbene (t) Raman spectrum (Fig. 15, bottom) is characterized by the prominent 1639 and 1596 cm^{-1} modes, respectively, assigned to motions dominated by central C=C stretch and phenyl rings stretch [155], in the S_1 state (t*) however, the same spectral range is dominated by a single mode at 1570 cm^{-1} and the accurate assignment of the t* vibrational modes remains challenging [149]. Multi-VCS therefore appears to be a powerful experimental approach for assessing the accuracy of state-of-the-art computational methodologies for excited-states modeling. The peak position of some t* Raman peaks (in particular the ~ 1570 cm^{-1} mode) are seen to shift on the 10-ps time scale by up to 5 cm^{-1} [147], depending on the excess vibrational energy [149], due to vibrational cooling

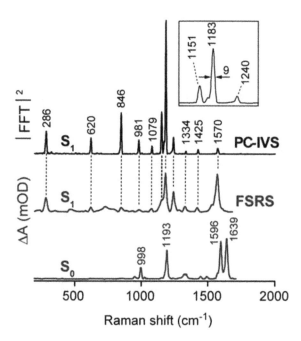

Fig. 15 *Trans*-stilbene Raman spectra measured in n-hexane, in the t* excited state (S_1) by PC-IVS (*top, in black*) or FSRS (*middle, in brown*), and in the t ground state (S_0) by FSRS (*bottom, in brown*). Both S_1 Raman spectra were recorded after actinic excitation at 325 or 326 nm. In the PC-IVS experiment [77], the vibrational coherence was subsequently induced with a 9-fs pulse centered at 550 nm and resonant with the t* excited state absorption (ESA), and probed in the time domain before Fourier transformation. In the FSRS experiment [149], a 645-nm, 2-ps-long Raman pump pulse is used, which is pre-resonant with the same ESA. The ground-state Raman spectrum is also acquired by FSRS, without actinic pulse (or at negative time delays) with the same 615-nm Raman pulse, which is off-resonant with respect to the ground state absorption Reprinted with permission from Ref. [77]. Copyright 2014 American Chemical Society https://pubs.acs.org/doi/abs/10.1021%2Fjp5075863

in the excited state. This was exploited to investigate the mechanism of solute–solvent energy dissipation [98]. Sub-ps decay kinetics of the t* Raman peak intensities reported with FSRS were eventually identified to be controlled by experimental conditions (Raman depletion effect) [46], and indeed not confirmed by population-controlled pump-IVS [77].

Multi-VCS of *cis*-stilbene was also carried out similarly, but its very short c* lifetime increases the experimental challenge [37, 148, 149], as portrayed above for the excited states of carotenoids. A seminal pump-IVS experiment demonstrated the ability of multi-VCS to reveal quantitative structural information along the ultrafast isomerization pathway from c* to the CInt [37]. Following actinic UV excitation, a 620-nm, 11-fs pulse (repump) on resonance with ESA and SE of c* was used to impulsively trigger and subsequently probe a S_1 vibrational wavepacket. A dominant oscillation is detected with a frequency around 240 cm^{-1}, which down-shifts by up to 30 cm^{-1} on the ps time scale as a function of the waiting time between actinic and repump pulses. This mode, previously observed in the form of a Franck–Condon activated vibrational wavepacket in a conventional pump-probe experiment [156] and by ps-resolved Raman spectroscopy [148], is attributed to a spectator mode whose frequency shift reveals the anharmonicity and gradual topography change of the excited state PES as the system evolves from the FC point towards the CInt. FSRS experiments performed with a Raman pump on resonance with the c* ESA band [149] revealed Raman peaks broadened by the very short c* lifetime, but the background subtraction (inherent to FSRS data, see above) is in this case particularly challenging, leading the authors to consider the data with caution.

While both c and t photoisomerizations have long been postulated to occur via the same transient perpendicular p state [141], its experimental evidence and spectroscopic characterization in condensed phase came four decades later [143]. Recently, further characterization of the p* state was made by investigating stilbene derivatives where the formation and decay time of the p* state vary by more than two orders of magnitude, while its characteristic UV absorption band is invariably observed at 350 ± 20 nm depending on the substitutions [111, 158–159]. The dependence of the p* lifetime on the solvent polarity is interpreted as the signature of its zwitterionic nature. Multi-VCS was performed on the *trans*- 1,1′-dicyanostilbene, which features a nearly 30-ps-long-lived p* state in *n*-hexane, thus facilitating the elucidation of some of its vibrational spectroscopic signatures thanks to a very sensitive FSRS set-up [111]. Three Raman peaks of a few μOD signal amplitude and characterized by a 20–30 ps decay kinetics were attributed to the transient p* species, including a 1558 cm^{-1} mode which was tentatively assigned to the phenyl quadrant vibrations, in the absence of any computational prediction of the p* vibrational modes.

The synthetic molecular rotary machines developed by Feringa's group are based on so-called crowded derivatives of stilbene and stiff-stilbene [139, 160], such as the fluorene-based compound displayed in Fig. 14. The photoisomerization of such compounds also involves a transient excited state species. For stiff-stilbene, a sub-ps, viscosity-independent excited state relaxation is observed but the UV absorption signature of a stilbene-like perpendicular "phantom" state is not detected [161]. While the S_0 Raman spectrum of stiff-stilbene also features the same double peak around 1600 cm^{-1} as in *trans* stilbene (see above), its S$_1$ Raman spectrum shows a dominating peak at 1500 cm^{-1}, i.e., downshifted by 70 cm^{-1} as compared to the 1570 cm^{-1} mode of the t* state of stilbene. For the fluorene-based rotary machine investigated by the Meech's group, the 0.1-ps, viscosity-independent formation of a transient excited state is also observed, which is characterized as a dark state, and proposed to result from the ultrafast relaxation along a volume-conserving coordinate, such as pyramidalization of one of the carbon atoms of the isomerizing bond [54, 162]. For both stiff-stilbene and the fluorene-based machine, the subsequent decay to the ground state is instead viscosity-dependent like it is for *trans*-stilbene. This is expected for the torsion around the central C=C double bond in these compounds, since it is a large amplitude motion displacing a significant volume of solvent. Recently, Meech's group applied multi-VCS to follow the structural relaxation of the transient dark state until the formation of the ground state isomer on the 1.6-ps time scale [119]. FSRS employing a Raman pump resonant with the dark state absorption around 550 nm revealed the Raman spectrum of the dark state dominated by 1345 and 1430 cm^{-1} Raman peaks, strongly downshifted with respect to the ground state Raman double peak at 1560 and 1585 cm^{-1} (see Fig. 16). These excited-state Raman signatures are discussed as being related to the isomerizing C=C double bond by analogy to the case of stiff-stilbene. They decay on the same time scale as the dark state population, to give rise to the Raman signature of the ground state photoproduct (so-called "unstable rotor", see Fig. 16) dominated by the C=C stretch peak pair at 1510 and 1550 cm^{-1}. The large frequency downshift of the C=C Raman signature of the dark state is tentatively attributed to the elongation of the central isomerizing bond in the excited state, in line with computational predictions [163].

Finally, another common feature of substituted *trans*-stilbene derivatives is the existence of ground state, sub-populations of rotamers resulting from the thermally activated phenyl ring rotations [164]. While the various rotamers may not easily be distinguished by their steady-state UV–Vis spectroscopic signatures, they may feature distinct photoreaction kinetics and vibrational signatures. Multi-VCS was recently applied to the investigation of a family of di-fluorinated stilbene compounds [165]. This work demonstrates the efficiency of multi-VCS at discriminating the vibrational signatures and photoreaction kinetics of distinct subpopulations. Chemical and/or structural heterogeneity is a common feature of complex molecular systems in condensed phase. One present challenge of time-resolved non-linear spectroscopy and of physical chemistry is to resolve such heterogeneity, and multi-VCS is a promising spectroscopic tool for that.

6 Conclusions

The detection of Raman spectra as a function of photoreaction time is one of the most natural ways to investigate structure changes and interactions at the molecular level. Multi-VCS has achieved this goal by detecting stimulated Raman scattering (SRS) after the interaction with an actinic pulse triggering the photoreaction of interest. SRS is a third–order non-linear spectroscopy technique which exploits

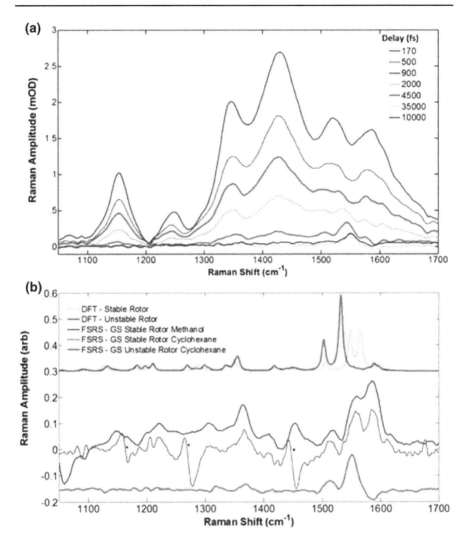

Fig. 16 **a** Time-resolved Raman spectra of the excited, so-called "dark" state of the fluorene-based molecular rotor displayed in Fig. 14 *right*, recorded in cyclohexane. The *blue spectrum* is attributed to the Raman signature of the dark state. It decays on the same time-scale as the dark state lifetime to produce the Raman spectrum (*yellow* and *brown spectra*) attributed to the photoproduct, called "unstable rotor". **b** Computed (*DFT*) and experimental (*FSRS*) ground state Raman spectra of the reactant ("*GS stable rotor*") and photoproduct ("*GS unstable rotor*") Adapted with permission from Ref. [54]. Copyright (2017) American Chemical Society

coherent light–matter interactions to prepare vibrational wavepackets in one (or two) electronic states and to probe their spectroscopic signatures either in the spectral domain or in the time domain. The examples discussed in this contribution portrayed how such stimulated Raman spectra are acquired by Multi-VCS and used to (1) map structural dynamics along a photoreaction, (2) identify transient molecular species, and (3) reveal chemical/structural heterogeneity of complex molecular systems in condensed phase. In particular, the ability to directly follow Raman shifts of only a few wavenumbers as well as Raman amplitudes during a photoreaction is a central feature of time- as well frequency-domain Multi-VCS methods.

The recent success of multi-VCS in clarifying the dynamics of a myriad of molecular system is mainly due to advances in key optical technologies of ultrashort pulse generation and the ability of tuning the excitation and probing spectra. On the one hand, multi-VCS in the time domain requires short pulses, i.e., below 15 fs to effectively induce high-frequency vibrational coherence. On the other hand, spectral tuneability has been shown to be a central piece when disentangling e.g. ground from excited-states vibrational dynamics as well as addressing different chromophores. These two aspects have become experimentally more accessible in recent years by the combination of commercial non-collinear optical parametric amplifiers (nc-OPA) and broadband chirped mirrors. As also shown in this contribution, this has enabled the application of multi-VCS to a wide range of different chromophores absorbing from the UV, over the visible and up to the near-infrared spectral region. Present experimental developments, e.g., towards microscopy, are the promise for turning multi-VCS into a so-called "high-content" analytical technique [167–168]. In particular, the combination of e.g., multi-VCS with super-resolution microscopy exemplifies the ability to identify chemical compounds with spatial resolutions beyond the diffraction limit without any chemical labeling [169].

In spite of the success of multi-VCS, there are still several ongoing research topics. One of these topics is the extraction of the pure excited state vibrational dynamics. The complexity of molecular signal extraction is present in all multi-VCS methods and can be due to different causes (spectral overlap between ground- and excited-state absorptions, signal distortion due to other optical signals contributions, etc.). The development of cautious data post-processing, for example, to isolate reliably the specific stimulated Raman spectra of interest is still an intense research topic.

Another ongoing research topic in multi-VCS is the calculation of the evolution of Raman spectra in excited molecular states. Compared to other multidimensional techniques like 2D electronic or 2D infrared spectroscopies, the application of theoretical methods to calculate the evolution of Raman spectra in excited molecular states still is in its infancy, and in the overwhelming majority of the experimental cases, analysis of multi-VCS spectra is still done in a very qualitative way. From the theoretical point-of-view, models (like sum-over-states (SOS) and multimode Brownian oscillator) and nonlinear signal calculation techniques are well known, but there have been very few examples where the experimental optical signal in multi-VCS has been completely numerically simulated [102]. A major challenge for accurate modeling of third-order non-linear spectroscopy signals (VCS or UV–Vis 2DES) obtained in complex molecular systems still resides in the accurate quantum chemical simulation of excited molecular states and of their photoreaction dynamics. The contribution by Segarra-Martin et al. in this collection reviews the present state-of-the-art theoretical developments for the simulation of 2DES spectroscopy signal in complex molecular systems. Rapid progress in theoretical development is opening new horizons towards quantitative modeling of experimental signal based on realistic models of complex systems.

In conclusion, multi-VCS is now established as one of the state-of-the-art non-linear spectroscopy techniques for the investigation of ultrafast photoreactivity in molecules. More specifically, it is a technique of choice for monitoring vibrational dynamics in excited states with unparalleled time resolution. It is our belief that the combination of multi-VCS methods with recent and rapid theoretical developments will further enhance the impact of multi-VCS in unraveling ultrafast photoreaction mechanisms.

References

1. Joo TH, Albrecht AC (1993) Vibrational frequencies and dephasing times in excited electronic states by femtosecond time-resolved 4-wave-mixing. Chem Phys 173(1):17–26
2. Weiner AM, Desilvestri S, Ippen EP (1985) 3-Pulse scattering for femtosecond dephasing studies—theory and experiment. J Opt Soc Am B 2(4):654–662
3. Hwang H, Rossky PJ (2004) Electronic decoherence induced by intramolecular vibrational motions in a betaine dye molecule. J Phys Chem B 108(21):6723–6732. https://doi.org/10.1021/jp037031b
4. Vos MH, Jones MR, Martin JL (1998) Vibrational coherence in bacterial reaction centers: spectroscopic characterisation of motions active during primary electron transfer. Chem Phys 233(2–3):179–190. https://doi.org/10.1016/S0301-0104(97)00355-8
5. Collini E, Wong CY, Wilk KE, Curmi PMG, Brumer P, Scholes GD (2010) Coherently wired light-harvesting in photosynthetic marine algae at ambient temperature. Nature 463(7281):U644–U669. https://doi.org/10.1038/nature08811
6. Fuller FD, Pan J, Gelzinis A, Butkus V, Senlik SS, Wilcox DE, Yocum CF, Valkunas L, Abramavicius D, Ogilvie JP (2014) Vibronic coherence in oxygenic photosynthesis. Nat Chem 6(8):706–711. https://doi.org/10.1038/nchem.2005
7. Huelga SF, Plenio MB (2013) Vibrations, quanta and biology. Contemp Phys 54(4):181–207. https://doi.org/10.1080/00405000.2013.829687
8. Romero E, Augulis R, Novoderezhkin VI, Ferretti M, Thieme J, Zigmantas D, van Grondelle R (2014) Quantum coherence in photosynthesis for efficient solar-energy conversion. Nat Phys 10(9):677–683. https://doi.org/10.1038/nphys3017
9. Scholes GD, Fleming GR, Olaya-Castro A, van Grondelle R (2011) Lessons from nature about solar light harvesting. Nat Chem 3(10):763–774. https://doi.org/10.1038/nchem.1145
10. Gueye M, Manathunga M, Agathangelou D, Orozco Y, Paolino M, Fusi S, Haacke S, Olivucci M, Leonard J (2018) Engineering the vibrational coherence of vision into a synthetic molecular device. Nat Commun 9:313. https://doi.org/10.1038/s41467-017-02668-w
11. Scholes GD, Fleming GR, Chen LX, Aspuru-Guzik A, Buchleitner A, Coker DF, Engel GS, van Grondelle R, Ishizaki A, Jonas DM, Lundeen JS, McCusker JK, Mukamel S, Ogilvie JP, Olaya-Castro A, Ratner MA, Spano FC, Whaley KB, Zhu XY (2017) Using coherence to enhance function in chemical and biophysical systems. Nature 543(7647):647–656. https://doi.org/10.1038/nature21425
12. Rose TS, Rosker MJ, Zewail AH (1989) Femtosecond real-time probing of reactions. IV. The reactions of alkali halides. J Chem Phys 91(12):7415–7436
13. Pollard WT, Fragnito HL, Bigot JY, Shank CV, Mathies RA (1990) Quantum-mechanical theory for 6 fs dynamic absorption spectroscopy and its application to Nile blue. Chem Phys Lett 168(3):239–245. https://doi.org/10.1016/0009-2614(90)85603-A
14. Fragnito HL, Bigot JY, Becker PC, Shank CV (1989) Evolution of the vibronic absorption spectrum in a molecule following impulsive excitation with a 6 fs optical pulse. Chem Phys Lett 160(2):101–104. https://doi.org/10.1016/0009-2614(89)87564-5
15. Dhar L, Rogers JA, Nelson KA (1994) Time-resolved vibrational spectroscopy in the impulsive limit. Chem Rev 94(1):157–193. https://doi.org/10.1021/cr00025a006
16. Ruhman S, Kosloff R (1990) Application of chirped ultrashort pulses for generating large-amplitude ground-state vibrational coherence: a computer simulation. J Opt Soc Am B 7(8):1748–1752. https://doi.org/10.1364/josab.7.001748

17. Bardeen CJ, Wang Q, Shank CV (1998) Femtosecond chirped pulse excitation of vibrational wave packets in LD690 and bacteriorhodopsin. J Phys Chem A 102(17):2759–2766. https://doi.org/10.1021/jp980346k

18. Malkmus S, Dürr R, Sobotta C, Pulvermacher H, Zinth W, Braun M (2005) Chirp dependence of wave packet motion in oxazine 1. J Phys Chem A 109(46):10488–10492. https://doi.org/10.1021/jp054462g

19. Kahan A, Nahmias O, Friedman N, Sheves M, Ruhman S (2007) Following photoinduced dynamics in bacteriorhodopsin with 7-fs impulsive vibrational spectroscopy. J Am Chem Soc 129(3):537–546. https://doi.org/10.1021/ja064910d

20. Kraack JP, Motzkus M, Buckup T (2011) Selective nonlinear response preparation using femtosecond spectrally resolved four-wave-mixing. J Chem Phys 135(22):224505

21. Wand A, Kallush S, Shoshanim O, Bismuth O, Kosloff R, Ruhman S (2010) Chirp effects on impulsive vibrational spectroscopy: a multimode perspective. Phys Chem Chem Phys 12(9):2149–2163

22. Chesnoy J, Mokhtari A (1988) Resonant impulsive-stimulated Raman scattering on malachite green. Phys Rev A 38(7):3566–3576

23. Pollard WT, Dexheimer SL, Wang Q, Peteanu LA, Shank CV, Mathies RA (1992) Theory of dynamic absorption spectroscopy of nonstationary states. 4. Application to 12-fs resonant impulsive Raman spectroscopy of bacteriorhodopsin. J Phys Chem 96(15):6147–6158. https://doi.org/10.1021/j100194a013

24. Ruhman S, Joly AG, Nelson KA (1988) Coherent molecular vibrational motion observed in the time domain through impulsive stimulated Raman scattering. IEEE J Quantum Electron 24(2):460–469. https://doi.org/10.1109/3.146

25. Mukamel S (1999) Principles of nonlinear optical spectroscopy, vol 6. Oxford University Press on Demand, Oxford

26. Pollard WT, Lee SY, Mathies RA (1990) Wave packet theory of dynamic absorption spectra in femtosecond pump–probe experiments. J Chem Phys 92(7):4012–4029. https://doi.org/10.1063/1.457815

27. Johnson AE, Myers AB (1996) A comparison of time- and frequency-domain resonance Raman spectroscopy in triiodide. J Chem Phys 104(7):2497–2507. https://doi.org/10.1063/1.470998

28. Liebel M, Schnedermann C, Wende T, Kukura P (2015) Principles and applications of broadband impulsive vibrational spectroscopy. J Phys Chem A 119(36):9506–9517. https://doi.org/10.1021/acs.jpca.5b05948

29. Tanimura Y, Mukamel S (1993) Temperature-dependence and non-condon effects in pump-probe spectroscopy in the condensed-phase. J Opt Soc Am B 10(12):2263–2268. https://doi.org/10.1364/josab.10.002263

30. Brazard J, Bizimana LA, Gellen T, Carbery WP, Turner DB (2016) Experimental detection of branching at a conical intersection in a highly fluorescent molecule. J Phys Chem Lett 7(1):14–19. https://doi.org/10.1021/acs.jpclett.5b02476

31. Hamm P, Zanni MT (2011) Concepts and methods of 2d infrared spectroscopy. Cambridge University Press, Cambridge

32. Siebert T, Schmitt M, Gräfe S, Engel V (2006) Ground state vibrational wave-packet and recovery dynamics studied by time-resolved CARS and pump-CARS spectroscopy. J Raman Spectrosc 37(1–3):397–403. https://doi.org/10.1002/jrs.1441

33. Motzkus M, Pedersen S, Zewail AH (1996) Femtosecond real-time probing of reactions. 19. Nonlinear (DFWM) techniques for probing transition states of uni- and bimolecular reactions. J Phys Chem 100(14):5620–5633. https://doi.org/10.1021/jp960265t

34. Dobryakov AL, Quick M, Ioffe IN, Granovsky AA, Ernsting NP, Kovalenko SA (2014) Excited-state Raman spectroscopy with and without actinic excitation: S1 Raman spectra of trans-azobenzene. J Chem Phys 140(18):184310. https://doi.org/10.1063/1.4874854

35. Sun Z, Lu J, Zhang DH, Lee S-Y (2008) Quantum theory of (femtosecond) time-resolved stimulated Raman scattering. J Chem Phys 128(14):144114. https://doi.org/10.1063/1.2888551

36. Payne SA, Hochstrasser RM (1986) Picosecond coherent anti-stokes Raman scattering from the excited states of stilbene and benzophenone. J Phys Chem 90(10):2068–2074

37. Takeuchi S, Ruhman S, Tsuneda T, Chiba M, Taketsugu T, Tahara T (2008) Spectroscopic tracking of structural evolution in ultrafast stilbene photoisomerization. Science 322(5904):1073–1077

38. Motzkus M, Pedersen S, Zewail AH (1996) Femtosecond real-time probing of reactions.19. Nonlinear (DFWM) techniques for probing transition states of uni- and bimolecular reactions. J Phys Chem Us 100(14):5620–5633

39. Hauer J, Buckup T, Motzkus M (2007) Pump-degenerate four wave mixing as a technique for analyzing structural and electronic evolution: multidimensional time-resolved dynamics near a conical intersection. J Phys Chem A 111(42):10517–10529

40. McCamant DW, Kukura P, Mathies RA (2003) Femtosecond broadband stimulated Raman: a new approach for high-performance vibrational spectroscopy. Appl Spectrosc 57(11):1317–1323

41. Mallick B, Lakhsmanna A, Umapathy S (2011) Ultrafast Raman loss spectroscopy (URLS): instrumentation and principle. J Raman Spectrosc 42(10):1883–1890. https://doi.org/10.1002/jrs.2996

42. Roy K, Kayal S, Kumar VR, Beeby A, Ariese F, Umapathy S (2017) Understanding ultrafast dynamics of conformation specific photo-excitation: a femtosecond transient absorption and ultrafast Raman loss study. J Phys Chem A 121(35):6538–6546. https://doi.org/10.1021/acs.jpca.7b03893

43. Kayal S, Roy K, Umapathy S (2018) Femtosecond coherent nuclear dynamics of excited tetraphenylethylene: ultrafast transient absorption and ultrafast Raman loss spectroscopic studies. J Chem Phys 148(2):024301. https://doi.org/10.1063/1.5008726

44. Tokmakoff A, Lang MJ, Larsen DS, Fleming GR, Chernyak V, Mukamel S (1997) Two-dimensional Raman spectroscopy of vibrational interactions in liquids. Phys Rev Lett 79(14):2702–2705. https://doi.org/10.1103/PhysRevLett.79.2702

45. Yoshizawa M, Hattori Y, Kobayashi T (1994) Femtosecond time-resolved resonance Raman gain spectroscopy in polydiacetylene. Phys Rev B 49(18):13259–13262. https://doi.org/10.1103/PhysRevB.49.13259

46. Kovalenko SA, Dobryakov AL, Ernsting NP (2011) An efficient setup for femtosecond stimulated Raman spectroscopy. Rev Sci Instrum 82(6):063102. https://doi.org/10.1063/1.3596453

47. Rhinehart JM, Challa JR, McCamant DW (2012) Multimode charge-transfer dynamics of 4-(dimethylamino)benzonitrile probed with ultraviolet femtosecond stimulated Raman spectroscopy. J Phys Chem B 116(35):10522–10534. https://doi.org/10.1021/jp3020645

48. Weigel A, Ernsting NP (2010) Excited stilbene: intramolecular vibrational redistribution and solvation studied by femtosecond stimulated Raman spectroscopy. J Phys Chem B 114(23):7879–7893. https://doi.org/10.1021/jp100181z

49. McCamant DW, Kukura P, Yoon S, Mathies RA (2004) Femtosecond broadband stimulated Raman spectroscopy: apparatus and methods. Rev Sci Instrum 75(11):4971–4980

50. Dietze DR, Mathies RA (2016) Femtosecond stimulated Raman spectroscopy. ChemPhysChem 17:1224–1251. https://doi.org/10.1002/cphc.201600104

51. Quick M, Dobryakov AL, Kovalenko SA, Ernsting NP (2015) Resonance femtosecond-stimulated Raman spectroscopy without actinic excitation showing low-frequency vibrational activity in the S-2 state of all-trans beta-carotene. J Phys Chem Lett 6(7):1216–1220. https://doi.org/10.1021/acs.jpclett.5b00243

52. Frobel S, Buschhaus L, Villnow T, Weingart O, Gilch P (2015) The photoformation of a phthalide: a ketene intermediate traced by FSRS. Phys Chem Chem Phys 17(1):376–386

53. Laimgruber S, Schachenmayr H, Schmidt B, Zinth W, Gilch P (2006) A femtosecond stimulated Raman spectrograph for the near ultraviolet. Appl Phys B Lasers Opt 85(4):557–564

54. Hall CR, Conyard J, Heisler IA, Jones G, Frost J, Browne WR, Feringa BL, Meech SR (2017) Ultrafast dynamics in light-driven molecular rotary motors probed by femtosecond stimulated Raman spectroscopy. J Am Chem Soc 139(21):7408–7414. https://doi.org/10.1021/jacs.7b03599

55. Kloz M, van Grondelle R, Kennis JTM (2012) Correction for the time dependent inner filter effect caused by transient absorption in femtosecond stimulated Raman experiment. Chem Phys Lett 544:94–101. https://doi.org/10.1016/j.cplett.2012.07.005

56. Kloz M, Weissenborn J, Polivka T, Frank HA, Kennis JTM (2016) Spectral watermarking in femtosecond stimulated Raman spectroscopy: resolving the nature of the carotenoid S-star state. Phys Chem Chem Phys 18(21):14619–14628. https://doi.org/10.1039/c6cp01464j

57. Kuramochi H, Takeuchi S, Tahara T (2016) Femtosecond time-resolved impulsive stimulated Raman spectroscopy using sub-7-fs pulses: apparatus and applications. Rev Sci Instrum 87(4):10. https://doi.org/10.1063/1.4945259

58. Kraack JP, Wand A, Buckup T, Motzkus M, Ruhman S (2013) Mapping multidimensional excited state dynamics using pump-impulsive-vibrational-spectroscopy and pump-degenerate-four-wave-mixing. Phys Chem Chem Phys 15(34):14487–14501

59. Cerullo G, De Silvestri S (2003) Ultrafast optical parametric amplifiers. Rev Sci Instrum 74(1):1–18. https://doi.org/10.1063/1.1523642

60. Riedle E, Beutter M, Lochbrunner S, Piel J, Schenkl S, Sporlein S, Zinth W (2000) Generation of 10–50 fs pulses tunable through all of the visible and the NIR. Appl Phys B Lasers Opt 71(3):457–465. https://doi.org/10.1007/s003400000351

61. Feng Y, Vinogradov I, Ge NH (2017) General noise suppression scheme with reference detection in heterodyne nonlinear spectroscopy. Opt Express 25(21):26262–26279

62. Lanzani G, Cerullo G, Brabec C, Sariciftci NS (2003) Time domain investigation of the intra-chain vibrational dynamics of a prototypical light-emitting conjugated polymer. Phys Rev Lett 90(4):047402

63. Nagasawa Y, Yoneda Y, Nambu S, Muramatsu M, Takeuchi E, Tsumori H, Miyasaka H (2014) Femtosecond degenerate four-wave-mixing measurements of coherent intramolecular vibrations in an ultrafast electron transfer system. Vib Spectrosc 70:58–62. https://doi.org/10.1016/j.vibspec.2013.11.006

64. Song Y, Hellmann C, Stingelin N, Scholes GD (2015) The separation of vibrational coherence from ground- and excited-electronic states in P3HT film. J Chem Phys 142(21):212410. https://doi.org/10.1063/1.4916325

65. Ruhman S, Joly AG, Nelson KA (1987) Time-resolved observations of coherent molecular vibrational motion and the general occurrence of impulsive stimulated scattering. J Chem Phys 86(11):6563–6565

66. Kumar ATN, Rosca F, Widom A, Champion PM (2001) Investigations of ultrafast nuclear response induced by resonant and nonresonant laser pulses. J Chem Phys 114(15):6795–6815

67. Kumar ATN, Rosca F, Widom A, Champion PM (2001) Investigations of amplitude and phase excitation profiles in femtosecond coherence spectroscopy. J Chem Phys 114(2):701–724

68. Cina JA, Kovac PA, Jumper CC, Dean JC, Scholes GD (2016) Ultrafast transient absorption revisited: phase-flips, spectral fingers, and other dynamical features. J Chem Phys 144(17):175102

69. Kraack JP, Buckup T, Motzkus M (2013) Coherent high-frequency vibrational dynamics in the excited electronic state of all-trans retinal derivatives. J Phys Chem Lett 4(3):383–387

70. Schnedermann C, Liebel M, Kukura P (2015) Mode-specificity of vibrationally coherent internal conversion in rhodopsin during the primary visual event. J Am Chem Soc 137(8):2886–2891. https://doi.org/10.1021/ja508941k

71. Kraack JP, Buckup T, Motzkus M (2011) Vibrational analysis of excited and ground electronic states of all-trans retinal protonated Schiff-bases. Phys Chem Chem Phys 13(48):21402–21410

72. Ohta K, Larsen DS, Yang M, Fleming GR (2001) Influence of intramolecular vibrations in third-order, time-domain resonant spectroscopies. II. Numerical calculations. J Chem Phys 114(18):8020–8039. https://doi.org/10.1063/1.1359241

73. Erland J, Balslev I (1993) Theory of quantum beat and polarization interference in four-wave mixing. Phys Rev A 48(3):R1765–R1768

74. Koch M, Feldmann J, von Plessen G, Göbel EO, Thomas P, Köhler K (1992) Quantum beats versus polarization interference: an experimental distinction. Phys Rev Lett 69(25):3631–3634

75. Faeder J, Pinkas I, Knopp G, Prior Y, Tannor DJ (2001) Vibrational polarization beats in femtosecond coherent anti-stokes Raman spectroscopy: a signature of dissociative pump–dump–pump wave packet dynamics. J Chem Phys 115(18):8440–8454. https://doi.org/10.1063/1.1412253

76. Weigel A, Sebesta A, Kukura P (2015) Shaped and feedback-controlled excitation of single molecules in the weak-field limit. J Phys Chem Lett 6(20):4032–4037

77. Wende T, Liebel M, Schnedermann C, Pethick RJ, Kukura P (2014) Population-controlled impulsive vibrational spectroscopy: background- and baseline-free Raman spectroscopy of excited electronic states. J Phys Chem A 118(43):9976–9984

78. Liebel M, Schnedermann C, Kukura P (2014) Vibrationally coherent crossing and coupling of electronic states during internal conversion in beta-carotene. Phys Rev Lett 112(19):198302

79. Wohlleben W, Buckup T, Hashimoto H, Cogdell RJ, Herek JL, Motzkus M (2004) Pump-deplete-probe spectroscopy and the puzzle of carotenoid dark states. J Phys Chem B 108(10):3320–3325

80. Buckup T, Savolainen J, Wohlleben W, Herek JL, Hashimoto H, Correia RRB, Motzkus M (2006) Pump-probe and pump-deplete-probe spectroscopies on carotenoids with $N = 9$–15 conjugated bonds. J Chem Phys 125(19):194505

81. Larsen DS, Papagiannakis E, van Stokkum IHM, Vengris M, Kennis JTM, van Grondelle R (2003) Excited state dynamics of beta-carotene explored with dispersed multi-pulse transient absorption. Chem Phys Lett 381(5–6):733–742. https://doi.org/10.1016/j.cplett.2003.10.016

82. Papagiannakis E, Vengris M, Larsen DS, van Stokkum IHM, Hiller RG, van Grondelle R (2006) Use of ultrafast dispersed pump-dump-probe and pump-repump-probe spectroscopies to explore

the light-induced dynamics of peridinin in solution. J Phys Chem B 110(1):512–521. https://doi.org/10.1021/jp053094d

83. Thaller A, Laenen R, Laubereau A (2006) The precursors of the solvated electron in methanol studied by femtosecond pump-repump-probe spectroscopy. J Chem Phys 124(2):024515

84. Draxler S, Brust T, Eicher J, Zinth W, Braun M (2010) Novel detection scheme for application in pump-repump-probe spectroscopy. Opt Commun 283(6):1050–1054

85. van Wilderen LJGW, Clark IP, Towrie M, van Thor JJ (2009) Mid-infrared picosecond pump-dump-probe and pump-repump-probe experiments to resolve a ground-state intermediate in cyanobacterial phytochrome cph1. J Phys Chem B 113(51):16354–16364

86. Bradler M, Werhahn JC, Hutzler D, Fuhrmann S, Heider R, Riedle E, Iglev H, Kienberger R (2013) A novel setup for femtosecond pump-repump-probe IR spectroscopy with few cycle CEP stable pulses. Opt Express 21(17):20145–20158

87. Buckup T, Weigel A, Hauer J, Motzkus M (2010) Ultrafast multiphoton transient absorption of beta-carotene. Chem Phys 373(1–2):38–44

88. Fujisawa T, Kuramochi H, Hosoi H, Takeuchi S, Tahara T (2016) Role of coherent low-frequency motion in excited-state proton transfer of green fluorescent protein studied by time-resolved impulsive stimulated Raman spectroscopy. J Am Chem Soc 138(12):3942–3945. https://doi.org/10.1021/jacs.5b11038

89. Hoffman DP, Mathies RA (2012) Photoexcited structural dynamics of an azobenzene analog 4-nitro-4[prime or minute]-dimethylamino-azobenzene from femtosecond stimulated Raman. Phys Chem Chem Phys 14:6298–6306. https://doi.org/10.1039/C2CP23468H

90. Gustafson TL, Roberts DM, Chernoff DA (1983) Picosecond transient Raman-spectroscopy—the photo-isomerization of trans-stilbene. J Chem Phys 79(4):1559–1564. https://doi.org/10.1063/1.446027

91. Gustafson TL, Roberts DM, Chernoff DA (1984) The structure of electronic excited-states in trans-stilbene—picosecond transient stokes and anti-stokes Raman-spectra. J Chem Phys 81(8):3438. https://doi.org/10.1063/1.448068

92. Hashimoto H, Koyama Y, Hirata Y, Mataga N (1991) S1 and T1 species of beta-carotene generated by direct photoexcitation from the all-trans, 9-cis, 13-cis, and 15-cis isomers as revealed by picosecond transient absorption and transient Raman spectroscopies. J Phys Chem Us 95(8):3072–3076. https://doi.org/10.1021/j100161a022

93. Weaver WL, Huston LA, Iwata K, Gustafson TL (1992) Solvent solute interactions probed by picosecond transient Raman-spectroscopy—mode-specific vibrational dynamics in S1 trans-stilbene. J Phys Chem Us 96(22):8956–8961. https://doi.org/10.1021/j100201a047

94. Laubereau A, Vonderli D, Kaiser W (1972) Direct measurement of vibrational lifetimes of molecules in liquids. Phys Rev Lett 28(18):1162. https://doi.org/10.1103/physrevlett.28.1162

95. Valley DT, Hoffman DP, Mathies RA (2015) Reactive and unreactive pathways in a photochemical ring opening reaction from 2D femtosecond stimulated Raman. Phys Chem Chem Phys 17(14):9231–9240. https://doi.org/10.1039/C4CP05323K

96. Ferrante C, Pontecorvo E, Cerullo G, Vos M, Scopigno T (2016) Direct observation of subpicosecond vibrational dynamics in photoexcited myoglobin. Nat Chem 8:1137

97. Weigel A, Dobryakov A, Klaumunzer B, Sajadi M, Saalfrank P, Ernsting N (2011) Femtosecond stimulated Raman spectroscopy of flavin after optical excitation. J Phys Chem B 115:3656–3680

98. Iwata K, H-o Hamaguchi (1997) Microscopic mechanism of solute- solvent energy dissipation probed by picosecond time-resolved Raman spectroscopy. J Phys Chem A 101:632–637

99. Kuramochi H, Takeuchi S, Yonezawa K, Kamikubo H, Kataoka M, Tahara T (2017) Probing the early stages of photoreception in photoactive yellow protein with ultrafast time-domain Raman spectroscopy. Nat Chem 9:660–666. https://doi.org/10.1038/nchem.2717

100. Namboodiri V, Scaria A, Namboodiri M, Materny A (2009) Investigation of molecular dynamics in beta-carotene using femtosecond pump-FWM spectroscopy. Laser Phys 19(2):154–161. https://doi.org/10.1134/s1054660x09020029

101. Buckup T, Hauer J, Mohring J, Motzkus M (2009) Multidimensional spectroscopy of beta-carotene: vibrational cooling in the excited state. Arch Biochem Biophys 483(2):219–223

102. Marek MS, Buckup T, Motzkus M (2011) Direct observation of a dark state in lycopene using pump-DFWM. J Phys Chem B 115(25):8328–8337

103. Liebel M, Kukura P (2013) Broad-band impulsive vibrational spectroscopy of excited electronic states in the time domain. J Phys Chem Lett 4(8):1358–1364

104. Miki T, Buckup T, Krause MS, Southall J, Cogdell RJ, Motzkus M (2016) Vibronic coupling in the excited-states of carotenoids. Phys Chem Chem Phys 18(16):11443–11453. https://doi.org/10.1039/c5cp07542d

105. Takaya T, Anan M, Iwata K (2018) Vibrational relaxation dynamics of b-carotene and its derivatives with substituents on terminal rings in electronically excited states as studied by femtosecond time-resolved stimulated Raman spectroscopy in the near-IR region. Phys Chem Chem Phys 20(5):3320–3327

106. Kraack JP, Buckup T, Hampp N, Motzkus M (2011) Ground- and excited-state vibrational coherence dynamics in bacteriorhodopsin probed with degenerate four-wave-mixing experiments. ChemPhysChem 12(10):1851–1859

107. Kraack JP, Buckup T, Motzkus M (2012) Evidence for the two-state-two-mode model in retinal protonated schiff-bases from pump degenerate four-wave-mixing experiments. Phys Chem Chem Phys 14(40):13979–13988

108. Liebel M, Schnedermann C, Bassolino G, Taylor G, Watts A, Kukura P (2014) Direct observation of the coherent nuclear response after the absorption of a photon. Phys Rev Lett 112(23):4. https://doi.org/10.1103/PhysRevLett.112.238301

109. Hoffman DP, Mathies RA (2016) Femtosecond stimulated Raman exposes the role of vibrational coherence in condensed-phase photoreactivity. Acc Chem Res 49:616–625. https://doi.org/10.1021/acs.accounts.5b00508

110. Barclay MS, Quincy TJ, Williams-Young DB, Caricato M, Elles CG (2017) Accurate assignments of excited-state resonance Raman spectra: a benchmark study combining experiment and theory. J Phys Chem A 121(41):7937–7946. https://doi.org/10.1021/acs.jpca.7b09467

111. Quick M, Dobryakov AL, Ioffe IN, Granovsky AA, Kovalenko SA, Ernsting NP (2016) Perpendicular state of an electronically excited stilbene: observation by femtosecond-stimulated Raman spectroscopy. J Phys Chem Lett 7(20):4047–4052. https://doi.org/10.1021/acs.jpclett.6b01923

112. Han FY, Liu WM, Zhu LD, Wang YL, Fang C (2016) Initial hydrogen-bonding dynamics of photoexcited coumarin in solution with femtosecond stimulated Raman spectroscopy. J Mater Chem C 4(14):2954–2963. https://doi.org/10.1039/c5tc03598h

113. Musser AJ, Liebel M, Schnedermann C, Wende T, Kehoe TB, Rao A, Kukura P (2015) Evidence for conical intersection dynamics mediating ultrafast singlet exciton fission. Nat Phys 11(4):352–357. https://doi.org/10.1038/nphys3241

114. Kuramochi H, Fujisawa T, Takeuchi S, Tahara T (2017) Broadband stimulated Raman spectroscopy in the deep ultraviolet region. Chem Phys Lett 683:543–546. https://doi.org/10.1016/j.cplett.2017.02.015

115. Harris MA, Mishra AK, Young RM, Brown KE, Wasielewski MR, Lewis FD (2016) Direct observation of the hole carriers in DNA photoinduced charge transport. J Am Chem Soc 138:5491–5494

116. Oscar BG, Liu WM, Zhao YX, Tang LT, Wang YL, Campbell RE, Fang C (2014) Excited-state structural dynamics of a dual-emission calmodulin-green fluorescent protein sensor for calcium ion imaging. Proc Natl Acad Sci USA 111(28):10191–10196. https://doi.org/10.1073/pnas.1403712111

117. Creelman M, Kumauchi M, Hoff WD, Mathies RA (2014) Chromophore dynamics in the PYP photocycle from femtosecond stimulated Raman spectroscopy. J Phys Chem B 118(3):659–667. https://doi.org/10.1021/jp408584v

118. Roy K, Kayal S, Ariese F, Beeby A, Umapathy S (2017) Mode specific excited state dynamics study of bis(phenylethynyl) benzene from ultrafast Raman loss spectroscopy. J Chem Phys 146(6):064303. https://doi.org/10.1063/1.4975174

119. Hall CR, Heisler IA, Jones GA, Frost JE, Gil AA, Tonge PJ, Meech SR (2017) Femtosecond stimulated Raman study of the photoactive flavoprotein AppA(BLUF). Chem Phys Lett 683:365–369. https://doi.org/10.1016/j.cplett.2017.03.030

120. Kukura P, McCamant DW, Yoon S, Wandschneider DB, Mathies RA (2005) Structural observation of the primary isomerization in vision with femtosecond-stimulated Raman. Science 310(5750):1006–1009

121. Polli D, Altoe P, Weingart O, Spillane KM, Manzoni C, Brida D, Tomasello G, Orlandi G, Kukura P, Mathies RA, Garavelli M, Cerullo G (2010) Conical intersection dynamics of the primary photoisomerization event in vision. Nature 467(7314):U440–U488

122. Frank HA (1999) The photochemistry of carotenoids, vol 8. Advances in photosynthesis. Kluwer Academic, Dordrecht

123. Polivka T, Sundstrom V (2004) Ultrafast dynamics of carotenoid excited states—from solution to natural and artificial systems. Chem Rev 104(4):2021–2071. https://doi.org/10.1021/cr020674n

124. Buckup T, Savolainen J, Wohlleben W, Herek JL, Hashimoto H, Correia RRB, Motzkus M (2006) Pump-probe and pump-deplete-probe spectroscopies on carotenoids with $N = 9$–15 conjugated bonds. J Chem Phys 125(19):194505. https://doi.org/10.1063/1.2388274

125. Jailaubekov AE, Song SH, Vengris M, Cogdell RJ, Larsen DS (2010) Using narrowband excitation to confirm that the S* state in carotenoids is not a vibrationally-excited ground state species. Chem Phys Lett 487(1–3):101–107. https://doi.org/10.1016/j.cplett.2010.01.014

126. Hauer J, Maiuri M, Viola D, Lukes V, Henry S, Carey AM, Cogdell RJ, Cerullo G, Polli D (2013) Explaining the temperature dependence of spirilloxanthin's S* signal by an inhomogeneous ground state model. J Phys Chem A 117(29):6303–6310. https://doi.org/10.1021/jp4011372

127. Ehlers F, Scholz M, Schimpfhauser J, Bienert J, Oum K, Lenzer T (2015) Collisional relaxation of apocarotenals: identifying the S* state with vibrationally excited molecules in the ground electronic state S-0*. Phys Chem Chem Phys 17(16):10478–10488. https://doi.org/10.1039/c4cp05600k

128. Balevicius V, Abramavicius D, Polivka T, Pour AG, Hauer J (2016) A unified picture of S* in carotenoids. J Phys Chem Lett 7(17):3347–3352. https://doi.org/10.1021/acs.jpclett.6b01455

129. Tavan P, Schulten K (1987) Electronic excitations in finite and infinite polyenes. Phys Rev B 36(8):4337–4358. https://doi.org/10.1103/PhysRevB.36.4337

130. Orlandi G, Zerbetto F (1986) Vibronic coupling in polyenes—the frequency increase of the active C=C Ag stretching mode in the absorption-spectra. Chem Phys 108(2):187–195. https://doi.org/10.1016/0301-0104(86)85040-6

131. Nagae H, Kakitani Y, Koyama Y (2009) Theoretical description of diabatic mixing and coherent excitation in singlet-excited states of carotenoids. Chem Phys Lett 474(4–6):342–351. https://doi.org/10.1016/j.cplett.2009.04.039

132. Yoshizawa M, Aoki H, Hashimoto H (2001) Vibrational relaxation of the 2A(g)(−) excited state in all-trans-beta-carotene obtained by femtosecond time-resolved Raman spectroscopy. Phys Rev B 63(18):180301. https://doi.org/10.1103/PhysRevB.63.180301

133. McCamant DW, Kukura P, Mathies RA (2003) Femtosecond time-resolved stimulated Raman spectroscopy: application to the ultrafast internal conversion in beta-carotene. J Phys Chem A 107(40):8208–8214

134. Yoshizawa M, Nakamura R, Yoshimatsu O, Abe K, Sakai S, Nakagawa K, Fujii R, Nango M, Hashimoto H (2012) Femtosecond stimulated Raman spectroscopy of the dark S-1 excited state of carotenoid in photosynthetic light harvesting complex. Acta Biochim Pol 59(1):49–52

135. Quick M, Kasper MA, Richter C, Mahrwald R, Dobryakov AL, Kovalenko SA, Ernsting NP (2015) Beta-carotene revisited by transient absorption and stimulated Raman spectroscopy. ChemPhysChem 16(18):3824–3835. https://doi.org/10.1002/cphc.201500586

136. Kukura P, McCamant DW, Mathies RA (2004) Femtosecond time-resolved stimulated Raman spectroscopy of the S-2 (1B(u)(+)) excited state of beta-carotene. J Phys Chem A 108(28):5921–5925

137. Kardas TM, Ratajska-Gadomska B, Lapini A, Ragnoni E, Righini R, Di Donato M, Foggi P, Gadomski W (2014) Dynamics of the time-resolved stimulated Raman scattering spectrum in presence of transient vibronic inversion of population on the example of optically excited trans-beta-apo-8′-carotenal. J Chem Phys. https://doi.org/10.1063/1.4879060

138. Shim S, Mathies RA (2008) Development of a tunable femtosecond stimulated Raman apparatus and its application to beta-carotene. J Phys Chem B 112(15):4826–4832

139. Koumura N, Zijlstra RWJ, van Delden RA, Harada N, Feringa BL (1999) Light-driven monodirectional molecular rotor. Nature 401:152

140. van Leeuwen T, Lubbe AS, Štacko P, Wezenberg SJ, Feringa BL (2017) Dynamic control of function by light-driven molecular motors. Nat Rev Chem 1:0096

141. Saltiel J (1967) Perdeuteriostilbene. The role of phantom states in the cis-trans photoisomerization of stilbenes. J Am Chem Soc 89:1036–1037

142. Sension RJ, Repinec ST, Szarka AZ, Hochstrasser RM (1993) Femtosecond laser studies of the cis–stilbene photoisomerization reactions. J Chem Phys 98:6291–6315. https://doi.org/10.1063/1.464824

143. Kovalenko SA, Dobryakov AL, Ioffe I, Ernsting NP (2010) Evidence for the phantom state in photoinduced cis–trans isomerization of stilbene. Chem Phys Lett 493:255–258. https://doi.org/10.1016/j.cplett.2010.05.022

144. Quenneville J, Martínez TJ (2003) Ab initio study of cis–trans photoisomerization in stilbene and ethylene. J Phys Chem A 107:829–837. https://doi.org/10.1021/jp021210w

145. Minezawa N, Gordon MS (2011) Photoisomerization of stilbene: a spin-flip density functional theory approach. J Phys Chem A 115:7901–7911. https://doi.org/10.1021/jp203803a

146. Harabuchi Y, Keipert K, Zahariev F, Taketsugu T, Gordon MS (2014) Dynamics simulations with spin-flip time-dependent density functional theory: photoisomerization and photocyclization mechanisms of cis-stilbene in $\pi\pi^*$ states. J Phys Chem A 118:11987–11998. https://doi.org/10.1021/jp5072428

147. Iwata K, H-o Hamaguchi (1992) Picosecond structural relaxation of S1 trans-stilbene in solution as revealed by time-resolved Raman spectroscopy. Chem Phys Lett 196:462–468. https://doi.org/10.1016/0009-2614(92)85721-L

148. Kwok WM, Ma C, Phillips D, Beeby A, Marder TB, Thomas RL, Tschuschke C, Baranović G, Matousek P, Towrie M, Parker AW (2003) Time-resolved resonance Raman study of S1 cis-stilbene and its deuterated isotopomers. J Raman Spectrosc 34:886–891. https://doi.org/10.1002/jrs.1070

149. Dobryakov AL, Ioffe I, Granovsky AA, Ernsting NP, Kovalenko SA (2012) Femtosecond Raman spectra of cis-stilbene and trans-stilbene with isotopomers in solution. J Chem Phys 137:244505. https://doi.org/10.1063/1.4769971

150. Sakamoto A, Tanaka F, Tasumi M, Torii H, Kawato K, Furuya K (2006) Comparison of the Raman spectrum of trans-stilbene in the S1 state calculated by the CIS method and the spectra observed under resonant and off-resonant conditions. A collection of papers presented at the 3rd international conference on advanced vibrational spectroscopy (ICAVS-3), Delavan, WI, USA, 14–19 August 2005—Part 1 42:176–182. https://doi.org/10.1016/j.vibspec.2006.04.001

151. Tsumura K, Furuya K, Sakamoto A, Tasumi M (2008) Vibrational analysis of trans-stilbene in the excited singlet state by time-dependent density functional theory: calculations of the Raman, infrared, and fluorescence excitation spectra. J Raman Spectrosc 39:1584–1591. https://doi.org/10.1002/jrs.2095

152. Angeli C, Improta R, Santoro F (2009) On the controversial nature of the 1 B1u and 2 B1u states of trans-stilbene: the n-electron valence state perturbation theory approach. J Chem Phys 130:174307. https://doi.org/10.1063/1.3131263

153. Ioffe IN, Granovsky AA (2013) Photoisomerization of stilbene: the detailed XMCQDPT2 treatment. J Chem Theory Comput 9:4973–4990. https://doi.org/10.1021/ct400647w

154. Orlandi G, Garavelli M, Zerbetto F (2017) Analysis of the vibronic structure of the trans-stilbene fluorescence and excitation spectra: the S 0 and S 1 PES along the C e [double bond, length as m-dash] C e and C e–C ph torsions. Phys Chem Chem Phys 19:25095–25104

155. Meić Z, Güsten H (1978) Vibrational studies of trans-stilbenes—I. Infrared and Raman spectra of trans-stilbene and deuterated trans-stilbenes. Spectrochim Acta Part A 34(1):101–111. https://doi.org/10.1016/0584-8539(78)80193-7

156. Ishii K, Takeuchi S, Tahara T (2004) A 40-fs time-resolved absorption study on cis-stilbene in solution: observation of wavepacket motion on the reactive excited state. Chem Phys Lett 398:400–406. https://doi.org/10.1016/j.cplett.2004.09.075

157. Berndt F, Dobryakov AL, Quick M, Mahrwald R, Ernsting NP, Lenoir D, Kovalenko SA (2012) Long-lived perpendicular conformation in the photoisomerization path of 1,1′-dimethylstilbene and 1,1′-diethylstilbene. Chem Phys Lett 544:39–42. https://doi.org/10.1016/j.cplett.2012.07.007

158. Dobryakov AL, Quick M, Lenoir D, Detert H, Ernsting NP, Kovalenko SA (2016) Time-resolved photoisomerization of 1,1′-di-tert-butylstilbene and 1,1′-dicyanostilbene. Chem Phys Lett 652:225–229. https://doi.org/10.1016/j.cplett.2016.04.060

159. Dobryakov AL, Quick M, Richter C, Knie C, Ioffe IN, Granovsky AA, Mahrwald R, Ernsting NP, Kovalenko SA (2017) Photoisomerization pathways and Raman activity of 1,1′-difluorostilbene. J Chem Phys 146:044501. https://doi.org/10.1063/1.4974357

160. Pollard MM, Meetsma A, Feringa BL (2008) A redesign of light-driven rotary molecular motors. Org Biomol Chem 6:507–512. https://doi.org/10.1039/B715652A

161. Quick M, Berndt F, Dobryakov AL, Ioffe IN, Granovsky AA, Knie C, Mahrwald R, Lenoir D, Ernsting NP, Kovalenko SA (2014) Photoisomerization dynamics of stiff-stilbene in solution. J Phys Chem B 118:1389–1402. https://doi.org/10.1021/jp411656x

162. Conyard J, Addison K, Heisler IA, Cnossen A, Browne WR, Feringa BL, Meech SR (2012) Ultrafast dynamics in the power stroke of a molecular rotary motor. Nat Chem 4:547–551. https://doi.org/10.1038/nchem.1343

163. Kazaryan A, Kistemaker JCM, Schäfer LV, Browne WR, Feringa BL, Filatov M (2010) Understanding the dynamics behind the photoisomerization of a light-driven fluorene molecular rotary motor. J Phys Chem A 114:5058–5067. https://doi.org/10.1021/jp100609m

164. Mazzucato U, Momicchioli F (1991) Rotational isomerism in trans-1,2-diarylethylenes. Chem Rev 91:1679–1719. https://doi.org/10.1021/cr00008a002

165. Quick M, Dobryakov AL, Ioffe IN, Berndt F, Mahrwald R, Ernsting NP, Kovalenko SA (2018) Rotamer-specific photoisomerization of difluorostilbenes from transient absorption and transient Raman spectroscopy. J Phys Chem B. https://doi.org/10.1021/acs.jpcb.7b09283

166. Ploetz E, Laimgruber S, Berner S, Zinth W, Gilch P (2007) Femtosecond stimulated Raman microscopy. Appl Phys B 87(3):389–393. https://doi.org/10.1007/s00340-007-2630-x

167. Schnedermann C, Lim JM, Wende T, Duarte AS, Ni L, Gu Q, Sadhanala A, Rao A, Kukura P (2016) Sub-10 fs time-resolved vibronic optical microscopy. J Phys Chem Lett 7(23):4854–4859. https://doi.org/10.1021/acs.jpclett.6b02387

168. Czerwinski L, Nixdorf J, Florio GD, Gilch P (2016) Broadband stimulated Raman microscopy with 0.1 ms pixel acquisition time. Opt Lett 41(13):3021–3024. https://doi.org/10.1364/OL.41.003021

169. Silva WR, Graefe CT, Frontiera RR (2016) Toward label-free super-resolution microscopy. ACS Photonics 3(1):79–86. https://doi.org/10.1021/acsphotonics.5b00467

170. Buckup T, Motzkus M (2014) Multidimensional time-resolved spectroscopy of vibrational coherence in biopolyenes. In: Johnson MA, Martinez TJ (eds) Annual review of physical chemistry, vol 65., pp 39–57. https://doi.org/10.1146/annurev-physchem-040513-103619

Top Curr Chem (Z) (2017) 375:87
https://doi.org/10.1007/s41061-017-0173-0

REVIEW

Two-Dimensional Resonance Raman Signatures of Vibronic Coherence Transfer in Chemical Reactions

Zhenkun Guo[1] · Brian P. Molesky[1] · Thomas P. Cheshire[1] ·
Andrew M. Moran[1]

Received: 11 April 2017 / Accepted: 2 October 2017 / Published online: 2 November 2017
© Springer International Publishing AG 2017

Abstract Two-dimensional resonance Raman (2DRR) spectroscopy has been developed for studies of photochemical reaction mechanisms and structural heterogeneity in condensed phase systems. 2DRR spectroscopy is motivated by knowledge of non-equilibrium effects that cannot be detected with traditional resonance Raman spectroscopy. For example, 2DRR spectra may reveal correlated distributions of reactant and product geometries in systems that undergo chemical reactions on the femtosecond time scale. Structural heterogeneity in an ensemble may also be reflected in the 2D spectroscopic line shapes of both reactive and non-reactive systems. In this chapter, these capabilities of 2DRR spectroscopy are discussed in the context of recent applications to the photodissociation reactions of triiodide. We show that signatures of "vibronic coherence transfer" in the photodissociation process can be targeted with particular 2DRR pulse sequences. Key differences between the signal generation mechanisms for 2DRR and off-resonant 2D Raman spectroscopy techniques are also addressed. Overall, recent experimental developments and applications of the 2DRR method suggest that it will be a valuable tool for elucidating ultrafast chemical reaction mechanisms.

Keywords Multidimensional spectroscopy · Raman spectroscopy · Ultrafast spectroscopy · Photodissociation · Coherence transfer

Chapter 6 was originally published as Guo, Z., Molesky, B. P., Cheshire, T. P. & Moran, A. M. Top Curr Chem (Z) (2017) 375: 87. https://doi.org/10.1007/s41061-017-0173-0.

✉ Andrew M. Moran
  ammoran@email.unc.edu

[1] Department of Chemistry, University of North Carolina at Chapel Hill, Chapel Hill, NC 27599, USA

1 Introduction

Time coincidence between chemical reactions and relaxation processes such as solvation and vibrational dephasing gives rise to interesting physical effects. Assumptions made in traditional kinetic models based on an equilibrium version of Fermi's golden rule often break down in the ultrafast regime [1–3]. For example, early femtosecond pump–probe experiments investigated the importance of coherent vibrational motions in electron transfer [4, 5], isomerization [6], and photodissociation [7–10] reactions. It was shown that recurrences of a photoexcited reactant at the transition state can give rise to periodic "bursts" in product formation [4, 5, 10–14]. Recent 2D photon echo studies have renewed interest in coherent photochemical reactions and inspired deeper thought about the significance of coherence in biological function [14–16]. To date, most studies of coherent reaction mechanisms have been conducted with pump–probe or 2D photon echo techniques [17, 18]. These methods possess a single delay time between laser pulses (i.e., a population time) during which the vibrational coherences of interest evolve. With only one population time, it is not possible to establish correlations between separate chemical species in a reaction that is initiated by light absorption; however, if laser pulses (and population times) are added to a traditional three-pulse experiment, then vibrational resonances of reactants and products can be displayed in separate dimensions of a 2D spectrum [19–21].

In this chapter, we discuss our development of two-dimensional resonance Raman (2DRR) spectroscopy and describe how it can be used to elucidate vibronic coherence transfer processes [22–26]. Our focus will be on ultrafast chemical reactions where 2DRR reveals non-trivial information; however, this is not the only way that 2DRR can be employed. For example, 2DRR can be used to uncover line-broadening mechanisms in non-reactive systems in the same manner as any other 2D vibrational spectroscopy technique. To begin, the basic sequence of events associated with a 2DRR experiment is outlined in Fig. 1. The experiment begins when a laser pulse initiates coherent vibrational motion on the ground state potential of a reactant by way of a stimulated Raman transition (these motions correspond to dimension #1). The second laser pulse promotes the system to the excited electronic state of the reactant which then undergoes an ultrafast transition to the product state. Most generally, the product can also exhibit coherent vibrational motion if the transition is fast compared to the vibrational period(s) of the system (these motions are displayed in dimension #2). The sequence outlined in Fig. 1 can be applied to a variety of ultrafast transitions (e.g., electron transfer, energy transfer, isomerization).

We use the term 2DRR to refer to a specific component of the fifth-order response in which the system evolves in a purely vibrational coherence in each of the two electronic population times. Of course, our implementation of 2DRR is preceded by a variety of other time-resolved vibrational spectroscopy techniques which can be adapted to provide similar information. The techniques most closely related to 2DRR are described at fifth-order in perturbation theory. Examples include femtosecond stimulated Raman spectroscopy (FSRS) [27, 28] and resonant

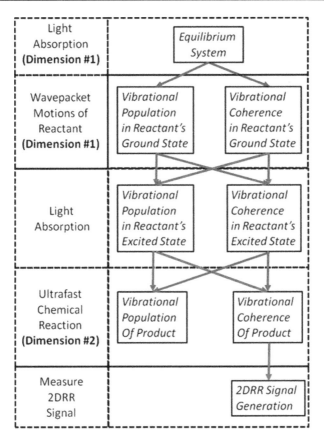

Fig. 1 The 2DRR technique can be used to detect correlations between reactants and products in ultrafast chemical reactions. The first dimension represents coherent vibrational motion of the reactant, whereas the second dimension corresponds to coherent vibrational motion of the product. This scheme applies to reactions in which the transition from the reactant to product is faster than the vibrational periods of the system. The 2DRR method effectively isolates the paths traced with red arrows, thereby facilitating study of reaction mechanisms involving vibronic coherence transfer

pump degenerate four-wave mixing (RP-D4WM) [29, 30]. These two techniques isolate the same components of the fifth-order response function but differ in how vibrational spectra are obtained. For example, data acquisition is generally faster in FSRS because only one delay time is scanned due to its use of frequency-domain detection. FSRS and RP-D4WM are usually conducted by measuring one-dimensional vibrational spectra of a product with respect to a variable population time. Vibrational coherences in two dimensions have been reported in recent FSRS applications (this is the same nonlinearity detected with 2DRR) [31, 32]. In addition, Harel and co-workers have developed a new fifth-order four-dimensional experiment that, like our version of 2DRR, yields a 2D vibrational spectrum for Raman active modes; however, this technique also possesses two dimensions associated with electronic resonance frequencies [33–35].

The information provided by 2DRR spectroscopy differs from that obtained in a traditional pump–probe measurement in that the reaction is initiated from a non-equilibrium state. That is, a system is usually at equilibrium before the (single) pump pulse is absorbed in a traditional pump–probe experiment. In the 2DRR technique, the chemical reaction is initiated by a second pump pulse, which arrives while the reactant is undergoing coherent vibrational motions. Such wave packet motions correspond to Franck–Condon active modes for the optical transition, whereas the vibrational motions in dimension #2 are Franck–Condon active for the non-radiative transition (i.e., the chemical reaction). Notably, the most prominent vibrational modes must also be Franck–Condon active for light absorption of the product in order for the final laser pulse to induce signal emission. For this reason, 2DRR spectroscopy can be used to distinguish modes that project onto the reaction coordinate from so-called spectator modes which are Franck–Condon active only for light absorption. Physically interesting effects are revealed when wave packet motions along the reaction coordinate in dimension #1 affect properties of the product in dimension #2. For example, the application to the photodissociation reaction of triiodide that we focus on in this chapter will show that coherent motion of the reactant in dimension #1 directly influences the distribution of vibrational quanta in the product [22, 23].

The extraordinary photodissociation mechanism of triiodide has motivated numerous ultrafast spectroscopic studies [7, 8, 36–41]. Light absorption in the ultraviolet spectral range induces photodissociation on a timescale that is shorter than [7, 37] or comparable to [40, 42] the ~ 300-fs vibrational period of the symmetric stretching mode. Therefore, the photodissociation process acts as an impulse that initiates vibrational coherence in the bond stretching coordinate of the diiodide product. Earlier work has shown that the oscillatory transient absorption response of this system reflects symmetry breaking in the excited state [38], whereas the "chirp" in the waveform of the vibrational coherence represents time evolution of the bond strength during the reaction [40]. It should be noted that the reaction is more complex than originally thought; relatively recent work shows that distinct populations of free solvated diiodide and a contact fragment pair (diiodide and iodine) are produced by photodissociation [43]. 2DRR spectroscopy is insensitive to the contact radical pair because its vibrational motion is known to be overdamped [42].

This chapter is organized as follows. In Sect. 2, we discuss theoretical aspects of the 2DRR response. Experimental approaches are detailed in Sect. 3. Our application to the photodissociation reaction of triiodide is then reviewed in Sect. 4. Finally, we conclude by summarizing key findings and discussing future directions in Sect. 5.

2 2DRR Signal Generation Mechanism for a Photoinduced Reaction

The 2DRR response function possesses a large number of components that can be selectively enhanced by tuning laser pulses into the electronic resonances of the reactant and/or product. Reaction mechanisms, line-broadening dynamics, and/or anharmonicity can be probed, depending on the laser pulse sequence [26]. In this section, we define signal components associated with a system in which the reactant and product absorb light in separate spectral regions. The 2DRR response is discussed in the context of the photodissociation reaction of triiodide to facilitate discussion of the measurements presented below; however, the principles generalize to other ultrafast reactions such as energy and electron transfer.

2.1 2DRR Response Function for the Photodissociation Reaction of Triiodide

Photodissociation of triiodide is initiated by light absorption in the UV spectral range; however, as shown in Fig. 2, light absorption by diiodide dominates the visible region of the spectrum. A relative shift between the electronic resonances of triiodide and diiodide is convenient, because this means that the two species are readily distinguished in a two-color 2DRR experiment. This will be shown by the model developed in this section. The electronic resonance frequencies extracted from the absorbance spectra can be used to parameterize effective Hamiltonians for triiodide [23],

$$H_{\text{triiodide}} = |r\rangle\langle r| \sum_{m=0}^{\infty} |m\rangle\langle m| [E_r + E_m] + |r*\rangle\langle r*| \sum_{n=0}^{\infty} |n\rangle\langle n| [E_{r*} + E_n], \quad (1)$$

and diiodide,

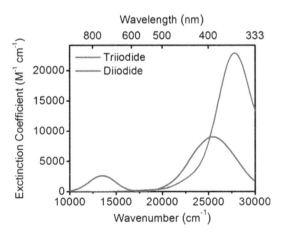

Fig. 2 Linear absorbance spectra of triiodide and diiodide in ethanol. The absorbance spectrum of triiodide is directly measured, whereas that of diiodide is derived from Ref. [44] because it is not stable in solution. The electronic resonance frequencies associated with this non-equilibrium state of diiodide are likely red-shifted from those displayed above. Displacement of the absorbance spectra of triiodide and diiodide facilitates detection of the pathway defined in Fig. 1 Reproduced from Guo et al. [23], with the permission of AIP Publishing

$$H_{\text{diiodide}} = |p\rangle\langle p| \sum_{m=0}^{\infty} |m\rangle\langle m| \left[E_p + E_m\right] + |p*\rangle\langle p*| \sum_{n=0}^{\infty} |n\rangle\langle n| \left[E_{p*} + E_n\right]. \qquad (2)$$

The indices r and p represent the ground electronic states of triiodide and diiodide, whereas an asterisk is used to denote the excited electronic state. The energies, E_r and E_r^* (E_p and E_p^*), correspond to the ground and excited states, respectively. The energy gaps, $E_r^* - E_r$ and $E_p^* - E_p$, govern the optical response and can be parameterized based on the linear absorbance spectrum in Fig. 2. The dummy indices, m and n, represent vibrational levels associated with the ground and excited electronic states.

The vibrational energy levels of the ground electronic state are described in the harmonic limit based on earlier spontaneous resonance Raman measurements [45]. The excited state potentials of both triiodide and diiodide are dissociative [8, 37]; however, the optical response is only sensitive to the gradient of the excited state potential at the Franck–Condon geometry. This is a general property of systems whose absorbance spectra do not exhibit vibronic progressions because of line broadening [46]. In the semiclassical perspective, this means that the wave packet initiated on the excited state potential energy surface does not return to the Franck–Condon geometry before electronic dephasing is complete (i.e., electronic dephasing is on the order of 10–20 fs in triiodide) [47]. Therefore, wave packet motions on the ground state potentials can be accurately simulated by introducing a bound excited state potential with a realistic slope at the Franck–Condon geometry. In Ref. [23], we used the cubic fitting parameters for the London–Eyring–Polanyi–Sato (LEPS) excited state potential energy surface of triiodide in ethanol [36, 45]. The potential energy minima of the excited state potentials (both triiodide and diiodide) are displaced to produce gradients consistent with models used in other work (see Appendix A) [42, 43].

Three types of 2DRR nonlinearities must be considered for signal interpretation: (1) both dimensions correspond to the triiodide reactant; (2) both dimensions correspond to the diiodide product; (3) the vibrational resonances of triiodide and diiodide appear in separate dimensions. These components of the response are understood by considering classes of terms in the fifth-order response function [1]. The Feynman diagrams presented in Fig. 3 show that the vibrational coherences detected in 2DRR spectra evolve in the two time intervals with even indices (t_2 and t_4). Vibrational levels associated with the electronic states ($r, r*, p, p*$) are specified by dummy indices (m, n, j, k, l, u, v, w). It is useful to consider that the experimentally controlled pulse delay times, τ_1 and τ_2, are good approximations to the time intervals between field-matter interactions, t_2 and t_4 (these time intervals are limited by vibrational dephasing). Electronic (or vibronic) coherences, which evolve in the intervals with odd indices (t_1, t_3, and t_5), dephase in 10–20 fs for solvated triiodide.

The first class of nonlinearities shown in Fig. 3 (terms 1–4) correspond to one-color (ultraviolet) experimental conditions and involve vibrational motions of only triiodide [22]. In Fig. 4, we illustrate how the Feynman diagram for term 1 can be viewed in an energy level representation. The Feynman diagrams associated with

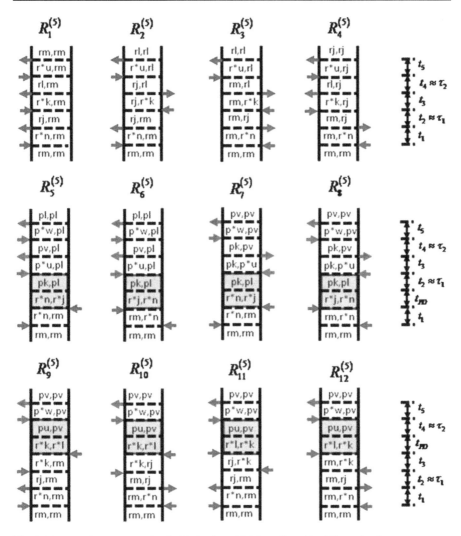

Fig. 3 Feynman diagrams associated with dominant 2DRR nonlinearities. Blue and red arrows represent pulses resonant with triiodide and diiodide, respectively (see Fig. 2). The indices r and r^* represent the ground and excited electronic states of the triiodide reactant, whereas p and p^* correspond to the diiodide photoproduct. Vibrational levels associated with these electronic states are specified by dummy indices (m, n, j, k, l, u, v, w). Each row represents a different class of terms: (1) both dimensions correspond to triiodide in terms 1–4; (2) both dimensions correspond to diiodide in terms 5–8; (3) vibrational resonances of triiodide and diiodide appear in separate dimensions in terms 9–12. The intervals shaded in blue represent a non-radiative transfer of vibronic coherence from triiodide to diiodide Reproduced from Guo et al. [23], with the permission of AIP Publishing

the other two classes of response functions incorporate both ultraviolet and visible laser pulses, thereby enabling observation of vibrational coherences in the diiodide product. The key point is that the photodissociation process transfers vibronic coherence from triiodide to diiodide either before (terms 5–8) or after (terms 9–12)

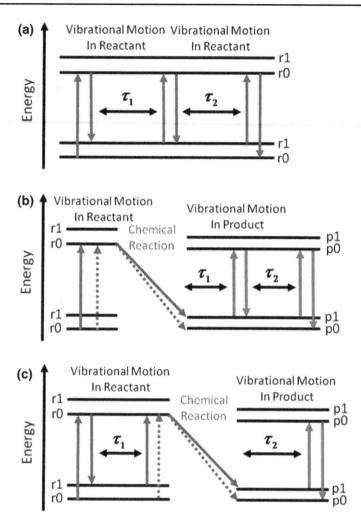

Fig. 4 Energy level representations associated with individual pathways in **a** term 1, **b** term 5, and **c** term 9. It is assumed that the chemical reaction is fast compared to the vibrational periods of the reactants and products. Solid and dashed arrows correspond to field-matter interactions with the ket and bra in the Feynman diagrams presented in Fig. 3

evolution of the vibrational coherence in t_2 (i.e., the delay time, τ_1). Most notably, vibrational resonances of triiodide and diiodide appear in separate dimensions of the 2DRR spectrum in terms 9–12; this is the pathway defined in Fig. 1.

In previous work [22, 23], response functions were written under the assumption that the laser pulses are short compared to the vibrational period but long compared to electronic dephasing (i.e., the snapshot limit) [23]. Both approximations are appropriate for the experiments we have conducted on triiodide. In these limits, the components of the fifth-order polarization consist of products of Lorentzian functions. Fourier transformation with respect to the experimentally controlled delay times, τ_1 and τ_2, yields a 2D spectrum in the conjugate dimensions, ω_1 and ω_2.

For example, the response function associated with the first term in Fig. 3 is given by

$$P_1^{(5)}(\omega_1, \omega_2) = -\frac{N\xi_{UV}^5|\mu_{r*r}|^6}{\hbar^5} \sum_{mnjklu} B_m \langle n \mid m \rangle \langle n \mid j \rangle \langle k \mid j \rangle \langle k \mid l \rangle \langle u \mid l \rangle \langle u \mid m \rangle,$$

$$\times L_{r*n,rm}(\omega_{UV})D_{rj,rm}(\omega_1)L_{r*k,rm}(\omega_{UV})D_{rl,rm}(\omega_2)L_{r*u,rm}(\omega_t) \tag{3}$$

where

$$L_{r*n,rm}(\omega) = \frac{1}{\omega - \omega_{r*r} - \omega_{nm} + i\Gamma_{r*r}}, \tag{4}$$

and

$$D_{rk,rm}(\omega) = \frac{2\Gamma_{vib} + 4\Lambda_{UV}}{\omega_{km}^2 + (\Gamma_{vib} + 2\Lambda_{UV})^2}\left(\frac{1}{\omega - \omega_{km} + i\Gamma_{vib}}\right). \tag{5}$$

The subscript of the electric field, UV, denotes an interaction with triiodide (VIS denotes an interaction with diiodide). The parameter N is the number density, B_m is a Boltzmann population, ξ_{UV} is the electric field amplitude, Λ_{UV} is the UV pulse width, μ_{r*r} is an electronic transition dipole for triiodide, ω_{r*r} is the electronic resonance frequency for triiodide, Γ_{r*r} is the electronic line width for triiodide, Γ_{vib} is the vibrational line width. The inner product, $\langle n \mid m \rangle$, represents a vibrational overlap integral, where the index on the left (right) represents the vibrational level of the excited (ground) electronic state [48]. The remaining 11 response functions are given in Appendix B.

The physical picture associated with the signal component of interest is illustrated in Fig. 5. The potential energy surfaces and wave packet widths in Fig. 5 are based on earlier work [36, 45]. The experiment begins when a laser pulse initiates vibrational motion in the ground electronic state of the triiodide reactant. This wave packet corresponds to the first dimension of the 2DRR spectrum. The second pulse promotes the wave packet to the excited state potential where asymmetric motion induces bond rupture. The photodissociation process is shorter than [7, 37] or comparable to [40, 42] the ~ 300-fs vibrational period of the bond stretching mode of diiodide. Therefore, the reaction initiates coherent wave packet motion in diiodide which can be detected in the second dimension of the 2DRR spectrum.

The possibility of observing vibrational resonances of reactants and products in separate dimensions of a 2DRR spectrum is the most interesting aspect of the model presented in this section. The ultrafast timescale of the reaction is the main prerequisite for observing correlations between reactants and products with 2DRR spectroscopy. The non-radiative transition (i.e., chemical reaction) can then impulsively excite vibrational motions in the product. In the language of a density matrix, the Feynman diagrams in Fig. 3 indicate that impulsive excitation transfers vibronic coherence from triiodide to diiodide (vibronic coherence transfer is highlighted in blue in Fig. 3). For example, in term 9, the density matrix element,

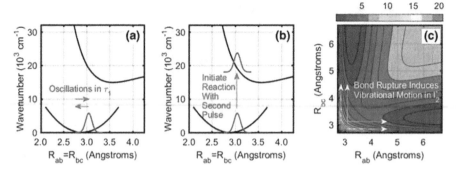

Fig. 5 The sequence of events associated with terms 9–12 and the pathway in Fig. 1. R_{ab} and R_{bc} denote the two bond lengths in triiodide and must be equal because wave packet motion in τ_1 occurs in the symmetric stretching coordinate. **a** The first pulse initiates a ground state wave packet in the symmetric stretching coordinate. Wave packet motion on the ground state potential energy surface is detected in the delay between the pump and repump laser pulses, τ_1. **b** Photodissociation of triiodide is initiated from a non-equilibrium geometry by the repump laser pulse, which separates the τ_1 and τ_2 delay times. The Raman spectrum of diiodide may then be detected by scanning the delay of the probe pulse, τ_2. **c** The repump pulse promotes the wave packet in triiodide to a steep portion of the excited state potential energy surface. Diiodide is produced by asymmetric motion on the excited state potential energy surface Adapted from Guo et al. [23], with the permission of AIP Publishing

$\rho_{r*k,r*l}$, is transformed into $\rho_{pu,pv}$ by the photodissociation process. Terms in the summation in which $u \neq v$ represent vibrational coherences of diiodide and appear in the second dimension of the 2DRR spectrum. Similarly, vibrational coherences of triiodide, $\rho_{rj,rm}$ ($j \neq m$), appear in the first dimension of the 2DRR spectrum. To further illustrate this point, we present an energy level representation associated with term 9 in Fig. 4.

Another unique aspect of terms 9–12 is that the third and fourth field-matter interactions occur on opposite sides of the density operator (see Fig. 3). Consequently, unlike terms 1–8, the two indices of the vibronic states in τ_2 differ from the two indices in τ_1. For example, in term 9, vibronic coherences associated with the first and second dimensions are $\rho_{rj,rm}$ ($j \neq m$) and $\rho_{pu,pv}$ ($u \neq v$), respectively. In contrast, the vibronic coherences associated with the first and second dimensions in term 1 are $\rho_{rj,rm}$ ($j \neq m$) and $\rho_{rl,rm}$ ($l \neq m$), respectively. Similarly, in term 5, the vibronic coherences associated with the first and second dimensions are $\rho_{pk,pl}$ ($k \neq l$) and $\rho_{pv,pl}$ ($v \neq l$), respectively. One index must remain unchanged in terms 1–8. The measurements and model calculations in Sect. 4 will show that retention of the indices m and l in terms 1 and 5 limits the vibrational resonances to two quadrants of the 2DRR spectrum. On the other hand, resonances can appear in all quadrants of the 2DRR spectrum for terms 9–12 because the vibronic indices associated with the two dimensions are independent. This is a convenient spectroscopic signature with which to establish vibronic coherence transfer.

2.2 Susceptibility to Third-Order Cascades

Cascades of third-order signals have been recognized as a serious experimental complication in off-resonant fifth-order Raman experiments conducted on pure liquids [49–54]. As indicated in Fig. 6, this artifact represents a process in which the four-wave mixing response on one molecule radiates a signal field that drives a four-wave mixing process on a second molecule. The second four-wave mixing signal (i.e., the cascaded signal) is radiated in the same direction as the desired fifth-order response and carries many of the same spectroscopic signatures. In contrast, all five field-matter interactions take place with an individual molecule in a genuine six-wave mixing process, thereby yielding a direct fifth-order signal field. Under off-resonant conditions, the cascaded response can be many orders of magnitude larger than the fifth-order 2D Raman signal. Success in measuring the off-resonant 2D Raman spectrum of CS_2 was finally achieved in 2002 when clever experimental geometries and detection schemes were developed to suppress the cascaded signal intensity [53, 55]. The possibility of conducting 2DRR spectroscopy without contributions from cascaded artifacts was first studied by our group in 2014 [22]. Cascades were more recently shown to be negligible in a related fifth-order method [33]. These investigations suggest potential for even higher-order nonlinear spectroscopy techniques such as the Raman echo [19, 56].

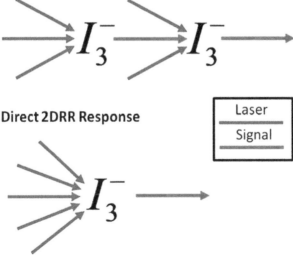

Fig. 6 Cascades of four-wave mixing signals generally dominate in off-resonant 2D Raman experiments. In the desired "direct" process, all 6 field-matter interactions take place with an individual molecule. The cascaded response is generally negligible in 2DRR spectroscopy because it involves 8 field-matter interactions, whereas the direct process only requires 6. Both types of nonlinearities are subject to the same selection rules under electronically resonant conditions

In our initial efforts, we ruled out cascaded signals with control experiments based on the signal phases, concentration dependence, and variation of the beam geometry [22, 24]. Our model calculations suggest that cascades dominate off-resonant experiments because the 2D Raman and cascaded responses are respectively forbidden and allowed for harmonic systems (i.e., they are subject to different selection rules). Of course, this is also why off-resonant 2D Raman experiments are useful for investigating anharmonicity in pure liquids [20]. The problem is that the cascaded signal is more intense because it relies on lower-order terms in the potential energy and/or polarizability. Tuning laser pulses into electronic resonance obviates such selection rules. Vibrational modes contribute to the 2DRR signals if they are Franck–Condon active (i.e., whether they are harmonic or not). Moreover, the 2DRR signal intensity is larger than that associated with cascades because cascades involve two more field-matter interactions (i.e., the cascade is higher-order in this sense). This can be proven by summing the interactions in Fig. 6.

While 2DRR spectroscopy is less susceptible to cascades than is off-resonant 2D Raman spectroscopy, the calculations presented in Fig. 7 suggest that it is still important to keep the optical density low (but not so low that the solvent response becomes comparable to that of the solute). Our model predicts that the 2DRR response will generally dominate in transmissive beam geometries; however, optically thick systems like molecular crystals, where a reflective geometry is required, will certainly be problematic. In Fig. 7, we present the ratio between the cascaded and 2DRR signal field magnitudes, $|E_{cas}(\omega_1, \omega_2)|/|E^{(5)}(\omega_1, \omega_2)|$, versus the dimensionless mode displacement, d, for the resonance at the fundamental frequency of the vibration. For our experimental conditions, the ratio is close to unity when the displacement is less than one but is extremely small for a

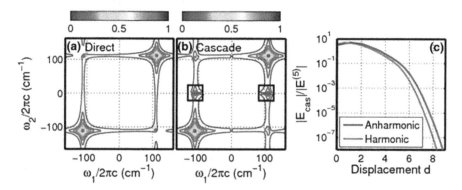

Fig. 7 Absolute values of the **a** direct fifth-order and **b** cascaded third-order signal magnitudes of triiodide at $\omega_1 = \omega_2 = \pm\ 112\ \mathrm{cm}^{-1}$ are computed with an empirical anharmonic excited state potential energy surface (see Appendix A). The ratio, $|E_{cas}(\omega_1, \omega_2)|/|E^{(5)}(\omega_1, \omega_2)|$, is computed using an empirical anharmonic model (blue) and a harmonic model (green) with equal ground and excited state frequencies ($112\ \mathrm{cm}^{-1}$) [45]. The features at $\omega_2 = 0\ \mathrm{cm}^{-1}$ (enclosed in boxes) in the cascaded signal spectrum represent imperfect subtraction of the non-oscillatory component of the signal (these are not vibrational resonances) Reproduced from Molesky et al [22], with the permission of AIP Publishing

displacement consistent with triiodide ($d = 7.0$). Reasons for this were discussed in Ref. [22]. The dependence on d is important to consider because Franck–Condon active modes in many larger molecules possess displacements that are much less than 1.0. Fortunately, the ratio, $|E_{cas}(\omega_1, \omega_2)|/|E^{(5)}(\omega_1, \omega_2)|$, also decreases linearly with the signal emission frequency.

The calculation in Fig. 7 employs a fairly large 1-mM concentration and deep ultraviolet detection wavelength of 267 nm. It is not necessary to carefully tune experimental parameters in triiodide because of its large displacement, d (Fig. 7c shows that the direct response dominates for displacements > 2). Larger polyatomic molecules generally possess much smaller displacements ($d < 1$). Fortunately, it is usually possible to detect signal emission at longer wavelengths for larger molecules, particularly for conjugated systems with delocalized excitations. For example, control experiments and model calculations show contributions from cascades to be negligible in 2DRR experiments conducted on myoglobin with a 0.2-mM concentration and 400-nm detection [24]. Harel and co-workers similarly found negligible contributions from cascades in a related 2D Raman experiment in which one of the pulses is pre-resonant with an electronic transition [34]. Our model calculations suggest that the cascaded signal intensity can reach and potentially exceed 10% of the total signal strength in a three-beam geometry with small crossing angles. For this reason, we recommend conducting control experiments based on the signal phase and sample concentration to rule out cascades when new systems are studied with all fifth-order Raman techniques (e.g., FSRS, RP-D4WM). Specialized beam geometries can be employed for challenging cases [53].

3 Experimental Methods

Approaches that we have used to conduct 2DRR spectroscopy differ in the bandwidths, frequencies, and geometries of the laser beams [22, 24, 56]. In this section, we describe how these aspects of a 2DRR pulse sequence can be used to selectively detect the three signal components defined in Sect. 2.1.

3.1 Pulse Sequences

In Fig. 8, we present pulse sequences that have been used to isolate each of the three signal components discussed in Sect. 2.1. The nonlinearity associated with terms 1–4 can be probed with a degenerate six-wave mixing configuration in which all pulses are resonant with the triiodide reactant. A resonant pump degenerate four-wave mixing approach is used to enhance signal components corresponding to terms 5–8. Finally, we find that terms 9–12 are most conveniently detected with a traditional pump–repump–probe setup. It should be noted that the fifth-order 2DRR nonlinearity is detected in each case even though the approaches differ in the number of laser pulses (i.e., more than one field-matter interaction can occur with each pulse).

Suppression of third-order cascades in electronically off-resonant 2D Raman experiments was achieved using carefully designed laser beam geometries in which

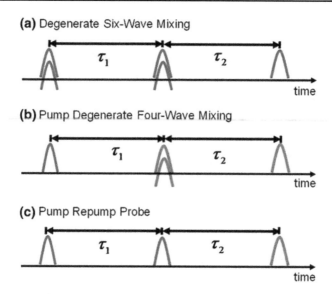

Fig. 8 Pulse sequences used to probe terms **a** 1–4, **b** 5–8, and **c** 9–12 in Fig. 3. In all cases, the signal is Fourier-transformed with respect to the delays, τ_1 and τ_2, to generate a 2D spectrum. Blue (deep- or near-ultraviolet) and red (visible) laser pulses represent resonance with triiodide and diiodide, respectively

phase-mismatch was induced in third-order signals [53, 55]. In contrast, the choice of 2DRR beam geometry is primarily governed by convenience because of negligible cascaded signal intensity. We conduct degenerate six-wave mixing 2DRR measurements using an interferometer originally developed by Mark Berg for a different type of experiment (see Figs. 8a, 9a) [57, 58]. In this approach, the first and second pulse-pairs, which are electronically resonant with triiodide, initiate vibrational coherences on the ground state potential which evolve during the delay times, τ_1 and τ_2. The advantage of the six-wave mixing geometry is that one field-matter interaction occurs with each of the incident beams. Therefore, the signal is generated in a background-free direction when the fifth pulse is diffracted from the holographic grating prepared by the first four pulses. The signal is weak but can be interferometrically detected using the sixth pulse as a reference field [59]. Experiments that take several hours can be conducted with the passive phase stabilization afforded by the diffractive optics approach [60].

The pump degenerate four-wave mixing pulse sequence is most convenient when the final four field-matter interactions are electronically resonant with the diiodide product (see Fig. 8b) [29, 30]. Inspired by Scherer and Blank [61–63], we have added a resonant pump to a third-order, diffractive optics-based transient grating setup (see Fig. 9b) [60, 64]. Interferometric detection is readily implemented in this design because the direction in which the signal is radiated is independent of the color of the pump pulse (pulse 1 in Fig. 8b) [59]. Because two field-matter interactions occur with pulse 1, a four-wave mixing background is produced by the three visible beams (beams 2–4 in Fig. 9b). Fortunately, this third-order background is negligible when the product of the reaction possesses a large extinction coefficient and the equilibrium sample is transparent at the detection wavelength. Fortunately,

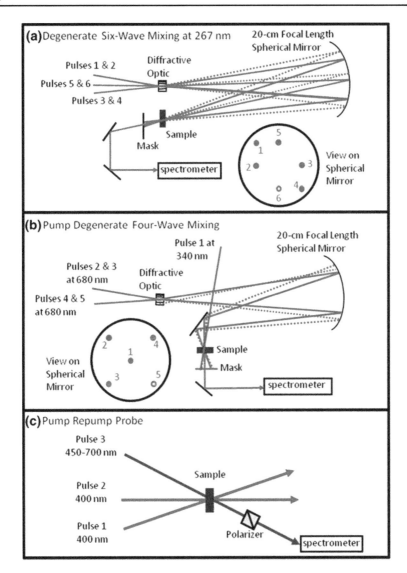

Fig. 9 Diffractive optics-based interferometers used for **a** degenerate six-wave mixing (with 267-nm beams) and **b** pump (340 nm) degenerate four-wave mixing (680 nm). The signal is radiated in the direction $k_s = k_1 - k_2 + k_3 - k_4 + k_5$ in panel (**a**). The wave vector of the signal is $k_s = k_2 - k_3 + k_4$ in panel (**b**) (i.e., the signal wave vector is independent of the direction and color of the 340-nm pump beam). In both cases, the passively phase-stabilized signal field is interferometrically detected using a local oscillator beam (beams 6 and 5 in panels (**a**) and (**b**), respectively). **c** Differential transmission of a probe pulse is detected in a pump–repump–probe geometry. The experimental setups in (**a**), (**b**), and (**c**) are used to implement the pulse sequences shown in Fig. 8a–c, respectively Adapted from Guo et al. [23] and Molesky et al [22], with the permission of AIP Publishing

triiodide does not absorb light in the visible spectral range at equilibrium, whereas diiodide possesses a significant extinction coefficient (see Fig. 2). Therefore, the 2DRR signal strength is at least an order of magnitude larger than the four-wave mixing background produced by the ethanol solvent when the visible pulses are tuned to wavelengths longer than 460 nm [23]. The experiment can therefore be conducted without a time-consuming differencing scheme.

Finally, we find that the traditional pump–repump–probe geometry shown in Fig. 9c is most convenient when the first four field-matter interactions are resonant with triiodide and the final two are resonant with diiodide (terms 9–12 in Fig. 3) [65]. The change in transmission of the continuum probe pulse induced by absorption of the pump and repump pulses is detected in this scheme. Both the pump and repump pulses must be chopped on a shot-to-shot basis to subtract a large background in this geometry, which makes this a fairly time consuming approach. Fortunately, the vibrational resonances of triiodide and diiodide are near 112 cm^{-1}, so a high density of points is not required. The 2D spectra in our published works were obtained by sampling a 40 by 40 grid of points with step sizes ranging from 60 to 75 fs [22, 23].

4 Application to the Photodissociation Reaction of Triiodide

Experiments conducted on the photodissociation reaction of triiodide show that it is indeed possible to isolate the three classes of nonlinearities defined in Sect. 2 [22, 56]. Triiodide is in many ways an ideal system with which to demonstrate such a decomposition of the response function. As discussed above, the electronic resonances of the reactant and product are well-separated, and the photodissociation reaction is faster than or comparable to the vibrational periods of the systems. In addition, the Franck–Condon active modes in the reactant and product possess extremely large displacements between ground and excited state potential energy minima. As a result, the 2DRR response dominates over the portion of the fifth-order nonlinearity associated with vibrational populations.

In Fig. 10, we present 2DRR spectra acquired using each of the three pulse sequences defined in Fig. 8. The all-UV pulse configuration yields vibrational resonances of triiodide in both dimensions, whereas the response of only diiodide is observed with the second pulse sequence. The 2DRR spectra in Fig. 10a, b differ slightly in both the resonance frequencies and line widths. At equilibrium, the vibrational mode frequencies of triiodide and diiodide differ by only a few wave numbers [7]; however, the resonance frequencies of diiodide detected by 2DRR reflect a highly non-equilibrium distribution of vibrational quanta. For this reason, the vibrational coherence frequency also depends on the detection wavelength due to anharmonicity [37]. A 2DRR spectrum in which resonances of triiodide and diiodide are displayed in separate dimensions is shown in Fig. 10c. The resonances are slightly off-diagonal at the 500-nm detection wavelength where we obtain the best signal-to-noise ratio. In agreement with earlier literature [37], we have also confirmed that the vibrational coherence frequency of diiodide decreases in ω_2 as

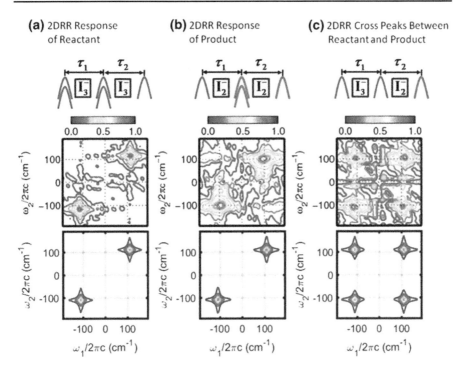

Fig. 10 Summary of 2DRR experiments conducted on triiodide: **a** the response of triiodide is detected in both dimensions (terms 1–4 in Fig. 3); **b** the response of the diiodide photoproduct is detected in both dimensions (terms 5–8 in Fig. 3); **c** the response of triiodide and diiodide are detected in separate dimensions (terms 9–12 in Fig. 3). Experimental and theoretical 2DRR spectra are presented in the second and third rows, respectively. Calculations in panels (**a**), (**b**), and (**c**) employ Eqs. (22), (23), and (24). Blue and red laser pulses represent wavelengths that are electronically resonant with triiodide and diiodide, respectively Adapted from Guo et al. [23], with the permission of AIP Publishing

the detection wavelength increases (i.e., the peaks move further off-diagonal at longer detection wavelengths) [23].

The measurement displayed in Fig. 10c is distinct from the other two in that resonances with equal intensities appear in all four quadrants of the 2DRR spectrum. The model calculations presented below the corresponding measurements in Fig. 10 suggest that this pattern of resonances originates in the sequences of field-matter interactions discussed in Sect. 2.1. Consider that the third and fourth field-matter interactions occur on the same sides of the Feynman diagrams in terms 1–8. Therefore, as discussed in Sect. 2.1, either the bra or ket must have the same vibrational state index in both τ_1 and τ_2. This constraint causes intensity to accumulate in the upper-right and lower-left quadrants of the spectrum [22, 23]. In contrast, the third and fourth field-matter interactions occur on opposite sides of the Feynman diagrams in terms 9–12; vibronic coherence transfer from triiodide to diiodide produces a coherence in τ_2 that is fully independent from that in τ_1. Consequently, peaks with equal intensities appear in all four quadrants. This unambiguous signature of vibronic coherence transfer should generalize to other photoinduced processes in which vibrational coherences are detected on the ground state potentials [23].

The observation of peaks in all four quadrants in Fig. 10c confirms that the process of interest has been isolated (i.e., the pathway defined in Fig. 1). We find that Fourier transforming the signal with respect to only τ_2 facilitates signal interpretation (see Fig. 11a) [23]. In this representation, oscillations of the signal in τ_1 represent recurrences of a vibrational wave packet on the ground state potential energy surface of the triiodide reactant, whereas the vibrational spectrum of diiodide is displayed in ω_2. The average vibrational frequency of diiodide is computed at each delay point using

$$\langle \omega_{\text{vib}}(\tau_1) \rangle = \frac{\int d\omega_2 S(\tau_1, \omega_2)\omega_2}{\int d\omega_2 S(\tau_1, \omega_2)}, \tag{6}$$

where $S(\tau_1, \omega_2)$ represents the absolute value of the signal displayed in Fig. 11a. The recurrences in $\langle \omega_{\text{vib}}(\tau_1) \rangle$ shown in Fig. 11b indicate that the frequency of wave packet motion of diiodide depends on the geometry of the triiodide at the "instant" the reaction is initiated by the repump laser pulse. Notably, this analysis is insensitive to the signal phase because it is carried out on the absolute value of the signal. Based on earlier work [37], we suggest that the non-equilibrium distribution of vibrational quanta in the diiodide product governs the vibrational coherence frequency of this anharmonic system.

Further insight is obtained by converting the delay time, τ_1, into the classical bond length of triiodide. This calculation makes use of the equilibrium bond length, the electronic dephasing time, the vibrational period of the symmetric stretching vibration, and the London–Eyring–Polanyi–Sato excited state potential energy surface of triiodide in ethanol [36, 45]. A classical view of the wave packet is justified by heterogeneity of the molecular geometry in the condensed phase environment. Wave packet dynamics can often be successfully described by solving

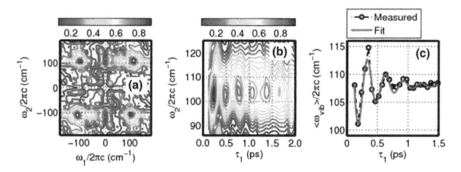

Fig. 11 **a** Fourier transforming the 2DRR signal with respect to only τ_2 reveals quantum beats of triiodide in the delay time, τ_1. **b** The average vibrational frequency of diiodide is computed at each delay point using the signal displayed in panel (**a**). **c** The delay, τ_1, is translated into the bond lengths of triiodide. The diagonal slant of the spiral suggests that a bond length displacement of 0.1 Å in triiodide induces a shift of approximately 6.8 cm^{-1} in the vibrational coherence frequency of diiodide Adapted from Guo et al. [23], with the permission of AIP Publishing

Schrödinger's equation using a Gaussian *ansatz* under such conditions [46, 47]. The trajectory shown in Fig. 11c indicates that turning points along triiodide's symmetric stretching coordinate are reached near delay times of 170 and 325 fs. Each revolution of the spiral represents a cycle of the symmetric stretching mode of triiodide, which possesses a 300-fs period. Vibrational dephasing causes the spiral to focus inward as time increases. Most importantly, the orientation of the spiral reflects positive correlation between the bond lengths of triiodide and the vibrational coherence frequency of diiodide.

The information provided by the 2DRR experiments is illustrated in Fig. 12. We conclude that the distribution of vibrational quanta in diiodide (as reflected by the vibrational coherence frequency) [37] depends on the bond lengths of triiodide at the "instant" the repump pulse induces photodissociation. Correlation between these two quantities can be understood within the framework of a traditional perturbative model in which vibrational overlap integrals weight state-specific paths from the initial (triiodide) to final (diiodide) states in the reaction [66]. In this perspective, the relative weights of the photodissociation channels are modulated by periodic changes in the bond lengths of the reactant in τ_1. That is, the distribution of vibrational quanta in diiodide reflects correlation between the amplitudes of state-specific photodissociation channels and the bond lengths of triiodide. 2DRR spectroscopy is specially equipped to provide such information about non-equilibrium processes because it possesses two electronic population times. In contrast, traditional third-order experiments such as transient absorption and photon echo spectroscopy must initiate reactions from equilibrium geometries [1].

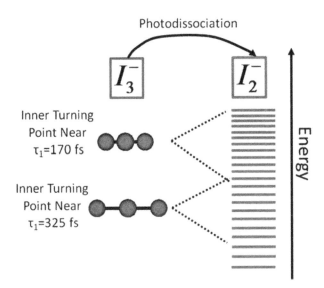

Fig. 12 2DRR experiments suggest correlation between the geometry of the wave packet in triiodide at the time of photodissociation and the distribution of vibrational quanta in the diiodide product. The vibrational coherence frequency of diiodide is smallest when the wave packet is at the inner turning point of the symmetric stretching mode. This information cannot be obtained from traditional pump–probe experiments where the reaction must be initiated from the equilibrium geometry of the system

5 General Applicability and Limitations of 2DRR Spectroscopy

The 2DRR signatures of vibronic coherence transfer discussed in this chapter are subject to several constraints. Firstly, the photoinduced reaction should generally be faster than the vibrational period (i.e., impulsive regime). It may be possible for a vibronic coherence transfer mechanism to promote phase coherence in an ensemble of systems where the reaction time is greater than or equal to the vibrational period; however, the prevalence of such mechanisms in chemical systems is not presently clear. Secondly, it is desirable to have reactants and products with non-overlapping electronic transitions. This constraint can be relaxed if the vibrational resonances of the reactant and product are well-separated and known beforehand, thereby allowing signal components to be assigned. Thirdly, the 2DRR method is sensitive to Franck–Condon active modes regardless of anharmonicity. Fourthly, the 2DRR method is limited to systems with modest optical densities because of the possibility of cascades. Optical densities less than 1.0 will usually be acceptable based on our calculations [25]. However, highly concentrated systems like molecular crystals will be problematic.

The constraints listed above are specific to studies of vibronic coherence transfer. In fact, 2DRR signals can be detected for any system with Franck–Condon active modes with relatively few constraints. Of course, cascades should be ruled out with control experiments as in any other fifth-order vibrational spectroscopy. Beyond that, the challenge is simply a matter of generating adequate signal strength. 2DRR experiments conducted on non-reactive systems yield information about line-broadening mechanisms. As in other 2D vibrational spectroscopy techniques, the 2DRR vibrational line shapes elongate with respect to the diagonal axis for inhomogeneously broadened transitions. We have observed such elongated 2DRR line shapes in the low-frequency vibrational modes of myoglobin [25, 26]. It should be noted that such 2DRR experiments are relatively insensitive to anharmonic couplings because Franck–Condon active modes contribute whether the modes are harmonic or not. However, in recent work, it has been predicted that sensitivity to anharmonic couplings can be achieved if the first pulse is pre-resonant with the electronic transition for a related 2D Raman technique [35].

The signatures of vibronic coherence transfer identified in this work generalize to any ultrafast process that can be photoinduced (e.g., energy or electron transfer). We suggest that electron transfer may be easier to study with 2DRR than is energy transfer. The reason is that one of the key requirements for fast energy transfer is spectral overlap between the donor's emission spectrum and the acceptor's absorption spectrum. It will not be possible to distinguish donor and acceptor modes by tuning incident laser beams in this situation, particularly if extremely broadband pulses are employed. Nonetheless, 2DRR studies of vibrational coherence transfer in systems like light harvesting proteins may be possible if the donor and acceptor possess readily distinguished vibrational mode frequencies. In contrast, applications to electron transfer reactions will be straightforward if the oxidized and/or reduced species have intense electronic transitions. For such systems, electron transfer can be investigated with an approach similar to that described here.

6 Concluding Remarks

In summary, 2DRR spectroscopy can be used to expose correlations between reactants and products in ultrafast photochemical reactions. These types of correlations cannot be revealed by traditional third-order methods such as transient absorption and photon echo spectroscopy in which reactions must be initiated from equilibrium conditions. Experimental tests show that 2DRR signals will generally be larger than cascaded nonlinearities for solutions with optical densities less than 1.0 [22, 24]. Model calculations suggest that Franck–Condon activity obviates the selection rules that promote artifacts under electronically off-resonant conditions [22]. Our experimental investigation of the photodissociation reaction of triiodide demonstrates that the information content of a 2DRR spectrum can be controlled by varying laser beam geometries and the colors of the pulses in the sequence. Results summarized in this article suggest that the bond length of the triiodide reactant is correlated to the distribution of vibrational quanta in the diiodide product [23].

The 2DRR technique can be used to probe the general vibronic coherence transfer process defined in Fig. 1 provided that the photoinduced reaction is faster than or comparable to the vibrational period(s). Thus, the 2DRR method can also be applied to vibronic coherence transfer in ultrafast energy and electron transfer transitions. For example, it may be interesting to apply 2DRR spectroscopy to ultrafast energy transfer processes in light harvesting proteins [14–16]. In these systems, the role played by vibronic effects in Frenkel exciton delocalization and energy transport has generated significant interest in recent years, and some issues remain unsettled [14, 67–72]. Notably, Harel and co-workers are making significant steps in this direction with a multidimensional method that is related to 2DRR spectroscopy [33, 34]. Electron transfer processes offer practical advantages that will facilitate decomposition of the 2DRR nonlinearity. Most importantly, the oxidized and/or reduced species in an electron transfer reaction often absorb light in different regions of the spectrum. Therefore, for some systems, the pathway defined in Fig. 1 may be readily isolated with multi-color pulse sequences (i.e., the same approach that we have taken for triiodide). In contrast, the electronic resonances of the donor and acceptor in an energy transfer transition usually possess significant overlap, thereby challenging an approach in which pathways are distinguished by tuning the wavelengths of the laser pulses.

Acknowledgements This work is supported by the National Science Foundation under CHE-0952439 and CHE-1504350.

Appendix A: Vibrational Hamiltonians

The present model assumes that both triiodide and diiodide possess two electronic levels and one nuclear coordinate with displaced ground and excited state potential energy minima. The anharmonic vibrational wave functions for the Franck–Condon active bond stretching mode of diiodide and the symmetric stretching coordinate of triiodide are generated using a Hamiltonian with the following form [73]

$$H_\alpha = \frac{\hbar\omega_{\alpha,\text{vib}}}{2}\left(2a^\dagger a + 1\right) + U_{3,\alpha}\left[a^\dagger a^\dagger a^\dagger + 3a^\dagger a^\dagger a + 3a^\dagger aa + aaa + 3a^\dagger + 3a\right],$$

(7)

where

$$U_{3,\alpha} = \frac{1}{3!\sqrt{2^3 m^3 \omega^3 \hbar^{-3}}}\left(\frac{d^3 V}{dq^3}\right)_0.$$

(8)

The wavefunctions are obtained by diagonalizing this Hamiltonian in a basis set of harmonic oscillators that includes states with up to the 40 vibrational quanta. Parameters of the vibrational Hamiltonian are given in Table 1. We use a notation in which α represents the molecule (r for triiodide or p for diiodide) and an asterisk indicates an electronically excited state.

The vibrational overlap integrals used to evaluate the response functions of diiodide are obtained using

$$\langle n \mid m \rangle = \sum_{jk} \varphi_{nk}\varphi_{mj}\langle k \mid j \rangle,$$

(9)

where φ_{nk} is the expansion coefficient for harmonic basis vector, k, and the anharmonic excited state vibrational wave function, n. Vibrational overlap integrals of triiodide are given by a different formula,

$$\langle n \mid m \rangle = \sum_{k} \varphi_{nk}\langle k \mid m \rangle,$$

(10)

because the ground and excited states are taken to be harmonic and anharmonic, respectively (see discussion in Sect. 2). In order to evaluate the overlap integrals, we assume a dimensionless displacement of 7.0 based on spontaneous Raman measurements for triiodide [45] and our earlier 2DRR study [22]. A displacement of 7.0 produces an excited state potential energy gradient of 225 eV/pm in diiodide which is identical to that associated with a previously employed exponential surface [42, 43].

Table 1 Parameters of Model Used to Compute 2DRR Spectra

Parameter[a]	Value
$\omega_{r*r}/2\pi c$	27,800 cm^{-1}
$\omega_{p*p}/2\pi c$[b]	13,300 cm^{-1} and 25,400 cm^{-1}
$\omega_{r,\text{vib}}/2\pi c = \omega_{r*,\text{vib}}/2\pi c$[c]	111 cm^{-1}
$\omega_{p,\text{vib}}/2\pi c = \omega_{p*,\text{vib}}/2\pi c$[c]	114 cm^{-1}
$U_{3,r}/hc$[c]	0 cm^{-1}
$U_{3,r*}/hc = U_{3,p}/hc = U_{3,p*}/hc$[c]	-1 cm^{-1}
$\Gamma_{r,\text{vib}}/c = \Gamma_{r*,\text{vib}}/c = \Gamma_{p,\text{vib}}/c = \Gamma_{p*,\text{vib}}/c$	10 cm^{-1}
$\Gamma_{r*r}/c = \Gamma_{p*p}/c$	2000 cm^{-1}
μ_{r*r}^{d}	2.3 D
μ_{p*p}^{d}	1.0 D
$\omega_{\text{UV}}/2\pi c$[e]	29,400 and 25,000 cm^{-1}
$\omega_{\text{VIS}}/2\pi c$[f]	14,705 and 18,900 cm^{-1}
$\omega_t/2\pi c$	$\omega_{\text{VIS}}/2\pi c$
$\Lambda_{\text{UV}}/c = \Lambda_{\text{VIS}}/c$	500 cm^{-1}

[a] The indices r and p represent triiodide and diiodide, respectively. Asterisks indicate the lowest-energy excited electronic states of the molecules

[b] The electronic resonance of diiodide that is probed depends on the experiment (see Sect. 3). In terms 1–4 and 9–12, the resonance is located at 25,400 cm^{-1}, whereas in terms 5–8 it is equal to 13,300 cm^{-1}

[c] Parameters of Eq. (7)

[d] Magnitudes of transition dipoles do not impact line shapes of simulated 2DRR spectra

[e] In Fig. 3, "pump" wave numbers are: 25,000 cm^{-1} for terms 1–4; 29,400 cm^{-1} for terms 5–8; 25,000 cm^{-1} for terms 9–12

[f] In Fig. 3, "probe" wave numbers are 14,705 cm^{-1} in terms 5–8 and 20,000 cm^{-1} in terms 9–12, respectively

Appendix B: Two-Dimensional Resonance Raman Signal Components

The Feynman diagrams presented in Fig. 3 include dummy indices for vibrational levels (m, n, j, k, l, u, v, w) associated with the ground and excited electronic states (r and $r*$ for triiodide, p and $p*$ for diiodide). Transition dipoles are written as product of integrals over electronic and nuclear degrees of freedom based on the Condon approximation. For example, an interaction that couples vibrational level m in the ground electronic state of the reactant and vibrational level n in the excited electronic state of the reactant contributes the product, $\mu_{r*r} \langle n \mid m \rangle$, to the response function, where μ_{r*r} is the electronic transition dipole and $\langle n \mid m \rangle$ is a vibrational overlap integral. We use a notation in which the excited state vibrational energy level is always written in the bra [48].

The first polarization component is given in Eq. (3). The remaining 11 polarization components are given by

$$P_2^{(5)}(\omega_1, \omega_2) = -\frac{N\xi_{\mathrm{UV}}^5 |\mu_{r*r}|^6}{\hbar^5} \sum_{mnjklu} B_m \langle n \mid m \rangle \langle n \mid j \rangle \langle k \mid m \rangle \langle k \mid l \rangle \langle u \mid j \rangle \langle u \mid l \rangle,$$
$$\times L_{r*n,rm}(\omega_{\mathrm{UV}}) D_{rj,rm}(\omega_1) L_{rj,r*k}(-\omega_{\mathrm{UV}}) D_{rj,rl}(\omega_2) L_{r*u,rl}(\omega_t)$$

(11)

$$P_3^{(5)}(\omega_1, \omega_2) = -\frac{N\xi_{\mathrm{UV}}^5 |\mu_{r*r}|^6}{\hbar^5} \sum_{mnjklu} B_m \langle n \mid m \rangle \langle n \mid j \rangle \langle k \mid j \rangle \langle k \mid l \rangle \langle u \mid m \rangle \langle u \mid l \rangle,$$
$$\times L_{rm,r*n}(-\omega_{\mathrm{UV}}) D_{rm,rj}(\omega_1) L_{rm,r*k}(-\omega_{\mathrm{UV}}) D_{rm,rl}(\omega_2) L_{r*u,rl}(\omega_t)$$

(12)

$$P_4^{(5)}(\omega_1, \omega_2) = -\frac{N\xi_{\mathrm{UV}}^5 |\mu_{r*r}|^6}{\hbar^5} \sum_{mnjklu} B_m \langle n \mid m \rangle \langle n \mid j \rangle \langle k \mid m \rangle \langle k \mid l \rangle \langle u \mid l \rangle \langle u \mid j \rangle,$$
$$\times L_{rm,r*n}(-\omega_{\mathrm{UV}}) D_{rm,rj}(\omega_1) L_{r*k,rj}(\omega_{\mathrm{UV}}) D_{rl,rj}(\omega_2) L_{r*u,rj}(\omega_t)$$

(13)

$$P_5^{(5)}(\omega_1, \omega_2) = -\frac{N\xi_{\mathrm{UV}}^2 \xi_{VIS}^3 |\mu_{r*r}|^2 |\mu_{p*p}|^4}{\hbar^5} \sum_{mnjkluvw} B_m \langle n \mid m \rangle \langle j \mid m \rangle \langle u \mid k \rangle \langle u \mid v \rangle \langle w \mid v \rangle \langle w \mid l \rangle,$$
$$\times L_{r*n,rm}(\omega_{\mathrm{UV}}) D_{pk,pl}(\omega_1) L_{p*u,pl}(\omega_{\mathrm{VIS}}) D_{pv,pl}(\omega_2) L_{p*w,pl}(\omega_t)$$

(14)

$$P_6^{(5)}(\omega_1, \omega_2) = -\frac{N\xi_{\mathrm{UV}}^2 \xi_{VIS}^3 |\mu_{r*r}|^2 |\mu_{p*p}|^4}{\hbar^5} \sum_{mnjkluvw} B_m \langle n \mid m \rangle \langle j \mid m \rangle \langle u \mid k \rangle \langle u \mid v \rangle \langle w \mid v \rangle \langle w \mid l \rangle,$$
$$\times L_{rm,r*n}(-\omega_{\mathrm{UV}}) D_{pk,pl}(\omega_1) L_{p*u,pl}(\omega_{\mathrm{VIS}}) D_{pv,pl}(\omega_2) L_{p*w,pl}(\omega_t)$$

(15)

$$P_7^{(5)}(\omega_1, \omega_2) = -\frac{N\xi_{\mathrm{UV}}^2 \xi_{VIS}^3 |\mu_{r*r}|^2 |\mu_{p*p}|^4}{\hbar^5} \sum_{mnjkluvw} B_m \langle n \mid m \rangle \langle j \mid m \rangle \langle u \mid l \rangle \langle u \mid v \rangle \langle w \mid k \rangle \langle w \mid v \rangle,$$
$$\times L_{r*n,rm}(\omega_{\mathrm{UV}}) D_{pk,pl}(\omega_1) L_{pk,p*u}(-\omega_{\mathrm{VIS}}) D_{pk,pv}(\omega_2) L_{p*w,pv}(\omega_t)$$

(16)

$$P_8^{(5)}(\omega_1, \omega_2) = -\frac{N\xi_{\mathrm{UV}}^2 \xi_{VIS}^3 |\mu_{r*r}|^2 |\mu_{p*p}|^4}{\hbar^5} \sum_{mnjkluvw} B_m \langle n \mid m \rangle \langle j \mid m \rangle \langle u \mid l \rangle \langle u \mid v \rangle \langle w \mid k \rangle \langle w \mid v \rangle,$$
$$\times L_{rm,r*n}(-\omega_{\mathrm{UV}}) D_{pk,pl}(\omega_1) L_{pk,p*u}(-\omega_{\mathrm{VIS}}) D_{pk,pv}(\omega_2) L_{p*w,pv}(\omega_t)$$

(17)

$$P_9^{(5)}(\omega_1,\omega_2) = -\frac{N\xi_{UV}^4\xi_{VIS}|\mu_{r*r}|^4|\mu_{p*p}|^2}{\hbar^5}\sum_{mnjkluvw}B_m\langle n\mid m\rangle\langle n\mid j\rangle\langle k\mid j\rangle\langle l\mid m\rangle\langle w\mid u\rangle\langle w\mid v\rangle,$$
$$\times L_{r*n,rm}(\omega_{UV})D_{rj,rm}(\omega_1)L_{r*k,rm}(\omega_{UV})D_{pu,pv}(\omega_2)L_{p*w,pv}(\omega_t)$$

(18)

$$P_{10}^{(5)}(\omega_1,\omega_2) = -\frac{N\xi_{UV}^4\xi_{VIS}|\mu_{r*r}|^4|\mu_{p*p}|^2}{\hbar^5}\sum_{mnjkluvw}B_m\langle n\mid m\rangle\langle n\mid j\rangle\langle k\mid m\rangle\langle l\mid j\rangle\langle w\mid u\rangle\langle w\mid v\rangle,$$
$$\times L_{rm,r*n}(-\omega_{UV})D_{rm,rj}(\omega_1)L_{r*k,rj}(\omega_{UV})D_{pu,pv}(\omega_2)L_{p*w,pv}(\omega_t)$$

(19)

$$P_{11}^{(5)}(\omega_1,\omega_2) = -\frac{N\xi_{UV}^4\xi_{VIS}|\mu_{r*r}|^4|\mu_{p*p}|^2}{\hbar^5}\sum_{mnjkluvw}B_m\langle n\mid m\rangle\langle n\mid j\rangle\langle k\mid m\rangle\langle l\mid j\rangle\langle w\mid u\rangle\langle w\mid v\rangle,$$
$$\times L_{r*n,rm}(\omega_{UV})D_{rj,rm}(\omega_1)L_{rj,r*k}(-\omega_{UV})D_{pu,pv}(\omega_2)L_{p*w,pv}(\omega_t)$$

(20)

$$P_{12}^{(5)}(\omega_1,\omega_2) = -\frac{N\xi_{UV}^4\xi_{VIS}|\mu_{r*r}|^4|\mu_{p*p}|^2}{\hbar^5}\sum_{mnjkluvw}B_m\langle n\mid m\rangle\langle n\mid j\rangle\langle k\mid j\rangle\langle l\mid m\rangle\langle w\mid u\rangle\langle w\mid v\rangle.$$
$$\times L_{rm,r*n}(-\omega_{UV})D_{rm,rj}(\omega_1)L_{rm,r*k}(-\omega_{UV})D_{pu,pv}(\omega_2)L_{p*w,pv}(\omega_t)$$

(21)

In the above polarization components, laser pulses with the subscripts UV and VIS are taken to interact with triiodide and diiodide, respectively.

For convenience, we further group the terms into three classes of signal fields under the assumption of perfect phase-matching conditions

$$E_{r,r}^{(5)}(\omega_1,\omega_2) = \left(\frac{i\omega_t l}{2\varepsilon_0 n(\omega_t)c}\right)\sum_{m=1}^{4}P_m^{(5)}(\omega_1,\omega_2),$$

(22)

$$E_{p,p}^{(5)}(\omega_1,\omega_2) = \left(\frac{i\omega_t l}{2\varepsilon_0 n(\omega_t)c}\right)\sum_{m=5}^{8}P_m^{(5)}(\omega_1,\omega_2),$$

(23)

and

$$E_{r,p}^{(5)}(\omega_1,\omega_2) = \left(\frac{i\omega_t l}{2\varepsilon_0 n(\omega_t)c}\right)\sum_{m=9}^{12}P_m^{(5)}(\omega_1,\omega_2).$$

(24)

Here, the two subscripts of the signal fields represent sensitivity to the triiodide reactant (subscript r) and diiodide product (subscript p) in the two frequency dimensions, ω_1 and ω_2.

References

1. Mukamel S (1995) Principles of nonlinear optical spectroscopy. Oxford University Press, New York
2. Nitzan A (2006) Chemical dynamics in condensed phases. Oxford University Press, Oxford
3. Valkunas L, Abramavicius D, Mančal T (2013) Molecular excitation dynamics and relaxation: quantum theory and spectroscopy. Wiley-VCH, Weinheim
4. Vos MH, Lambry J-C, Robles SJ, Youvan DC, Breton J, Martin J-L (1991) Proc Natl Acad Sci 88:8885
5. Vos MH, Rappaport F, Lambry J-C, Breton J, Martin J-L (1993) Nature 363:320
6. Peteanu LA, Schoenlein RW, Wang H, Mathies RA, Shank CV (1993) Proc Natl Acad Sci 90:11762
7. Banin U, Kosloff R, Ruhman S (1993) Isr J Chem 33:141
8. Benjamin I, Banin U, Ruhman S (1993) J. Chem. Phys. 98:8337
9. Zhu L, Sage JT, Champion PM (1994) Science 266:629
10. Zewail AH (2000) J Phys Chem A 104:5560
11. Bixon M, Jortner J (1997) J Chem Phys 107:1470
12. Wynne K, Reid GD, Hochstrasser RM (1996) J Chem Phys 105:2287
13. Song Y, Clafton SN, Pensack RD, Kee TW, Scholes GD (2014) Nat Commun 5:4933
14. Fuller FD, Pan J, Gelzinis A, Butkus V, Senlik SS, Wilcox DE, Yocum CF, Valkunas L, Abramavicius D, Ogilvie JP (2014) Nat Chem 6:706
15. Mohseni M, Omar Y, Engel GS, Plenio MB (2014) Quantum effects in biology. Cambridge University Press, Cambridge
16. Chenu A, Scholes GD (2015) Annu Rev Phys Chem 66:69
17. Jonas DM (2003) Annu Rev Phys Chem 54:425
18. Ogilvie JP, Kubarych KJ (2009) Adv At Mol Opt Phys 57:249
19. Loring RF, Mukamel S (1985) J Chem Phys 83:2116
20. Tanimura Y, Mukamel S (1993) J Chem Phys 99:9496
21. Tanimura Y, Okumura K (1996) J Chem Phys 106:2078
22. Molesky BP, Giokas PG, Guo Z, Moran AM (2014) J Chem Phys 114:114202
23. Guo Z, Molesky BM, Cheshire TP, Moran AM (2015) J Chem Phys 143:124202
24. Molesky BM, Guo Z, Moran AM (2015) J Chem Phys 142:212405
25. Molesky BM, Guo Z, Cheshire TP, Moran AM (2016) J Chem Phys 145:034203
26. Molesky BM, Guo Z, Cheshire TP, Moran AM (2016) J Chem Phys 145:180901
27. McCamant DW, Kukura P, Yoon S, Mathies RA (2004) Rev Sci Instrum 75:4971
28. Kukura P, McCamant DW, Mathies RA (2007) Annu Rev Phys Chem 58:461
29. Takeuchi S, Ruhman S, Tsuneda T, Chiba M, Taketsugu T, Tahara T (2008) Science 322:1073
30. Kraack JP, Wand A, Buckup T, Motzkus M, Ruhman S (2013) Phys Chem Chem Phys 15:14487
31. Wang Y, Liu W, Tang L, Oscar B, Han F, Fang C (2013) J Phys Chem A 117:6024
32. Hoffman DP, Mathies RA (2016) Acc Chem Res 49:616
33. Hutson WO, Spencer AP, Harel E (2016) J Phys Chem Lett 7:3636
34. Spencer AP, Hutson WO, Harel E (2017) Nat Commun 8:14732
35. Harel E (2017) J Chem Phys 146:154201
36. Banin U, Kosloff R, Ruhman S (1994) Chem Phys 183:289
37. Kühne T, Vöhringer P (1996) J Chem Phys 105:10788
38. Gershgoren E, Gordon E, Ruhman S (1997) J Chem Phys 106:4806
39. Kühne T, Küster R, Vöhringer P (1998) Chem Phys 233:161
40. Hess S, Bürsing H, Vöhringer P (1999) J Chem Phys 111:5461
41. Yang T-S, Chang M-S, Hayashi M, Lin SH, Vöhringer P, Dietz W, Scherer NF (1999) J Chem Phys 110:12070
42. Nishiyama Y, Terazima M, Kimura Y (2012) J Phys Chem B 116:9023
43. Baratz A, Ruhman S (2008) Chem Phys Lett 461:211
44. Herrmann V, Krebs P (1995) J Phys Chem 99:6794
45. Johnson AE, Myers AB (1995) J Chem Phys 104:3519
46. Myers AB (1995) In: Myers AB, Rizzo TR (eds) Laser techniques in chemistry, vol 23. John Wiley & Sons, New York
47. Heller EJ (1981) Acc Chem Res 14:368
48. Myers AB, Mathies RA, Tannor DJ, Heller EJ (1982) J Chem Phys 77:3857
49. Ivanecky JE III, Wright JC (1993) Chem Phys Lett 206:437

50. Ulness DJ, Kirkwood JC, Albrecht AC (1998) J Chem Phys 108:3897
51. Jansen TIC, Snijders JG, Duppen K (2001) J Chem Phys 114:109210
52. Blank DA, Kaufman LJ, Fleming GR (1999) J Chem Phys 111:3105
53. Kubarych KJ, Milne CJ, Lin S, Astinov V, Miller RJD (2002) J Chem Phys 116:2016
54. Mehlenbacher R, Lyons B, Wilson KC, Du Y, McCamant DW (2009) J Chem Phys 131:244512
55. Kaufman LJ, Heo J, Ziegler LD, Fleming GR (2002) Phys Rev Lett 88(207402):1
56. Berg M, Vanden Bout DA (1997) Acc Chem Res 30:65
57. van Veldhoven E, Khurmi C, Zhang X, Berg MA (2007) Chem Phys Chem 8:1761
58. Berg MA (2012) Adv Chem Phys 150:1
59. Lepetit L, Chériaux G, Joffre M (1995) J Opt Soc Am B 12:2467
60. Goodno GD, Dadusc G, Miller RJD (1998) J Opt Soc Am B 15:1791
61. Underwood DF, Blank DA (2005) J Phys Chem A 109:3295
62. Moran AM, Nome RA, Scherer NF (2007) J Chem Phys 127(184505):1
63. Park S, Kim J, Scherer NF (2012) Phys Chem Chem Phys 14:8116
64. Moran AM, Maddox JB, Hong JW, Kim J, Nome RA, Bazan GC, Mukamel S, Scherer NF (2006) J Chem Phys 124(194904):1
65. Busby E, Carroll EC, Chinn EM, Chang L, Moulé AJ, Larsen DS (2011) J Phys Chem Lett 2:2764
66. Band YB, Freed KF (1977) J Chem Phys 67:1462
67. Hennebicq E, Beljonne D, Curutchet C, Scholes GD, Silbey RJ (2009) J Chem Phys 130(214505):1
68. Roden J, Schulz G, Eisfeld A, Briggs J (2009) J Chem Phys 131:044909
69. Womick JM, Moran AM (2011) J Phys Chem B 115:1347
70. Chenu A, Chrisensson N, Kauffmann HF, Mančal T (2013) Sci Rep 3:2029
71. O'Reilly EJ, Olaya-Castro A (2014) Nat Commun 5:3012
72. Singh VP, Westberg M, Wang C, Dahlberg PD, Gellen T, Gardiner AT, Cogdell RJ, Engel GS (2015) J Chem Phys 142:212446
73. Moran AM, Dreyer J, Mukamel S (2003) J Chem Phys 118:1347

Top Curr Chem (Z) (2018) 376:6
https://doi.org/10.1007/s41061-018-0185-4

REVIEW

Two-Dimensional Spectroscopy at Terahertz Frequencies

Jian Lu[1] · **Xian Li**[1] · **Yaqing Zhang**[1] · **Harold Y. Hwang**[1] ·
Benjamin K. Ofori-Okai[1] · **Keith A. Nelson**[1]

Received: 31 August 2017 / Accepted: 5 January 2018 / Published online: 23 January 2018
© Springer International Publishing AG, part of Springer Nature 2018

Abstract Multidimensional spectroscopy in the visible and infrared spectral ranges has become a powerful technique to retrieve dynamic correlations and couplings in wide-ranging systems by utilizing multiple correlated light-matter interactions. Its extension to the terahertz (THz) regime of the electromagnetic spectrum, where rich material degrees of freedom reside, however, has been progressing slowly. This chapter reviews some of the THz-frequency two-dimensional (2D) spectroscopy techniques and experimental results realized in recent years. Examples include gas molecule rotations, spin precessions in magnetic systems, and liquid molecular dynamics studied by 2D THz or hybrid 2D THz-Raman spectroscopy techniques. The methodology shows promising applications to different THz-frequency degrees of freedom in various chemical systems and processes.

Keywords Terahertz · Gas-phase molecular rotations · Liquid-phase molecular dynamics · Magnetic resonances

1 Introduction

Recent years have witnessed increasing interest in two-dimensional spectroscopy in many regions of the electromagnetic spectrum. Two-dimensional infrared (2D IR) vibrational spectroscopy [1–4] is a powerful technique for studying structural dynamics and correlations between coupled molecular motions in chemical and

---

Chapter 7 was originally published as Lu, J., Li, X., Zhang, Y., Hwang, H. Y., Ofori-Okai, B. K. & Nelson, K. A. Top Curr Chem (Z) (2018) 376: 6. https://doi.org/10.1007/s41061-018-0185-4.

✉ Keith A. Nelson
kanelson@mit.edu

[1] Department of Chemistry, Massachusetts Institute of Technology, Cambridge, MA 02139, USA

biologic systems such as water [5], proteins [6], and DNA [7]. With visible light, multidimensional electronic spectroscopies [8] have been applied to probe the high-order correlations of excitons in quantum wells [9, 10] and light-harvesting complexes [11–13]. Recently, 2D electronic spectroscopy has been extended across the visible spectrum and into the ultraviolet regime [8, 14]. In the most common 2D spectroscopy experiments, three resonant interactions between incident light fields and the sample generate a nonlinear signal field that is fully characterized through heterodyne mixing with a reference or local-oscillator field and either time- or frequency-resolved measurement of the superposition. By varying the relative time delays and relative phases between incident fields and measuring the effects on the signal field, dynamical information about sample coherences and populations, couplings between different modes, and a great deal more can be learned. Variation of the time interval between two phase-coherent incident fields provides a second dimension along which the nonlinear signal can be measured, usually displayed along a second frequency axis after Fourier transformation of the signal as a function of the relative (inter-pulse) delay. The results often reveal features that remain hidden in conventional 1D linear spectra.

The terahertz (THz) regime of the electromagnetic spectrum is overlapped with a rich variety of material degrees of freedom including rotations of polar gas molecules, lattice vibrations in solids, molecular dynamics of liquids, spin dynamics in materials with magnetic order and/or high magnetic anisotropy, and many others [15]. Linear THz time-domain spectroscopy allows access to both the amplitude and phase of the THz electric field after its interaction with the sample, from which the real and imaginary parts of the dielectric function can be retrieved simultaneously without referring to the Kramers-Kronig relation [15]. Linear THz time-domain spectroscopy has found wide-ranging applications in the characterization of molecular and material systems relevant to chemistry, biology, and physics. In such experiments, THz pulses are typically generated by femtosecond laser pulses that either undergo optical rectification (OR) in nonlinear optical crystals, such as zinc telluride (ZnTe), gallium phosphate (GaP), or gallium arsenide (GaAs), or that generate ultrafast photocurrent in biased photoconductive antennas [15]. The achievable electric field strength is generally less than 10 kV/cm [16], which in most samples precludes any significant nonlinear light-matter interactions. The absence of strong tabletop THz sources placed early THz spectroscopy exclusively in the linear response region [17–19].

With the advent of strong tabletop THz sources, the subfield of nonlinear THz spectroscopy has been growing rapidly in the past decade. Nowadays, intense single-cycle THz pulses spanning the low THz region between 0.1 and 3 THz with pulse energies of several μJ and electric field strengths as high as 1 MV/cm (corresponding to magnetic field strengths of 0.33 T) have been routinely generated by OR of Ti:sapphire laser pulses in lithium niobate ($LiNbO_3$) crystals [20, 21]. THz pulses of 10 MV/cm electric fields (corresponding to multi-tesla magnetic fields), higher center frequencies, and broader bandwidths have been recently demonstrated by OR of near-IR fs pulses with center wavelengths between 1.2 and 1.5 μm in novel organic nonlinear optical crystals [22–24]. The rapid development of intense THz sources has enabled coherent THz spectroscopy and control over electronic, orbital, lattice, and spin degrees of freedom in various systems [25, 26].

Compared to the well-established research areas of multidimensional spectroscopy in the visible and IR frequency ranges, the methodology and application of 2D THz spectroscopy are still in a nascent stage. In recent years, there have been developments of 2D THz techniques that led to the first demonstrations of 2D multi-THz and 2D THz spectroscopies in the study of mostly nonresonant electronic and some lattice vibrational responses with strong nonlinearities in condensed-matter systems. Examples include carrier dynamics in graphene [27], correlations between carriers and phonons in quantum wells [27, 28], THz nonlinear frequency mixing in LiNbO$_3$ [29], and coherent cyclotron resonance nonlinear mixing in a 2D electron gas [30]. Demonstrations of 2D THz spectroscopy more directly relevant to chemistry include 2D THz rotational spectroscopy of molecules in the gas phase [31, 32], 2D THz vibrational spectroscopy of phonons in a semiconductor [33, 34], and 2D THz magnetic resonance spectroscopy of magnons [35, 36]. In the first of these, THz photon echo signals have been revealed for the first time in the gas phase, and the full set of THz third-order ($\chi^{(3)}$) nonlinear responses has been mapped into 2D spectra. In the second example, the first demonstration of 2D THz spectroscopy using three separate pulses has been realized, which reveals strong nonresonant nonlinearity beyond $\chi^{(3)}$ responses associated with the two-phonon quantum coherences in the system. The last example, though conducted on magnons in an antiferromagnetic crystal, is directly relevant for the development of 2D THz electron paramagnetic resonance (EPR) spectroscopy and the study of chemical and biologic systems that have spin resonances in the THz regime. THz pulses combined with optical excitation and detection through nonresonant Raman processes have enabled demonstrations of hybrid 2D THz-Raman spectroscopies. Hybrid 2D THz-Raman studies have revealed couplings among the intramolecular vibrational modes in halogenated liquids [37, 38], and the intermolecular dynamics of the hydrogen-bond network in water [39] and aqueous salt solutions [40] through THz photon echoes.

This chapter is organized as follows. In Sect. 2, the basic techniques for intense THz pulse generation and time-domain signal detection are reviewed. The methods involving 2D THz spectroscopy in collinear geometry and typical pulse sequences are discussed. As there are key differences between the techniques used in 2D THz and 2D IR or visible spectroscopies, we will elaborate the 2D THz methods in detail and frequently make comparisons with their analogs in the IR and visible. In what follows, we will review 2D THz rotational spectroscopy, 2D THz vibrational spectroscopy, 2D THz-Raman spectroscopies and 2D THz magnetic resonance spectroscopy in detail. Lastly, a summary and a brief outlook of the directions to which 2D THz spectroscopy can lead are presented.

2 Methods

2.1 THz Pulse Generation and Signal Detection

As 2D spectroscopies involve nonlinear signal generation and detection, experiments require strong excitation sources and sensitive detection schemes. Intense THz pulse generation mainly relies on OR of strong fs laser pulses in a nonlinear optical crystal. The typical picosecond (ps) to sub-ps durations of THz pulses allow detection of their electric field profiles in the time domain, gated by fs laser pulses.

2.1.1 Intense THz Pulse Generation Methods

OR is a second-order ($\chi^{(2)}$) nonlinear process. Typically, a broadband fs laser pulse at 800 nm from a Ti:sapphire amplifier is used to pump a nonlinear optical crystal in collinear geometry. Field components at nearby frequencies, $E(\omega_1)$ and $E(\omega_2)$, within the laser pulse bandwidth undergo difference-frequency mixing and generate electromagnetic radiation centered at $|\omega_1 - \omega_2|$, which is in the THz frequency range. In the perturbative regime, the THz-frequency fields are radiated by a $\chi^{(2)}$ polarization $P^{(2)}$ [41] given by

$$P^{(2)}(\omega_{THz}) = \chi^{(2)}(\omega_{THz};\omega_1, -\omega_2)E(\omega_1)E^*(\omega_2). \tag{1}$$

Similar to other nonlinear processes [41], the $\chi^{(2)}$ nonlinear coefficient of the crystal and the phase matching of the optical pump and the generated THz pulses in the crystal are crucial for efficient generation of the THz fields. The phase-matching condition for THz generation by OR with collinear optical and THz wavevectors is given by a scalar equation,

$$k(\omega_{THz}) = k(\omega_1) - k(\omega_2) \approx k(\omega_2) + dk(\omega_2)/d\omega \cdot \omega_{THz} - k(\omega_2), \tag{2}$$

where k is the wavevector magnitude, ω the angular frequency, and $\omega_{THz} = |\omega_1 - \omega_2|$ is satisfied. From the perspective of photons, $\omega_{THz} = |\omega_1 - \omega_2|$ dictates photon energy conservation and Eq. (2) dictates photon quasi-momentum conservation. The index matching condition can be obtained from Eq. (2) and is given by

$$n_{THz} = n_{op}^g, \tag{3}$$

where $n_{THz} = c \cdot k(\omega_{THz})/\omega_{THz}$ is the THz refractive index and $n_{op}^g = c \cdot dk(\omega_2)/d\omega$ the optical group index in the nonlinear optical crystal (c is the speed of light in vacuum).

Inorganic nonlinear crystals including ZnTe, GaP, and GaAs have optimal index matching between 800 nm and THz pulses. Phase matching can be satisfied in a simple collinear geometry, and the experimental implementation is hence straightforward. But these crystals have relatively small $\chi^{(2)}$ nonlinear coefficients and large-area crystals are typically required for strong THz pulse generation [16]. LiNbO$_3$ has a large $\chi^{(2)}$ nonlinear coefficient, but the large phase mismatch between 800 nm and THz pulses in LiNbO$_3$ leads to inefficient THz generation in collinear geometry. To circumvent the phase-matching problem and utilize the large $\chi^{(2)}$ coefficient in LiNbO$_3$, a non-collinear phase-matching scheme involving tilting of the 800-nm pulse intensity front to match the THz wavefront has been developed [20, 21, 42, 43], which enabled the rapid proliferation of tabletop nonlinear THz spectroscopy experiments [25].

The geometry for the tilted-pulse-front method is shown schematically in Fig. 1a. The broadband fs pump pulses from the Ti:sapphire laser with a flat intensity front are incident onto a grating. The first-order diffraction, which has a tilted intensity front, is collected and imaged into a specially cut LiNbO$_3$ crystal. The

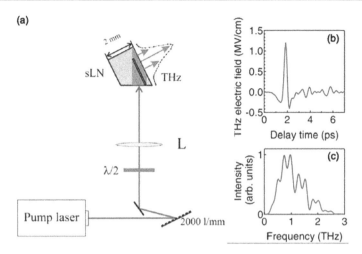

Fig. 1 **a** Schematic illustration of the setup for THz generation in a stoichiometric LiNbO$_3$ (sLN) crystal using the tilted-pulse-front method. Single-cycle THz electric field waveform (**b**) and spectrum (**c**) generated in LiNbO$_3$. From [21, 43]

demagnification of the imaging lens adjusts the tilt angle of the intensity front for optimal phase matching. Once achieved, the wave front of the generated THz pulses is parallel to the intensity front of the pump pulses. They travel at the same velocity along the direction normal to the pump intensity front, and the THz field amplitude becomes coherently enhanced. As a result of this non-collinear phase matching, strong THz pulses with μJ energies can be achieved at the output facet of the LiNbO$_3$ crystal. Typical pump-to-THz energy conversion efficiency in LiNbO$_3$ using the tilted-pulse-front method is on the order of 0.1% at room temperature. The generated THz pulses are collimated and focused to a submillimeter spot that can result in single-cycle THz electric fields of 1 MV/cm strength and bandwidth ranging from 0.1 to 3 THz [21]. The focused THz fields are used to excite and interrogate the sample.

Strong THz pulses can also be generated from organic nonlinear optical crystals with high $\chi^{(2)}$ nonlinear coefficients such as 2-(3-(4-hydroxystyryl)-5,5-dimethylcyclohex-2-enylidene)malononitrile (OH1) and 4-*N*,*N*-dimethylamino-4′-*N*′-methylstilbazolium 2,4,6-trimethylbenzenesulfonate (DSTMS) [23]. Collinear phase matching between THz and pump pulses is achieved when the crystals are pumped by near-IR pulses centered between 1.1 and 2 μm, typically from a near-IR optical parametric amplifier (OPA). THz pulses with μJ energies have been demonstrated in such organic crystals [44]. Due to the higher center frequencies of the generated THz pulses compared to those from LiNbO$_3$, the THz pulses can be focused to smaller diffraction-limited spots, thus resulting in electric fields on the order of tens of MV/cm as demonstrated recently [22, 44]. Examples of intense THz pulses generated in organic crystals DSTMS and OH1 are shown in Fig. 2.

In addition, gallium selenide (GaSe) is a nonlinear medium that allows the generation of multi-cycle THz radiation at higher frequencies (e.g., around 20 THz) by

Fig. 2 THz electric field profiles (**a**) and spectra (**b**) generated by OR in organic crystals DSTMS and OH1. From [22]

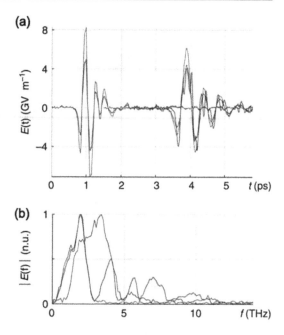

OR with collinear phase matching [27, 34]. These three THz generation methods are used in the works reviewed in this chapter. There are also other methods for strong and broadband THz pulse generation using either fs lasers in tabletop setups [45, 46] or relativistic electron bunches in accelerators [47, 48], which will not be discussed here. We note that the THz electric field profile is determined from the optical pump pulse intensity profile, so the THz fields used for 2D THz spectroscopy are inherently carrier-envelope phase-stable. This simplifies measurement of the full THz field as described below.

2.1.2 THz Time-Domain Detection by Electro-Optic Sampling

THz pulses generated as described above typically have sub-ps durations and can be sampled in the time domain using optical gate pulses of considerably shorter duration. For that purpose, a weak portion of the fs pulse used for THz generation is used to detect the THz electric field profile in a nonlinear optical crystal such as ZnTe and GaP via the electro-optic (EO) Pockels effect, i.e., electro-optic sampling (EOS) [49, 50]. As the THz generation and detection both originate from the same laser pulse, timing jitter between the gate and THz pulses is minimal. The same measurement method is used for the nonlinear THz signals in 2D THz spectroscopy measurements. Because the time-dependent THz electric field profile is measured in this manner, there is no need for heterodyne detection of the signal as typically used in 2D IR and visible spectroscopies.

The geometry for EOS is shown schematically in Fig. 3. The THz pulse and optical gate pulse, both with linear polarization, are focused into the EO crystal. Without the presence of the THz electric field, the gate pulse does not experience any

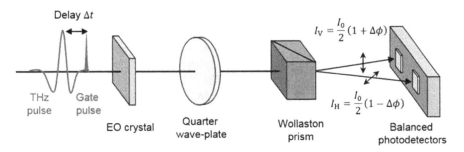

Fig. 3 Schematic illustration of the optical setup for THz electric field profile characterization via EOS. The phase retardation $\Delta\phi$ of the optical gate pulse in the EO crystal is converted to intensity modulations of the horizontal and vertical polarization components I_V and I_H by a quarter wave-plate and a Wollaston prism. The difference between I_V and I_H is detected by the balanced photodetectors. Sweeping the delay Δt between the THz and gate pulses, the THz electric field waveform is mapped out

birefringence in the EO crystal and remains linearly polarized. It is subsequently transmitted through a quarter wave-plate and becomes circular polarized. The horizontal and vertical polarization components (I_H and I_V) of the circularly polarized gate pulse are separated by a Wollaston prism (polarizing beamsplitter) and the difference (zero without the THz electric field) in their intensities is detected by a pair of balanced photodetectors. When the THz pulse is overlapped with the gate pulse in time, the THz electric field biases the EO crystal, causing a rotation $\Delta\phi$ of the index ellipsoid of the crystal and resulting in a transient birefringence. After transmission through the EO crystal and the quarter wave-plate, the gate pulse becomes elliptically polarized. The difference between I_H and I_V is linearly proportional to the THz electric field in the limit of small $\Delta\phi$ and is measured by the balanced photodetectors. The polarity of the THz electric field determines the sign of $\Delta\phi$ and hence the sign of the detected signal. Sweeping the time delay Δt between the THz pulse and the gate pulse, one can map out the electric field profile of the THz pulse. A numerical Fourier transformation of the THz field yields the complex THz spectrum.

In THz transmission measurements, the THz fields transmitted through the sample are collected and re-focused into the EO crystal for detection. If the sample under study has a resonant absorption in the THz excitation bandwidth, it absorbs and radiates THz radiation at its resonant frequency, manifested as a free-induction decay (FID) signal in the time domain. A numerical Fourier transformation of the FID signals yields the absorption or emission spectrum of the sample resonant with the excitation THz pulse. In a 2D THz spectroscopy measurement, three THz field interactions with the sample produce the emitted THz field that is measured through EOS.

2.1.3 Optical Detection Methods

In many THz pump-probe experiments, a weak optical pulse is used as the probe pulse. It is time-delayed relative to and spatially overlapped at the sample with the THz pump pulse(s). The optical responses of the sample induced by the THz field, such as absorption [51], frequency shifts [52], optical harmonic generation [53], birefringence [54], and polarization rotation [55], are detected. Examples of THz pump-optical birefringence probe spectroscopy include the THz Kerr effect in liquid molecules [37, 54, 56] and ferroelectric crystals [57, 58] and THz-induced dipolar alignment of polar gas molecules [59, 60]. Here, THz electric fields orient or align the molecules and the orientational diffusion in the liquid or alignment revivals in the gas phase are monitored by optical birefringence through the anisotropic polar-izability. Raman-active vibrational modes have also been observed using the THz pump-optical birefringence probe method, where anharmonic couplings between THz-driven THz-active modes and optically detected Raman-active modes are believed to play a role [37, 56, 58].

2.2 Nonlinear 2D THz Spectroscopy Methods

2.2.1 Pulse Sequences in 2D THz and 2D THz-Raman Spectroscopies

In 2D IR and visible spectroscopies, three pulses with controlled time delay between two neighboring pulses are usually used to conduct the experiments. In 2D THz spectroscopy, there is only one example to date using three time-delayed THz pulses [33, 34]. This is limited in part by experimental difficulties in the generation of mul-tiple THz pulses and the recombination of the pulses at the sample. Most impor-tantly, a very long data acquisition time is usually required, as all the inter-pulse delays and the time-domain signal detection time need to be scanned by mechanical delay stages. In this chapter, we mainly consider three types of pulse sequences and relevant light-matter interactions represented by the Feynman diagrams (details on the Feynman diagrams can be found in Refs. [1, 61, 62], etc.) shown in Fig. 4, which are typically used in different types of 2D THz spectroscopies. In the Feynman dia-grams, the notations of the states may indicate the number of quanta in a particular degree of freedom in some cases, and in others they denote various levels of differ-ent modes. In the hybrid THz-Raman spectroscopies of liquid molecules, the transi-tions induced through THz or Raman excitations can in general include overtone and combination band transitions that are usually allowed in liquids where the intra- and intermolecular vibrations largely have strong anharmonicity [63].

In the THz-THz-THz sequence shown in Fig. 4a, all field-matter interactions involve THz fields. An example of the $\chi^{(3)}$ interaction pathways following the pulse sequence is illustrated here with examples of typical nonrephasing (NR) and rephas-ing (R) pathways shown by the Feynman diagrams. As shown in Fig. 4a, THz field E_A generates a first-order coherence that evolves during time period τ. THz field E_B interacts with the sample twice, generating in succession a second-order population and a third-order coherence. The third-order coherence evolves during time period t and radiates the signal, which is detected by EOS as a function of t. If we index the possible time delays by which field interaction they follow, then inter-pulse delay τ is the coherence time t_1 and population time $t_2 = 0$ (field interactions 2 and 3 are time-coincident), and the detection time t corresponds to t_3. E_A can also interact twice to generate in succession a first-order coherence and a second-order population, the

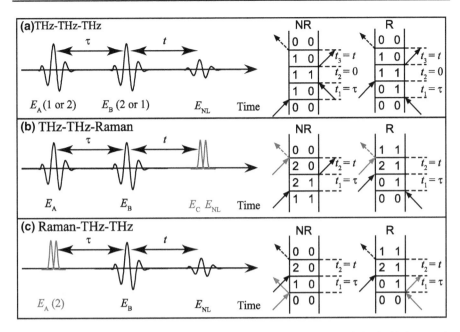

Fig. 4 Pulse sequences of 2D THz and 2D THz-Raman spectroscopies. The fields in black are THz fields and those in red are optical fields. Each field has one interaction with the sample unless indicated otherwise in parentheses. **a** THz-THz-THz sequence where all field-matter interactions arise from THz fields. The final signal emission is a THz field E_{NL} detected by EOS. Typical nonrephasing (NR) and rephasing (R) pathways following this sequence are described by the Feynman diagrams shown. **b** THz-THz-Raman sequence where each THz field interacts with the sample once and a final Raman (optical) field interacts with the sample once (E_C) to generate the nonlinear signal field E_{NL}, which is detected optically through heterodyne mixing with the Raman pulse field. Typical NR and R pathways following this sequence are described by the Feynman diagrams shown. **c** Raman-THz-THz sequence where field-matter interactions involve a second-order Raman interaction by E_A and a THz interaction by E_B. The final signal emission is a THz field E_{NL}, detected by EOS. Typical NR and R pathways following this sequence are described by the Feynman diagrams shown

latter of which evolves during time period τ. E_B interacts once to generate a third-order coherence radiating the nonlinear signal. In this case, we have coherence time $t_1 = 0$ (first two interactions time-coincident) and population time $t_2 = \tau$; the detection again, and generally, is $t = t_3$.

In the THz-THz-Raman sequence and typical Feynman diagrams shown in Fig. 4b, THz fields E_A and E_B each interact with the sample once and together result in a second-order coherence or population. Inter-pulse delay τ corresponds to the coherence time t_1, and time period t corresponds to either a coherence time or a population time. The Raman pulse converts the second-order coherence or population into a Raman coherence, which is detected optically. In the birefringence detection, the detection time t_3 is integrated by the photodetector.

In the Raman-THz-THz sequence and typical Feynman diagrams shown in Fig. 4c, a second-order interaction of the Raman pulse E_A generates a Raman coherence that evolves during τ. Via one THz interaction, THz pulse E_B converts the

Raman coherence into a second-order THz-active coherence radiating the signal field E_{NL}, which is detected by EOS as a function of t. In this case, we have $t_1 = \tau$ and $t_2 = t$.

These three pulse sequences, as well as the pulse sequence in 2D Raman spectroscopy [64, 65], are complementary to each other. They allow one to study all THz-active modes, all Raman-active modes, or coupled THz- and Raman-active modes with flexibility.

2.2.2 Collinear Phase Matching and Differential Chopping Detection

As the wavelength of THz pulses is typically comparable to the spot size of a focused THz beam, the THz wavevector is not well defined at the focus. The non-collinear FWM method with phase matching satisfied by the BOXCARS geometry, which has been routinely used in 2D IR and visible spectroscopies, cannot directly apply to THz fields. However, phase-resolved time-domain THz field detection and optical birefringence signal detection methods can allow nonlinear THz spectroscopy to be conducted with collinear phase matching, i.e., through wavevector-degenerate FWM.

To separate the nonlinear signals induced by both THz pulses from the signals induced by each THz pulse individually, a differential chopping detection method is usually used. Each THz pulse is modulated at a sub-harmonic frequency of the laser repetition rate by an optical chopper. One example of the differential chopping detection method is shown in Fig. 5. The laser repetition rate is 1 kHz. THz pulses A and B are both modulated at 250 Hz. In four successive laser shots, one can detect the signal emerging from the sample in response to both pulses together, pulse A only and pulse B only, and no incident pulses (i.e., the background noise). The nonlinear signal field $S_{NL}(t, \tau)$ that is measured as a function of inter-pulse delay τ and detection time t is given by

$$S_{NL}(t, \tau) = S_{AB}(t, \tau) - S_A(t, \tau) - S_B(t), \tag{4}$$

where $S_{AB}(t, \tau)$ is the signal field with both THz pulses present, and $S_A(t, \tau)$ and $S_B(t)$ are the signal fields with either pulse A or B present individually. With collinear phase matching and the differential chopping detection method, the measured signal

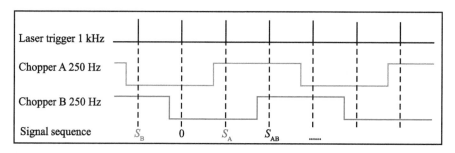

Fig. 5 Schematic representation of the differential chopping detection method

$S_{NL}(t, \tau)$ is not restricted to $\chi^{(3)}$ signals, as nonlinear signals of other orders are also collinear. Numerical 2D Fourier transformation of $S_{NL}(t, \tau)$ with respect to t and τ yields the 2D spectrum as a function of detection and excitation frequencies denoted as f and ν, respectively. Different contributions to the signal field $S_{NL}(t, \tau)$ associated with different interaction pathways have different characteristics and phase accumulation as functions of t and τ, so spectral peaks of different types can be separated in the 2D spectrum, which we will elaborate with examples in the subsequent sections.

Differential chopping detection in experiments that have three separate THz pulses denoted by A, B and C becomes more complicated, as the signals that originate from each individual pulse and each pair of pulses need to be accounted for. The nonlinear signal field is described by the following equation:

$$\begin{aligned} S_{NL}(t, \tau, T_w) = S_{ABC}(t, \tau, T_w) &- S_{AB}(t, \tau, T_w) - S_{BC}(t, T_w) - S_{CA}(t, \tau, T_w) \\ &+ S_A(t, \tau, T_w) + S_B(t, T_w) + S_C(t), \end{aligned} \tag{5}$$

where τ is the coherence time, T_w is the population time or waiting time, and t is the detection time. Details about the nonlinear signal detection method in experiments that involve three separate THz pulses can be found in references [33, 34].

2.3 Experimental Setups for 2D THz and 2D THz-Raman Spectroscopies

An example of the experimental setup for 2D THz spectroscopy is shown in Fig. 6. Two time-delayed THz pulses are generated in a LiNbO$_3$ crystal by OR of two time-delayed optical pulses recombined at the LiNbO$_3$ crystal using the tilted-pulse-front

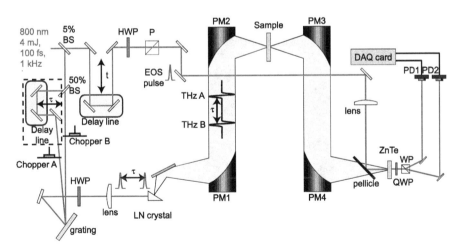

Fig. 6 An example of 2D THz spectroscopy experimental setup. Two time-delayed THz pulses are generated by two time-delayed optical pulses using the tilted-pulse-front method in a LiNbO$_3$ crystal. The THz signals are detected by EOS in a ZnTe crystal. *BS* beamsplitter, *HWP* half wave-plate, *QWP* quarter wave-plate, *P* polarizer, *WP* Wollaston prism, *PM* parabolic mirror, *LN* lithium niobate, *PD* photodiode, *DAQ card* data acquisition card. From [31]

method. The collinearly propagating THz pulses are collimated and focused into the sample by a pair of parabolic mirrors. The peak THz electric field strength is typically larger than 0.3 MV/cm for each pulse. THz signals transmitted through the sample are collected and re-focused into an EO crystal for detection. Two choppers are used to modulate the optical beams that generate the THz pulse pair. The differential nonlinear signal E_{NL} is processed by a data acquisition (DAQ) card. The total data acquisition time depending on the time windows required for τ and t as well as data averaging typically ranges from 1 day to 1 week.

An example of the experimental setup with three separate THz pulses for 2D THz spectroscopy is shown in Fig. 7. Three multi-cycle THz pulses centered at around 20 THz, which are time delayed with respect to each other, are generated by OR in three separate GaSe crystals. Due to the relatively high frequency content of the pulses, the output THz radiation is nearly collimated with small divergence. The three THz beam paths are focused and recombined at the sample by one parabolic mirror. The resulting nonlinear signals are collected and focused onto a ZnTe crystal by a pair of parabolic mirrors. An ultrashort optical pulse from the laser oscillator is time-delayed and overlapped with the THz signals in a ZnTe crystal for high-bandwidth signal detection by EOS. Three choppers at 1/2, 1/4, and 1/8 of the laser repetition rate are used for differential chopping detection.

An example of the experimental setup for 2D THz-THz-Raman spectroscopy is shown in Fig. 8. Two DSTMS crystals are pumped by two time-delayed near-IR pulses from the signal and idler of an OPA. The time-delayed THz pulses, generated

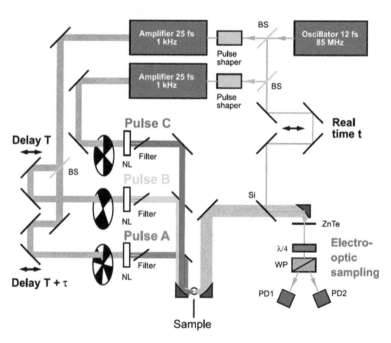

Fig. 7 An example of the experimental setup for 2D THz spectroscopy with three separate THz pulses. From [34]

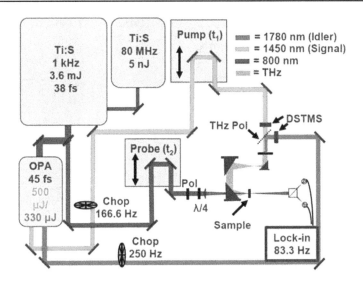

Fig. 8 An example of 2D THz-THz-Raman spectroscopy experimental setup. From [38]

with orthogonal polarizations, are recombined by a THz polarizer and focused onto the sample by a pair of parabolic mirrors. A weak portion of the Ti:sapphire laser pulse is time delayed with respect to the THz pulses and incident onto the sample through a hole in the parabolic mirror for birefringence measurements. Differential chopping detection is realized by chopping the signal and idler at 250 and 166.6 Hz, respectively, and detecting the differential signal at 83.3 Hz by a lock-in amplifier.

An example of the experimental setup for 2D Raman-THz-THz is shown in Fig. 9, which is essentially adapted from an optical pump-THz probe setup. Optical pulses from the Ti:sapphire amplifier are split into three paths. One path is used for THz generation by OR in a GaP crystal. The generated THz pulses are focused onto the sample by an elliptical mirror. The THz signals transmitted through the sample are collected by another elliptical mirror and focused onto another GaP crystal for detection. The second portion of the optical pulse is time delayed with respect to the THz pulse and incident through a hole in the elliptical mirror onto the sample where it is used as a Raman pump. The third path is time delayed and recombined with the THz signal at the GaP crystal for THz detection by EOS. Differential chopping detection is realized similar to Fig. 6. The GaP crystals used for THz generation and detection provide broad bandwidth, but the THz field strength generated is not as strong as in the former cases. As a result, a high repetition laser (5 kHz in Fig. 9), a sensitive detection scheme (Brewster windows and large numerical aperture optics shown in Fig. 9) and a long data averaging time (2 weeks in [39]) are required to observe weak nonlinear signals, such as the THz photon echo signals from water observed with this setup [39].

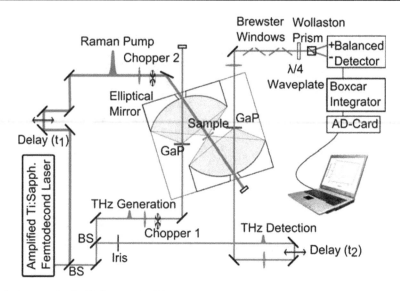

Fig. 9 An example of a 2D Raman-THz-THz spectroscopy experimental setup. From [39]

3 2D THz Rotational Spectroscopy of Gas-Phase Molecules

3.1 Molecular Orientation Induced by THz Pulses

Rotational dynamics of polar molecules have been the subject of extensive efforts in coherent spectroscopy and coherent control. Motivated by interests in rotational angular momentum, energy relaxation processes, and high-order optical interactions with multilevel quantum systems, rotational excitations were examined by optical, microwave, and THz spectroscopies. Due to their unique energy level structure and their inherently quantum mechanical behavior, molecular rotations show dramatic differences from other degrees of freedom such as vibrations and spin precessions. We first discuss the linear responses of polar molecules to THz excitation.

For small linear molecules with a permanent dipole moment, the Hamiltonian considering linear THz field-dipole interactions is given by

$$H = H_0 + H_1 = \hat{J}^2/2I - \boldsymbol{\mu} \cdot \mathbf{E}_{\mathrm{THz}}(t). \tag{6}$$

The static Hamiltonian H_0 is assumed to be a rigid rotor Hamiltonian, which accounts for the rotational dynamics of linear molecules. In H_0, $\hat{J}$ is the angular momentum operator and I the moment of inertia, given by $I = h/8\pi^2 cB$ where h is the Planck constant, c is the speed of light, and B is the rotational constant of the molecule. $\hat{H}_0$ in this form has an analytical solution. The eigenvectors are constructed by spherical harmonics, and the eigenenergies are given by $E_J = 2hBcJ(J + 1)$ as shown in Fig. 10a. Here, each eigenstate is denoted by the angular momentum quantum number J, and the energy difference between two adjacent J states is given by $\Delta E_{J,J+1} = 2hBc(J + 1)$. The field-dipole interaction Hamiltonian is $H_1 = -\boldsymbol{\mu} \cdot \mathbf{E}_{\mathrm{THz}}(t) = -\mu E_{\mathrm{THz}}(t) \cos \theta$, where θ is the angle between the permanent

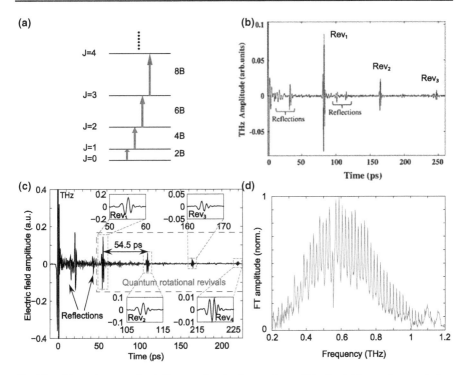

Fig. 10 **a** Schematic energy level diagram of molecular rotations within the rigid rotor framework. **b, c** FID signals from carbonyl sulfide (OCS) and acetonitrile (CH$_3$CN), respectively. The superposition of all the rotational coherences results in bursts of THz emission called rotational revivals, labeled Rev$_1$, Rev$_2$, etc. **d** The rotational spectrum of CH$_3$CN, which results from a numerical Fourier transformation of the FID signals in **c** and which shows each rotational transition as a sharp peak. From [31, 59]

dipole μ and the polarization of $\mathbf{E}_{THz}$. A THz field couples adjacent rotational states J and $J + 1$ (i.e., the selection rule is $\Delta J = \pm 1$) yielding 1-quantum coherences (1QCs) between adjacent states of the thermal ensemble. From the viewpoint of classical field-dipole interactions, the THz electric field exerts a torque on the dipoles, and the dipoles rotate at discrete 1QC frequencies $f_{J,J+1} = 2Bc(J + 1)$. The time-dependent THz emission from the rotating dipoles, given by the net dipolar orientation ⟨cos θ⟩, is the rotational free-induction decay (FID).

The THz pulse induces a net orientation of the molecular dipoles, i.e., nonzero cos θ, but since the rotational coherences are at many different frequencies (determined by the different initial rotational levels J, many of which are thermally populated at ordinary temperatures), they rapidly go out of phase and the net orientation is lost. However, since the 1QC frequencies are all integer multiples of the lowest frequency $2Bc$, the dipoles go back in phase and there is a short-lived periodic "revival" of net dipole orientation cos θ with the quantum rotational revival period [66] given by $T_{rev} = (2Bc)^{-1}$. Upon each revival, the net dipolar orientation results in a macroscopic polarization in the sample, which emits a burst of coherent THz-frequency radiation. The FID signal thus consists of a sequence of such bursts separated by T_{rev}. Examples of the FID signals from carbonyl sulfide (OCS) [59] and acetonitrile (CH$_3$CN) [31] gases excited by one single-cycle THz pulse generated from LiNbO$_3$ and detected by EOS are shown in Fig. 10b, c, respectively. The revival periods are $T_{rev} = 82$ ps for OCS and $T_{rev} = 54.5$ ps for CH$_3$CN, both consistent with literature values of their rotational constants B. Fourier transformation of the periodic FID signal in Fig. 10c yields the linear rotational spectrum (the absorption spectrum) of CH$_3$CN consisting of equally spaced peaks separated by $2Bc$ as shown in Fig. 10d. Each rotational transition between adjacent rotational states is resolved as a sharp peak.

3.2 Molecular Alignment Induced by THz Pulses and Two-Quantum THz-THz-Optical Rotational Spectroscopy

The molecular orientational 1QCs discussed in Sect. 3.1 are the linear responses induced by one THz field-dipole interaction. Second-order nonlinear rotational responses resulting from two successive THz field-dipole interactions, manifested as 2-quantum coherences (2QCs) and non-thermal populations, have also been demonstrated. The second-order field-dipole interaction Hamiltonian takes the approximate form $H_2 = (-\mu \cdot \mathbf{E}_{THz}(t))^2 = \mu^2 E_{THz}(t)^2 \cos^2 \theta$, where we are neglecting other excitation pathways including stimulated scattering. 2QCs and the excited rotational population correspond to net alignment of molecular dipoles, described by the alignment factor $\langle\cos^2 \theta\rangle$, with no net molecular orientation or associated macroscopic polarization because the dipoles may be antiparallel as well as parallel. Optical detection methods such as optical birefringence can be used to measure the alignment since it leads to anisotropy in the optical refractive indices between the direction of alignment (THz electric field polarization direction) and the direction normal to the alignment direction. The experimentally measured birefringence is proportional to $\cos^2 \theta$.

Rotational 2QCs result from two successive THz field-dipole interactions that generate coherent superpositions of first J and $J + 1$ levels and then $J, J + 1$ and $J + 2$ levels through successive transitions with the selection rule $\Delta J = \pm 1$. The 2QCs between levels J and $J + 2$ evolve in time at frequencies of $f_{J,J+2} = 2Bc(2J + 3)$, and their superposition leads to periodic, short-lived revivals of constructive interference similar to those of the 1QCs but with a period of $T_{rev}/2 = (4Bc)^{-1}$. The measurement results from OCS in Fig. 11a show that the 2QC revivals appear as expected. Fourier transformation of the birefringence signal yields the rotational spectrum in Fig. 11b. Note that the successive resonant THz field interactions with the dipoles are completely different from nonresonant optical excitation of 2QCs through stimulated rotational Raman interactions with the molecular polarizabilities, with the selection rule $\Delta J = \pm 2$ [67].

The periodic 2QC signals are superimposed on the steady-state second-order population response that is also induced by successive THz field interactions that produce first the 1QCs between adjacent levels J and $J + 1$ and then excited-state populations in levels $J + 1$. The excited rotational populations are anisotropic because each thermally populated level J includes equal populations in all of

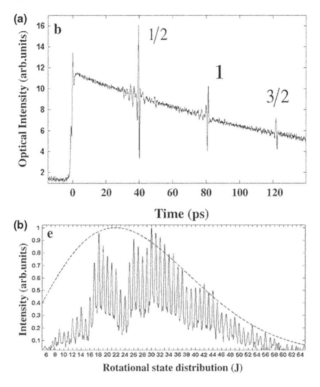

Fig. 11 **a** Alignment-induced birefringence signal from OCS. **b** Rotational spectrum of OCS resulting from a numerical Fourier transformation of the signal in **a**. From [59]

the $2J + 1$ degenerate rotational sublevels labeled by quantum number M, where $M = -J, -J + 1, ..., J - 1, J$, while the excited population in level $J + 1$ includes only the same sublevels and not the two sublevels $-(J + 1)$ and $(J + 1)$ because of the $\Delta M = 0$ selection rule.

As the molecular alignment involves two THz field-dipole interactions, one can separate these two interactions and control the amplitudes of 2QCs using two time-delayed THz pulses. Molecular alignment in response to two time-delayed THz pulses has been measured [60], yielding the results shown in Fig. 12. The pulse sequence used in the experiment is shown in the inset of Fig. 12a, which is the same as the THz-THz-Raman sequence shown in Fig. 4b. In Fig. 12a, the birefringence signal induced by two THz pulses with an inter-pulse delay of τ is separated from the second THz pulse by $\tau/2$. The 2QC signal level from time-delayed field-dipole interactions is far higher than that induced by two field-dipole interactions from either individual THz pulse. In Fig. 12b, the birefringence signals are measured at several different inter-pulse delays, and the 2QCs signals achieve the maximum amplitude when the delay τ is close to $T_{rev}/2 = 41$ ps. This is because at $\tau = T_{rev}/2$, the 1QCs induced by the first THz pulse with odd-number J are in phase with each other, and those with even-number J are also in phase with each other, and the alignment factor

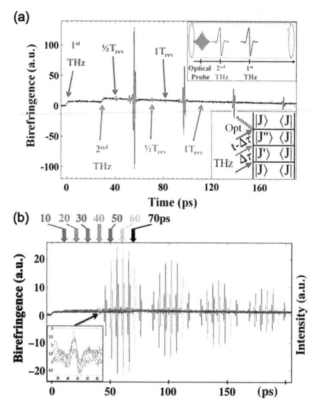

Fig. 12 **a** Optical birefringence signal from OCS in response to two THz pulses with a relative delay of 29 ps. Insets show the experimental geometry and the double-sided THz-THz-Raman Feynman diagram describing the 2QC excitation and detection pathway. **b** Optical birefringence signals from OCS in response to THz pulse pairs with variable relative delays color-coded as shown. The inset shows the far smaller birefringence signals induced by two successive interactions from one THz pulse. From [60]

of the molecular dipoles is maximized. After the interaction with the second THz field, the generated 2QCs are in phase at $t = T_{rev}/4$ ($T_{rev}/4$ after the second THz pulse). The constructive interference of the 2QCs is optimal at $\tau = T_{rev}/2$, which leads to the maximum coherent enhancement of the 2QC signal amplitudes [60].

This experiment demonstrates 2D two-quantum rotational spectroscopy in the time domain. In Fig. 13, we show the two-quantum signals at different time delays in a 2D plot, which clearly shows that the maximum signal is achieved at $\tau = T_{rev}/2$ (41 ps for OCS). The 2D Fourier transformation of the 2D time-domain signal with respect to the inter-pulse delay τ and the detection time t would yield a 2D rotational spectrum where 2-quantum (2Q) signals appear at J-resolved positions along the frequency diagonal of $\nu = f/2$, where the excitation and detection frequencies ν and f are conjugate to the inter-pulse delay τ and the detection time t. This is because the phase accumulation of the 2QCs during t is approximately twice as fast as the phase accumulation of the 1QCs during τ.

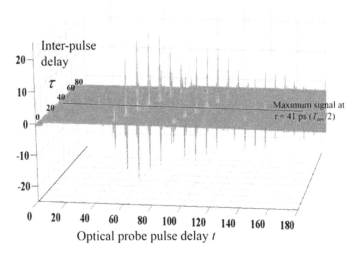

Fig. 13 The 2D time-domain plot of the optical birefringence signals from OCS in response to THz pulse pair

3.3 2D THz Rotational Spectroscopy of Acetonitrile

In molecular rotations, there have been theoretical works that predict the possibility to observe THz photon echoes using both classical calculation of the rigid-rotor angular distribution [68] and quantum mechanical calculation of the dipolar orientation [26] in response to pairs of time-delayed THz pulses. The characteristic frequency patterns and the narrow linewidths typically observed in molecular rotational spectroscopy make it an ideal testbed for measurement of rephasing (R or photon echo) and other contributions to 2D THz spectra. The 2D THz rotational spectra were observed for the first time in CH_3CN [31].

The experimental setup is the same as shown in Fig. 6, where a static pressure gas cell with CH_3CN is placed at the sample position. Nonlinear signal traces E_{NL} as a function of EOS detection time t at various inter-pulse delays τ are measured by the differential chopping detection technique. The nonlinear signals E_{NL} are shown in Fig. 14. THz pulse B is fixed at $t = 0$, and a positive inter-pulse delay τ is incremented, which means THz pulse A appears earlier than pulse B. At each delay τ shown in the insets, a burst of THz signal appears at $t = \tau$, which confirms that the observed signals are photon echoes. As THz signal emission results from the collective polarization formed during the orientation of the molecular dipoles, the photon echo signals also show periodic revivals, of the same form as the linear FID signals. In this process, pulse A interacts with the dipoles once and generates first-order 1QCs that evolve and accumulate phase during τ. Pulse B interacts with the dipoles twice, generating second-order populations and third-order 1QCs with reversed phase accumulation with respect to the first-order 1QCs. The third-order 1QCs are in phase at $t = \tau$ and at $t = \tau + nT_{rev}$, $(n = 1, 2, 3...)$, at which times they form collective polarizations that radiate the photon echo signals. At the revivals of THz pulse

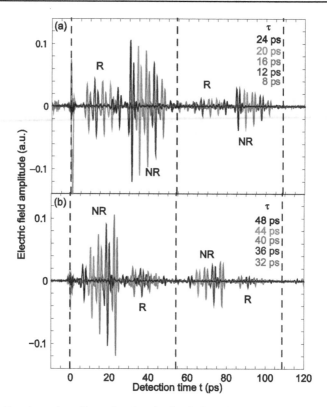

Fig. 14 a, b Experimental nonlinear time-domain signals E_{NL} showing photon echoes (R) and non-rephasing signals (NR) at various inter-pulse delays τ indicated in the inset. The dashed lines are separated by $T_{rev} = 54.5$ ps. From [31]

A, additional nonlinear signals are observed, which are due to field-induced changes in the amplitudes of the revival signals generated by pulse A. These are identified as nonrephasing (NR) signals. In this signal pathway, the first two interactions are the same as those in the R pathway, and the second interaction with pulse B results in third-order 1QCs that are not phase-reversed with respect to the first-order 1QCs. The NR third-order 1QCs are hence in phase at the same time as the FID revivals induced by pulse A.

The 2D THz rotational spectrum $E_{NL}(f, \nu)$ of CH_3CN results from 2D numerical Fourier transformation of the nonlinear trace $E_{NL}(t, \tau)$ recorded by measurement of t-dependent EOS signals at each inter-pulse delay τ. The 2D magnitude spectrum is shown in Fig. 15. The spectrum is separated into the NR and R quadrants with a difference in the sign of the excitation frequency ν. The NR quadrant consists of NR and 2Q signals that do not have reversed phase during time periods τ and t, while the R quadrant consists of R and also weaker fifth-order 2-quantum rephasing (2Q-R) signals that have reversed phase during τ and t. Along $\nu = 0$, there are pump-probe (PP) signals that do not have phase accumulation during τ. Each type of signal

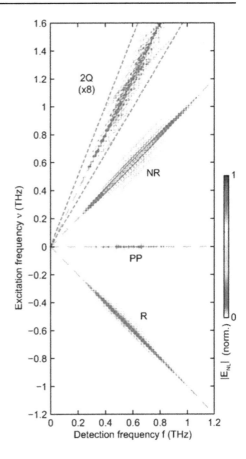

Fig. 15 Normalized 2D THz rotational spectrum of CH₃CN. *NR* nonrephasing, *R* rephasing, *PP* pump probe, *2Q* 2-quantum, magnified ×8. From [31]

consists of *J*-resolved spectral peaks located at the frequencies of rotational 1QCs or 2QCs as will be discussed below.

The NR and R quadrants of the 2D magnitude spectrum of CH₃CN are plotted separately in Fig. 16, with the excitation frequency ν of the R quadrant made positive. Enlarged spectra at the centers of each quadrant are shown and plotted as a function of rotational *J* quantum number. Each quadrant is normalized to its maximum amplitude. Weak spectral features of the fifth-order 2Q-R signals appear along the frequency diagonal $\nu = 2f$ when they are magnified. The THz field-dipole interaction pathways for the different types of signal are discussed as follows.

In the NR and R signals, *J*-state-resolved diagonal peaks are clearly observed along the frequency diagonals $\nu = f$ in Fig. 16a, b. For these diagonal peaks, THz pulse *A* interacts once with the molecular dipoles to induce first-order 1QCs described by density matrix elements $|J\rangle\langle J + 1|$. After inter-pulse delay τ, pulse *B* interacts twice with the dipoles to induce in succession second-order rotational populations $|J + 1\rangle\langle J + 1|$ and third-order 1QCs $|J\rangle\langle J + 1|$ (NR) or $|J + 1\rangle\langle J|$ (R). The NR and R signals are radiated by these third-order 1QCs and are measured during the detection time *t*. The third-order nonlinear signal field $E_{NL}(t, \tau)$ shows

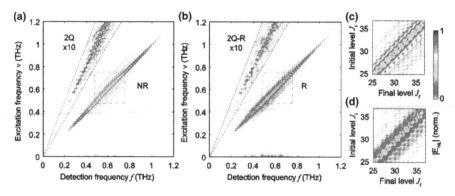

Fig. 16 a NR and **b** R (excitation frequency shown as positive) quadrants of the 2D spectrum of CH_3CN. Spectral amplitudes inside the red dashed area are magnified ×10 to bring out the 2Q (in **a**) and 2Q-R (in **b**) signals. The dashed boxes cover rotational transition frequencies $f_{J,J}$, from $f_{25,26}$ to $f_{37,38}$. **c, d** Enlarged views of the NR (**c**) and R (**d**) spectra within the dashed boxes in the NR quadrant (in **a**) and R quadrant (in **b**) as functions of initial and final J quantum numbers along the vertical and horizontal axes, respectively. Third- and fifth-order off-diagonal peaks are separated from the diagonal peaks at J-resolved positions. All of the spectra are normalized and plotted based on the color map shown. From [31]

oscillations at the 1QC frequencies along the two time variables, and in this example the frequencies are the same, such that Fourier transformation of the signal with respect to both time variables yields J-state-resolved diagonal peaks at frequencies $v = f = f_{J,J+1} = 2Bc(J + 1)$ in the 2D spectrum. In addition, for each excitation frequency $v = f_{J,J+1} = 2Bc(J + 1)$, off-diagonal peaks are observed at detection frequencies $f = f_{J-1,J} = 2BcJ$ and $f = f_{J+1,J+2} = 2Bc(J + 2)$. The off-diagonal peaks indicate that the first-order $|J\rangle\langle J + 1|$ coherences induced by THz pulse A are correlated not only to the third-order coherences $|J\rangle\langle J + 1|$ discussed above but also to the third-order coherences $|J - 1\rangle\langle J|$ and $|J + 1\rangle\langle J + 2|$ involving two neighboring J levels induced by pulse B. These spectral peaks are located at J-resolved positions as shown in the 2D J-number map plotted as a function of initial and final rotational level J_i and J_f (related to frequencies variables by $v = 2Bc(J_i + 1)$ and $f = 2Bc(J_f + 1)$) in Fig. 16c, d.

For 2Q and PP signals, pulse A interacts twice with the molecular dipoles to produce either 2QCs $|J\rangle\langle J + 2|$ or populations $|J + 1\rangle\langle J + 1|$, respectively. After inter-pulse delay τ, pulse B interacts once with the dipoles to produce third-order 1QCs $|J\rangle\langle J + 1|$ that radiate the measured signals during time t. The 2Q signal field $E_{NL}(\tau, t)$ shows oscillations as a function of τ at the 2QC frequencies and oscillations as a function of t at the 1QC frequencies, giving rise to J-state-resolved peaks at $v \cong 2f$. For PP signals, there is no coherence evolution during the inter-pulse delay so the signal appears in the 2D spectrum at zero frequency along v and at J-resolved positions along f.

In addition to these third-order signals, fifth-order spectral peaks including 2Q-R signals and off-diagonal NR and R signals coupling J and $J \pm 2$ levels are also observed. They originate from five THz field-dipole interactions, which are elaborated as follows. The 2Q-R peaks arise from two field interactions with pulse A

to create 2QCs $|J\rangle\langle J + 2|$ and three field interactions with pulse B. The first two interactions with pulse B create a population $|J + 2\rangle\langle J + 2|$ and the third induces a rephased 1QC $|J + 3\rangle\langle J + 2|$ or $|J + 2\rangle\langle J + 1|$, which radiates the nonlinear signals during t. The 2Q-R signals give rise to peaks along $f = 2\nu$ in the R quadrant shown in Fig. 16b. For the fifth-order NR and R signals shown in Fig. 16c, d, THz pulse A induces 1QCs $|J\rangle\langle J + 1|$ evolving during τ, and pulse B promotes them via four field-dipole interactions to fifth-order 1QCs with final rotational level, J_f, two quanta away from the initial level J_i (namely, $|J_f - J_i| = 2$) at $|J + 3\rangle\langle J + 2|$ (R) or $|J + 2\rangle\langle J + 3|$ (NR). The fifth-order 1QCs then radiate signals correlated to the 1QCs induced during τ.

Example pathways of the THz field-dipole interactions described here are further elaborated by the double-sided Feynman diagrams presented in Fig. 17.

3.4 Extensions of 2D THz Rotational Spectroscopy

The initial demonstrations of 2D THz rotational spectroscopy have provided illustrations of multiple-pulse THz coherent control over rotational dynamics and a detailed

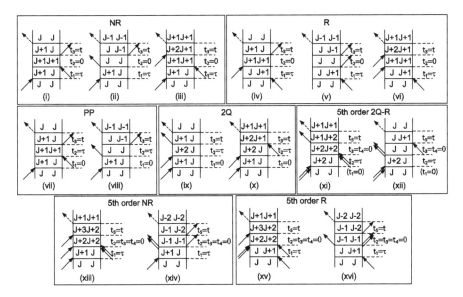

Fig. 17 Double-sided Feynman diagrams describing the THz field-dipole interactions. Diagram (i)–(iii) describes the third-order diagonal and off-diagonal peaks in the NR quadrant of the 2D spectrum. Diagrams (iv)–(vi) describe the third-order diagonal and off-diagonal peaks in the R quadrant. Diagrams (vii)–(viii) describe two excitation pathways that lead to the PP peaks. Diagrams (ix)–(xii) describe typical excitation pathways leading to the 2Q and 2Q-R peaks. Diagrams (xii)–(xiv) and (xv)–(xvi) describe typical excitation pathways leading to the fifth-order NR and R off-diagonal peaks. The bra and ket symbols for the density matrix elements $|J\rangle\langle J'|$ are assumed in all the diagrams. The time subscripts denote the number of preceding field interactions. The nonlinear signal emission time period t corresponds to t_3 in all third-order processes and t_5 for the fifth-order processes. The inter-pulse delay time τ corresponds to t_1 for the R and NR signals (for third- and fifth-order signals) and to t_2 for the PP, 2Q and 2Q-R signals. From [31]

elaboration of the different signal contributions to 2D rotational spectra at third and fifth order. The results suggest possible experimental advances and a wide range of new information that can be uncovered. For example, by including a third THz pulse at a controlled delay with respect to the second pulse [33] such that the rotational population time can be varied, 2D rotational spectroscopy with three THz pulses can enable measurements of the time-dependent evolution of off-diagonal spectral peaks and spectral diffusion as in other 2D spectroscopies [69, 70]. Such measurements may reveal specific rotational energy transfer and relaxation pathways (due to dipole-dipole interactions among the molecules under study, collisions with other species, etc.) by comparing the strengths and line shapes of the 2D spectral peaks for different population times. Independent control of the THz pulse polarizations could reveal the dynamics of relaxation among the M sublevels of the rotational J levels. The use of stronger THz pulses will allow measurement of still higher-order signal contributions, which will reveal additional correlations among rotational transitions and will allow the method to be used on molecules with moderate or small dipole moments. In addition to THz excitation pulses, optical excitation can be used to achieve enhanced control over rotational dynamics in linear and nonlinear molecules [71, 72]. Thus, a wide range of experimental refinements is possible, offering prospects for new insights into molecular rotational dynamics and the molecular interactions that mediate them.

4 2D THz and Hybrid 2D THz-Raman Vibrational Spectroscopies

Multidimensional IR vibrational spectroscopy has proved to be a powerful tool to study complex liquid-state vibrational dynamics, for example, the hydrogen bonds in water [5, 73] and intramolecular vibrations in proteins and DNA [6, 7]. Multidimensional THz vibrational spectroscopy of liquids could extend the range to low-frequency molecular vibrations and intermolecular motions that may provide additional insights, for example, into the structural dynamics of water, proteins, and DNA and to their chemical properties. However, 2D THz vibrational spectroscopy has so far been realized only on lattice vibrations in semiconductors. On the other hand, hybrid 2D spectroscopy combining THz and optical excitation or detection methods has enabled 2D THz-Raman variations that are complementary to 2D THz and 2D Raman spectroscopies. As dipole and polarizability interactions are both involved, the hybrid methods can provide enhanced sensitivity to study some vibrational modes and their interactions. In this section, we discuss some of the 2D THz and 2D THz-Raman vibrational spectroscopies and their applications to the study of lattice vibrations in semiconductors, molecular vibrations in halogenated liquids, and intermolecular dynamics of the hydrogen-bond networks in water and ionic aqueous solutions.

4.1 2D THz Spectroscopy of Phonons in Semiconductors

The first demonstration of 2D THz spectroscopy with three THz pulses has been realized very recently in the study of phonon nonlinearity in the semiconductor indium antimonide (InSb) crystal [33, 34]. The experimental setup is shown in Fig. 7. Multi-cycle THz pulses centered at 20 THz with 6-THz bandwidth are generated by OR in GaSe crystals and are incident onto the sample. The waveforms of the generated THz pulse and the three pulses transmitted through the sample denoted as A, B and C are shown in Fig. 18. Comparing the spectra of the incident THz pulse and the pulses transmitted through the sample, a spectral peak at 10 THz emerges, which is identified as a two-phonon coherence in InSb. As there is nearly no spectral amplitude at 10 THz in the incident THz spectrum, the two-phonon coherence is generated nonresonantly by impulsive excitation.

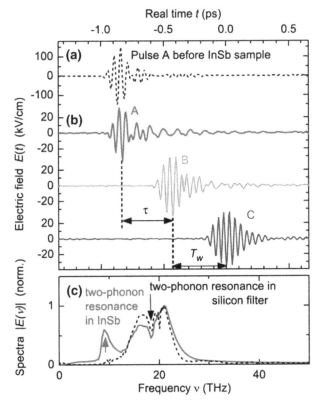

Fig. 18 a Waveform of THz pulse A before the InSb sample as a function of EOS detection ("real") time t. **b** Waveforms of pulses A, B, and C transmitted through the InSb sample. Each pulse develops oscillatory features emerging from the sample. The delays between pulses A and B and between pulses B and C are respectively the coherence time and waiting time denoted by τ and T_w. **c** Fourier transform spectra of pulse A before the sample (dashed line) and transmitted through the sample (solid line). The dip at around 18 THz is due to a two-phonon resonance in the silicon filter in the beam path, while the peak at around 10 THz emerges from the two-phonon resonance in the InSb sample. From [34]

To study the nonlinear responses of the two-phonon coherences in detail, 2D THz spectroscopy with three pulses was conducted. The pulse sequence is shown in Fig. 18. The coherence time and waiting (population) time are denoted by τ and T_w, respectively, while the "real" time is the EOS detection time with its zero at the peak of pulse C. The 2D time-domain signals emerging from the sample in response to the three THz pulses and the 2D THz spectra at two different waiting times are shown in Fig. 19. The 2D time-domain data shown in Fig. 19a are the signal $S_{ABC}(t, \tau, T_w)$ in response to all three pulses. Pulses B and C have fixed relative delay, i.e., waiting time, T_w, while pulse A is delayed with respect to pulse B. The data shown in Fig. 19b show the nonlinear signal $S_{NL}(t, \tau, T_w)$ extracted according to Eq. (6) at a selected waiting time of 827 fs. It is evident that there are oscillatory signals following the pulse that arrives latest (pulse A for $\tau < -850$ fs and pulse C for $\tau > -850$ fs). The nonlinear signals exhibit interference patterns due to the presence of several signal contributions following different interaction pathways. The 2D Fourier transformation of the signal in Fig. 19b gives rise to the 2D spectrum $S_{NL}(\nu_t, \nu_\tau, T_w)$ where ν_t and ν_τ are detection and excitation frequencies at the selected waiting time T_w. The experimental 2D magnitude spectra at $T_w = 35$ and 827 fs and calculated 2D magnitude spectrum at a $T_w = 750$ fs are shown in Fig. 19c–e. There

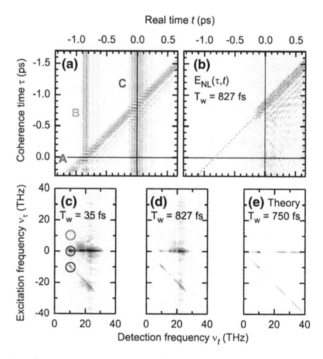

Fig. 19 **a** 2D plot of $S_{ABC}(t, \tau, T_w)$, which is superposition of the electric fields of all three pulses transmitted through the sample as a function of coherence time τ and real-time t. **b** Nonlinear signal field $S_{NL}(t, \tau, T_w)$ extracted according to Eq. (6) for a waiting time $T_w = 827$fs. The dashed line indicates the center of pulse A. **c, d** 2D magnitude spectra $|S_{NL}(\nu_t, \nu_\tau, T_w)|$ for $T_w = 35$ and 827 fs. The circles in **c** mark the spectral peaks resulting from two-phonon resonances. **e** Calculated 2D spectrum for $T_w = 750$ fs. From [33]

is very good agreement between the experimental data and calculation. The strong features at $\nu_t = 22$ THz are mainly due to field-induced interband tunneling of carriers and interband two-photon absorption, which are in the strongly nonperturbative regime due to the extremely large transition dipole between the valence and conduction bands [33]. The weak features at $\nu_t = 10$ THz originate from the two-phonon resonances in InSb, which are isolated from the 2D spectra and analyzed in the time domain in detail in the following.

The 2D time-domain signals as functions of coherence time and real time resulting from inverse Fourier transformation of selected isolated spectral peaks in Fig. 19c–e are shown in Fig. 20. There is good agreement between the experimental data and calculations. The signal in Fig. 20b is assigned as a rephasing signal as the phase front of the signal is perpendicular to the phase front of pulse A (orange dashed line), i.e., the nonlinear signal is phase-reversed with respect to the coherences generated by pulse A. Due to the lack of even-order signals in the data and the nonresonant nature of the two-phonon coherence generation, the seventh-order ($\chi^{(7)}$) pathway shown by diagram (i) in Fig. 21 was proposed [33] as the lowest-order pathway to describe the light-matter interactions. According to the diagrams, pulse A induces a two-phonon coherence via impulsive excitation with two interactions, and pulse B projects the two-phonon coherence back to the ground state also impulsively with two interactions. Finally, pulse C generates a rephased two-phonon coherence of seventh order also via impulsive excitation, but with three interactions. This excitation process is shown separately in diagram (i). The two-phonon coherence radiates the nonlinear signal shown in Fig. 20b.

The origins of the nonlinear signals in Fig. 20a, c were assigned to an eleventh-order ($\chi^{(11)}$) NR pathway, which was proposed as the lowest-order pathway that can lead to these signals. It is described by the ladder diagram (iii) in Fig. 21 and analyzed as follows. Each pulse has to interact at least once to lead to the nonlinear signal. Due to the observation that the phase front of the signal is independent of that of pulse A, pulse A should interact an even number of times. A third-order process hence is not sufficient to explain the signal origin. As shown in Fig. 4 of Ref. [33], the nonlinear signal at $T_w = 827$ fs is found to be a direct continuation in amplitude and phase of that at $T_w = 35$ fs. Besides, its phase is independent of the pulse sequences. Hence, the nonlinear two-phonon coherence signal follows the phase of pulse B only, while pulses A and C create long-lived electronic excitations in InSb whose bandgap is about 41 THz, twice the frequency of the THz pulses. The ladder diagram (iii) shown in Fig. 21 can be read as follows. Pulse A excites an electronic population via four interactions, pulse B impulsively induces a two-phonon coherence via three interactions, pulse C promotes this two-phonon coherence from the first electronic excited state to the second one, and finally the two-phonon coherence at the second electronic excited state radiates the nonlinear signal.

In this experiment, the exceptionally high-order nonlinear interactions were proposed on the basis of the huge transition dipole moments associated with the electronic and vibronic transitions in the InSb sample. In pure vibrational systems where nonlinearity is not as dramatic, 2D spectra resulting from all THz interactions have so far not been available. Hybrid 2D THz-Raman methods can provide better sensitivity because of simultaneous interactions via dipole and polarizability. Several

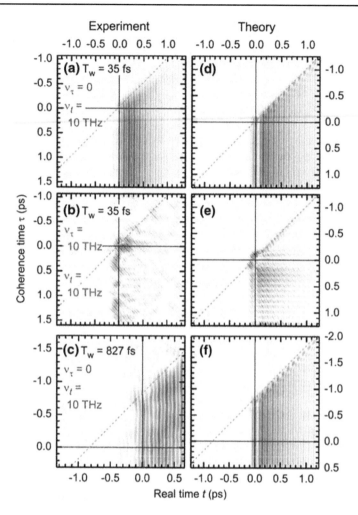

Fig. 20 Left column, experimental 2D time-domain signals from the spectral peaks at $(v_\tau, v_t, T_w) = (0, 10\,\text{THz}, 35\,\text{fs})$ **a,** at $(v_\tau, v_t, T_w) = (10\,\text{THz}, 10\,\text{THz}, 35\,\text{fs})$ **b,** and at $(v_\tau, v_t, T_w) = (0, 10\,\text{THz}, 827\,\text{fs})$. Right column, calculated 2D time-domain signals resulting from the corresponding spectral peaks. The black solid lines mark the zeroes of coherence time and real time while the orange dashed line marks the center of pulse A. From [33]

examples of 2D THz-Raman vibrational spectroscopies are discussed in the following sections.

4.2 2D THz-THz-Raman Spectroscopy of Intramolecular Vibrations in Liquids

A demonstration of 2D THz-THz-Raman spectroscopy has successfully revealed the anharmonic couplings between intramolecular vibrational modes in halogenated liquids [37]. The experimental setup used is the same as in Fig. 8. Two THz pulses with

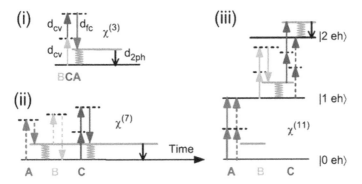

Fig. 21 Ladder diagrams describing relevant light-matter interactions for impulsive generation of two-phonon coherences (i), experimentally observed R signal (ii), and NR signal (iii). Time evolution is rightward. The red, green, and blue solid and dashed arrows represent interactions among pulses A, B, and C and the system on the ket and bra sides, respectively. The yellow lines represent two-phonon states, and the yellow wavy lines represent two-phonon coherences. The black arrows pointing downward denote final nonlinear signal emission. From [33]

orthogonal polarizations and a relative time delay τ are generated from two DSTMS crystals and are focused into the sample to excite the liquid. An optical probe pulse that is delayed from the second THz pulse by time t and polarized parallel to the first THz pulse is incident onto the sample to probe the THz-induced response. The birefringence of the probe pulse is measured as a function of THz inter-pulse delay τ and measurement time t.

Let us first consider the experiment with a single THz excitation pulse, i.e., a THz pump-optical birefringence probe measurement [56]. The excitation band-width spanning 1–5 THz includes THz-active vibrational modes in many liquids. The transient birefringence of the probe pulse is measured as a function of the delay t between the THz and optical pulses. The results are shown in Fig. 22. During the THz pulse, an instantaneous birefringence signal that scales with the square of the THz electric field emerges because of the THz nonresonant electronic Kerr effect [54, 56]. It is followed by a slow decay due to the THz-induced orientation of the molecules, i.e., the THz rotational Kerr effect, and subsequent diffusion to

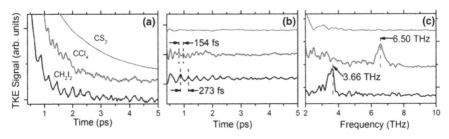

Fig. 22 **a** Orientational diffusion and vibrational coherences observed in THz Kerr effect measurements of halogenated liquids. **b** Vibrational signals extracted from **a**. **c** Fourier transforms of the oscillatory signals in **b** reveal the vibrational spectra. From [56]

an isotropic distribution. On top of the orientational signal, oscillatory signals corresponding to Raman-active vibrational modes are observed. These Raman-active modes are excited via two THz interactions, and the excitation pathways are revealed in 2D THz-THz-Raman spectroscopy as discussed below.

In the 2D THz-THz-Raman spectroscopy experiment, differential chopping detection ensures that the nonlinear signal S_{NL} results from one interaction with each THz pulse, as the signals resulting from two interactions with either individual THz pulse are excluded. An example of the THz field-matter interaction pathways is illustrated following the THz-THz-Raman pulse sequence shown in Fig. 4b. Let us consider a three-level system consisting of three states $|a\rangle$, $|b\rangle$, and $|c\rangle$, which denote the combination bands $|1, 0\rangle$, $|0, 1\rangle$, and $|2, 0\rangle$ of two vibrational modes. The transitions between each pair of states can be both THz- and Raman-allowed. The typical pathways for different types of signals are described by the Feynman diagrams shown in Fig. 23. We elaborate one possible pathway as shown in diagram (i) describing the NR signals observed in Ref. [37] as follows. The system starts with a population $|1, 0\rangle\langle 1, 0|$. THz pulse A interacts once to generate a first-order coherence $|2, 0\rangle$ $\langle 0, 1|$, which evolves at frequency ω_{bc} during the time period τ (or t_1 in [37]). THz pulse B interacts once to induce a second-order coherence $|2, 0\rangle\langle 1, 0|$, which evolves at frequency ω_{ac} during the time period t (or t_2 in [37]). As this coherence is correlated to the amplitude of $|2, 0\rangle\langle 0, 1|$, the amplitude of $|2, 0\rangle\langle 1, 0|$ is modulated as the delay τ is varied. The optical probe pulse detects $|2, 0\rangle\langle 1, 0|$ at each t point through a Raman transition between $|2, 0\rangle$ and $|1, 0\rangle$. The transient birefringence that the probe pulse experiences measures the amplitude of $|2, 0\rangle\langle 1, 0|$. The interaction pathway is described by a third-order response function $R^{(3)}(\tau, t)$ as follows:

$$R^{(3)}(t, \tau) \propto \Pi_{ca} \mu_{ba} e^{-i\omega_{ac} t} \mu_{bc} e^{-i\omega_{bc} \tau} \rho_{eq}, \qquad (7)$$

Fig. 23 Feynman diagrams that describe the typical nonrephasing and rephasing pathways in THz-THz-Raman experiments

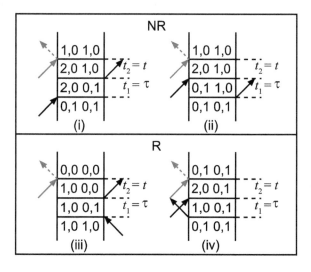

where Π_{xy} and μ_{xy} are the polarizability and dipole transition moments and ρ_{eq} is the density matrix at equilibrium. The 1D signals shown in Fig. 22 essentially arise from the same response function with τ set to zero. The 2D Fourier transformation of the response function with respect to τ and t gives rise to an off-diagonal peak located at $(f, \nu) = (\omega_{ac}, \omega_{bc})/2\pi$ showing the coupling between the two vibrational modes.

The 2D THz spectra of several simple liquids are shown in Fig. 24. Off-diagonal peaks from different NR pathways are observed, which clearly show the anharmonic vibrational couplings between different modes in these simple liquids. The experimental data agree well with simulations using the full set of response functions including couplings between different states and all possible pathways as outlined above. The 2D spectra can guide theoretical and computational efforts to elaborate the rich insights into the microscopic dynamics of thermally populated liquid-state vibrational dynamics.

Fig. 24 The 2D spectra of several simple liquids showing off-diagonal NR peaks originating from anharmonic couplings among various intramolecular vibrations. From [37]

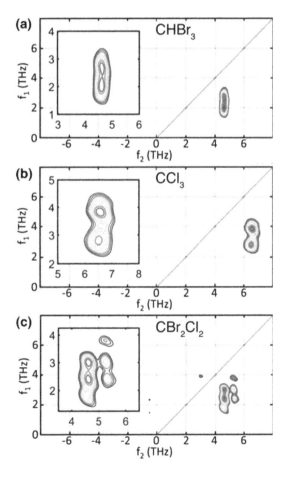

Diagonal peaks that involve vibrational population pathways were not observed in this experiment because of the orthogonal polarizations of the input THz pulse pair and the limited THz bandwidth. Moreover, rephasing signals were not observed because of the limited THz bandwidth. As shown by the example R pathway in diagram (iii) of Fig. 23, it requires THz excitation of the transition from $|0\rangle$ to $|2\rangle$, the frequency of which is beyond the available THz bandwidth in Ref. [37]. In a follow-up work with improved THz bandwidth, R signals were observed [38]. The experimental 2D spectrum of $CHBr_3$ is shown in Fig. 25. In comparison with the spectrum in Fig. 24a, in addition to the two NR peaks, several NR peaks are observed at higher frequencies in the first quadrant while several R peaks are observed in the second quadrant. The 2D spectrum reveals the rich anharmonic couplings among different vibrational modes in this simple liquid. The relevant light-matter interactions can be described by the Feynman pathways shown in Fig. 23.

4.3 2D Raman-THz Spectroscopy of Water and Aqueous Salt Solutions

The hydrogen-bond dynamics in water have been explored extensively by 2D IR spectroscopy, which focuses on the intramolecular vibrations, for example, the OH stretch mode [73, 74]. These observations are rather indirect, as information is inferred from the IR-frequency spectator modes that are sensitive probes of the strength of hydrogen bonding to the environment, but may not necessarily provide a full picture of the complex collective intermolecular motions of the hydrogen-bond network. The possibility to use 2D Raman-THz spectroscopy to investigate the hydrogen bond in water has been examined theoretically recently [75, 76] as an alternative. THz-frequency intermolecular dynamics of the hydrogen bond, including the librational motion (~ 20 THz), and the stretch and bend vibrations (< 10 THz) [38], may reveal direct information about the molecular dynamics related to the hydrogen bond of water under ambient conditions. Specifically, a photon echo signal investigating the heterogeneity of the hydrogen-bond network would allow one to infer

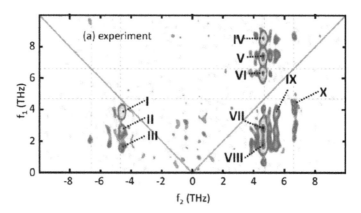

Fig. 25 Experimental 2D THz spectrum from $CHBr_3$. The spectral peaks are labeled. The peaks in the first quadrant are NR signals while those in the second quadrant are R signals. Adapted with permission from [38]

the time scale of a stable network pattern and thus provide mechanistic insights into the microscopic structure of the hydrogen bond in neat water and aqueous solutions [39].

The first 2D Raman-THz spectroscopy experiments have been realized in water and aqueous solutions recently [38], yielding 2D time-domain signals including that shown in Fig. 26. The observation of photon echoes in neat water and various aqueous salt solutions yields the hydrogen-bond relaxation times for these systems. The experimental setup was the same as shown in Fig. 9. The relevant light-matter interaction is elaborated by the pulse sequence shown in Fig. 4c and the Feynman diagrams describing typical rephasing pathways shown in Fig. 26a. For example, in diagram (i), the initial state of the system is a population $|0\rangle\langle0|$. The Raman excitation pulse induces a first-order vibrational coherence $|0\rangle\langle1|$ that evolves during time period τ (or t_1 in Fig. 26b). This 1QC originates presumably from the broad range of collective intermolecular modes involving the hydrogen bond bend and stretch vibrations and shows rapid decay. The THz field interacts with the system once, generating a second-order rephasing coherence $|2\rangle\langle1|$ through a two-quantum transition, for example, to a combination band. The rephasing coherence evolves and emits the nonlinear THz-frequency signals during t (or t_2 in Fig. 20b). The nonlinear signals are detected by EOS. The photon echo signal lies along the diagonal $\tau = t$ (or $t_1 = t_2$ in Fig. 26b) in the 2D time-domain trace, i.e., as usual the echo signal emerges after a delay t equal to the inter-pulse delay τ. The 2D Raman-THz-THz time-domain signal from neat water is shown in Fig. 26b in comparison to the instrument response function (IRF) in Fig. 26c, which assumes the sample response to behave as a δ-function. The difference between them shows evidence of a photon echo signal along the diagonal $\tau = t$ with an average relaxation time of less than 100 fs [39].

In a follow-up study [40], several aqueous salt solutions were investigated using the same 2D Raman-THz-THz method. The experimental 2D Raman-THz-THz time-domain signals from these solutions are shown in Fig. 27. In these solutions, the presence of different cations changes the viscosity of water presumably by

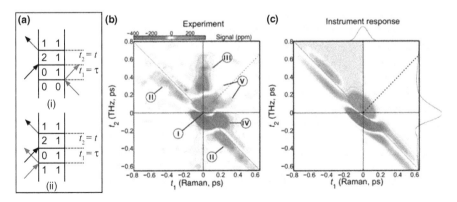

Fig. 26 **a** Feynman diagrams shows the typical rephasing pathways that describe the Raman-THz-THz photon echo signals. The states may denote different states from two intermolecular vibrational modes. **b** 2D Raman-THz time-domain signal of neat water and **c** the instrument response function. From [39]

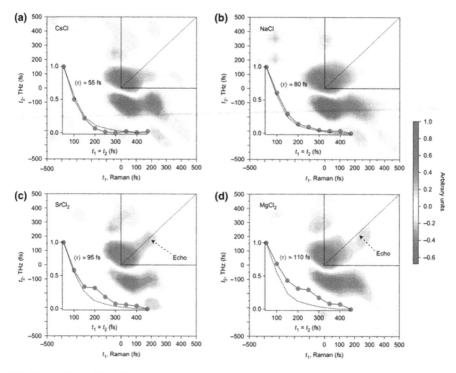

Fig. 27 2D Raman-THz-THz responses of various aqueous salt solutions. The photon echoes lie on the diagonals (dashed lines). The insets show the photon echoe signals as 1D cuts along the diagonals, which give average relaxation times from different solutions. From [40]

modulating the hydrogen-bond network via ion–water interactions. The 2D Raman-THz-THz echoes report on the inhomogeneity of the hydrogen-bond network and can therefore provide insight into the range of water–ion interactions. The decays of the photon echo signals in these solutions show subtle differences. The average relaxation times extracted from the photon echo signals show a strong correlation with the effects of the corresponding cations on the viscosity of water [40].

4.4 Extensions of 2D THz and 2D THz-Raman Vibrational Spectroscopies

The 2D spectroscopies utilizing high-frequency THz pulses and hybrid sequences of THz and Raman pulses have proved successful in measuring the nonresonant two-phonon coherences in semiconductors and intramolecular and intermolecular dynamics in liquids including simple halogenated liquids, water, and aqueous salt solutions. The methodology provides new information about low-frequency motions that are important in a wide range of systems including crystalline solids, simple liquids, biomolecules, polymers, glasses, and other materials. All-THz 2D spectroscopy also should prove useful for further progress. Improvements in the excitation THz field strength and bandwidth [22, 46] as well as data acquisition methods utilizing single-shot signal detection [77] could accelerate the use of these approaches for study of condensed phase molecular dynamics.

5 2D THz Magnetic Resonance Spectroscopy

The examples discussed in the above two sections all utilize the THz electric field for 2D spectroscopy of molecular rotational and vibrational degrees of freedom. It is also possible to use the THz magnetic field to drive the spin degrees of freedom in molecular and condensed matter systems and to conduct 2D THz magnetic resonance spectroscopy, which is directly relevant to chemistry and biology. In this section, we will discuss 2D THz magnetic resonance spectroscopy and show the first example of its application to collective spin waves (magnons) in a magnetic material.

5.1 Background and Motivation

Nonlinear manipulation of spins is the basis for all advanced methods in magnetic resonance including multidimensional nuclear magnetic resonance and electron paramagnetic resonance (EPR) spectroscopies [78, 79], magnetic resonance imaging, and, in recent years, quantum control over individual spins [80]. The methodology is facilitated by the ease with which the strong-field regime can be reached for radiofrequency or microwave magnetic fields that drive nuclear or electron spins, respectively, typified by sequences of magnetic pulses that control the magnetic moment directions [78–80]. The capabilities meet a bottleneck, however, for far-infrared magnetic resonances, which are characteristic of molecular complexes including molecular magnets [81] and metalloproteins [82] containing high-spin transition-metal or rare-earth ions. In these systems, zero-field splittings (ZFSs) due to high magnetic anisotropy and/or spin-spin interactions result in transition frequencies in the THz frequency region even in the absence of external magnetic fields. Measurements of the spin resonances originating from ZFSs can provide mechanistic insight into molecular magnetic properties and protein catalytic function as the ZFSs show exquisite sensitivity to ligand geometries and transition metal electronic structure. With strong applied magnetic fields (~ 10 T), resonances of unpaired electron spins in molecular complexes and metalloproteins can be shifted from the usual microwave regime into the THz range, thereby drastically improving the resolution of spectral splittings [79, 83, 84].

Despite the critical importance of THz-frequency EPR spectroscopy, current EPR technology remains limited at THz frequencies because the weak sources used only permit measurements of linear responses, i.e., 1QC or free-induction decay (FID) signals. Utilization of the strong THz generation techniques discussed in Sect. 2.1.1 can circumvent this limitation. Nonlinear and 2D THz spectroscopy methods can allow the extension of established, commercially available multidimensional EPR spectroscopy from the microwave to the THz frequency range. To date, the only available example of 2D THz spectroscopy of the spin degree of freedom was conducted on magnons in a magnetic material [36]. We review it here in the expectation that the methodology will be extended to molecular and biomolecular samples.

5.2 2D THz Spectroscopy of Collective Spin Waves

Magnons are the elementary excitations in material systems with spin order such as ferromagnetic (FM) and antiferromagnetic (AFM) phases. In these systems, the high magnetic anisotropy and strong spin-spin interactions result in an intrinsic internal magnetic field that is commonly on the order of 10 T. As a result, magnon resonances are usually found in the THz range. Some of these materials have been studied with continuous-wave and pulsed THz fields, revealing the magnon frequencies through their FID signals [85, 86] and demonstrating linear superposition in the responses to time-delayed pulse pairs [55, 85]. So far, there are very limited examples of nonlinear THz driving of spins [87–89]. As in other types of 2D spectroscopy, 2D magnetic resonance allows the distinct nonlinear responses to be separated from each other and from linear responses.

An initial demonstration of 2D THz spectroscopy using THz magnetic fields was conducted on magnons in yttrium orthoferrite ($YFeO_3$ or YFO), which has canted AFM order, as shown in Fig. 22a. The static spin Hamiltonian describing the two sublattice spins is given by [90, 91],

$$H_0 = -J\mathbf{S}_1 \cdot \mathbf{S}_2 + \mathbf{D} \cdot (\mathbf{S}_1 \times \mathbf{S}_2) - \sum_{i=1}^{2}(K_a S_{ia}^2 + K_c S_{ic}^2). \tag{8}$$

The first term describes the AFM coupling between neighboring spins $\mathbf{S}_1$ and $\mathbf{S}_2$ with a positive exchange constant J. The second term derives from the Dzyaloshinskii-Moria (DM) spin-spin interaction with the antisymmetric exchange parameter $\mathbf{D}$, a vector along the crystal b-axis. As the first term favoring AFM order of the spins is much larger than the second term favoring orthogonal orientation between the two spins, the interplay between them results in the canted AFM order shown in Fig. 28a with a canting angle of about 0.45° [90, 91]. A net magnetization $\mathbf{M}$ is formed along the crystal c axis because of the canting. The third term accounts for the orthorhombic magnetic anisotropy, which is manifested as the ZFS of the unpaired spins of the high-spin Fe^{3+}, with K_a and K_c the magnetic anisotropy parameters along the crystal a and c axes.

The Zeeman interaction between the THz magnetic field $\mathbf{B}_{THz}$ and the sublattice spins $\mathbf{S}_1$ and $\mathbf{S}_2$ describes the light-matter interactions. The interaction Hamiltonian H_1 is given by

$$H_1 = \gamma \mathbf{B}_{THz} \cdot \sum_{i=1}^{2} \mathbf{S}_i, \tag{9}$$

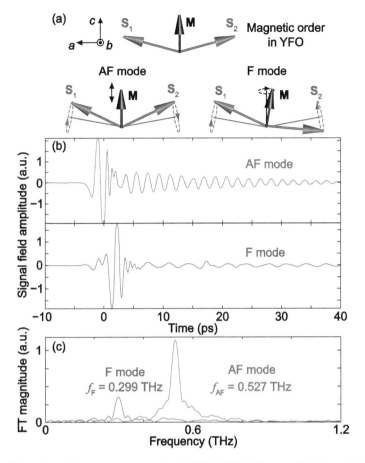

Fig. 28 **a** Spin order and the two magnon modes in YFO. **b** THz fields transmitted through the sample followed by FID signals of each magnon mode. **c** FT magnitude spectra of both magnon modes, resulting from a numerical Fourier transformation of the FID signals in **b**. From [36]

where γ is the gyromagnetic ratio. In the linear response region, two THz-active magnon modes, the quasi-AFM (AF) and quasi-FM (F) modes, can be constructed based respectively on the out-of-phase and in-phase cooperative precessional motions of the sublattice spins. Macroscopically, the AF mode corresponds to oscillation of the net magnetization amplitude and the F mode to precession about the net magnetization direction as shown in Fig. 28a. In YFO at room temperature, the AF mode at $f_{AF} = 0.527$ THz and the F mode at $f_F = 0.299$ THz can be selectively excited by THz pulses with magnetic field polarizations parallel and perpendicular respectively to the net magnetization direction. Experimentally, this is done by rotating the single-crystal sample while maintaining the polarization directions of the THz fields. In response to linear THz excitation, magnons radiate FID signals $B(t)$ at their resonance frequencies, revealing the linear-response spin dynamics. The FID signal fields are measured by EOS, and each magnon mode is resolved in the spectrum from Fourier transformation of the FID signals, as shown in Fig. 28b, c.

The experimental setup used is identical to that shown in Fig. 6. The sample is an a-cut single crystal of YFO. In response to time-delayed strong THz pulse pair excitation, nonlinear magnon responses are readily revealed in the time-domain signal of B_{NL}. An example of time-domain nonlinear signal measurements of the AF mode is shown in Fig. 29. In this case, the delay τ is selected to be 3.7 ps, which is twice the AF mode period. The magnon responses B_A and B_B induced by each THz pulse individually are in phase as shown in Fig. 29a. As a result, the AF mode response B_{AB} induced by both THz pulses shows a coherent enhancement in the signal amplitude as shown in Fig. 29b [55, 85]. The nonlinear signal B_{NL} is detected via the differential chopping detection method following $B_{NL} = B_{AB} - B_A - B_B$. In the trace of B_{NL}, weak oscillations that have a phase shift of $3\pi/2$ relative to B_{AB} and slight asymmetric distortions are observed. The shifted phase is that of a third-order response function, and the asymmetry indicates a second-harmonic generation (SHG) signal. Fourier transformation of the nonlinear signal reveals the nonlinear spectrum

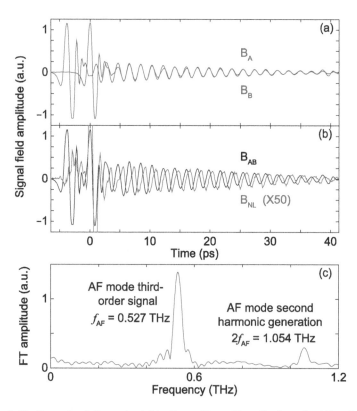

Fig. 29 **a, b** Nonlinear signal B_{NL}, extracted by $B_{NL} = B_{AB} - B_A - B_B$, from the AF mode magnons excited with a fixed inter-pulse delay of $\tau = 3.7$ ps. **c** Fourier transformation of the oscillatory signals in B_{NL} reveals the third- and second-order nonlinear signals at f_{AF} and $2f_{AF}$, respectively. From [36]

as shown in Fig. 29c. A third-order spectral peak is located at the fundamental AF mode frequency f_{AF}. The peak arises from a sum of the third-order signals resulting from different excitation pathways as detailed below. Magnon SHG signal is located at the second-harmonic AF mode frequency $2f_{AF}$.

The 2D B_{NL} trace of the AF mode was recorded as a function of the inter-pulse delay τ and detection time t by incrementing τ with a small step size and recording the t-dependent signal B_{NL} at each τ. Rotating the sample about the crystal a axis allowed excitation of the F mode, and the 2D B_{NL} trace of the F mode was recorded in the same manner. The 2D numerical Fourier transformation of the $B_{NL}(t, \tau)$ traces with respect to t and τ generated the 2D complex spectrum of each magnon mode. The 2D magnitude spectra of the two magnon modes in YFO are shown in Fig. 30. As in the 2D rotational spectra, the 2D magnetic resonance spectra are separated into NR and R quadrants because of the phase evolution of the magnon coherences during τ and t in the two different excitation pathways. In each magnon mode, the full set of $\chi^{(3)}$ signals is observed, and the R, NR, 2Q and PP signals appear at easily distinguished locations in the spectra. The relevant THz field-spin interaction pathways are elaborated as follows.

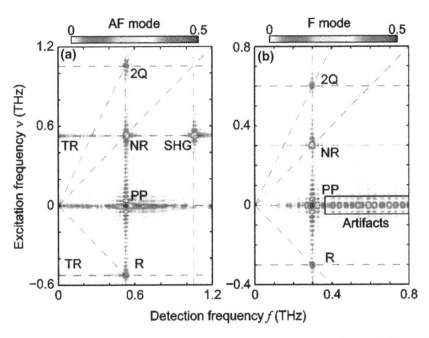

Fig. 30 2D THz magnetic resonance spectra of magnons in YFO. **a** Experimental AF mode 2D magnitude spectrum. Third-order spectral peaks include pump-probe (PP), non-rephasing (NR), rephasing (R), and 2-quantum (2Q) peaks. Second-order peaks include second harmonic generation (SHG) and THz rectification (TR) peaks. **b** Experimental F mode 2D magnitude spectrum showing the full set of third-order peaks. The artifacts are due to signal double-reflections in the sample. Both spectra are normalized and plotted according to the color scale shown. From [36]

The R (spin echo) peak and NR peak each result from a single field interaction during pulse A that creates a first-order magnon 1QC and, after delay τ, two field interactions during pulse B that generate a magnon population and then a third-order magnon 1QC (either phase-reversed or not relative to the first-order 1QC) that radiates the nonlinear signal. The 2Q peak arises from two field interactions during pulse A that create a 2QC that accumulates phase at twice the magnon frequency and, after time τ, one field interaction during pulse B that induces transitions to a third-order 1QC that radiates the signal. The 2Q signal reveals correlations between pairs of zone-center magnons [92], which are distinct from zone-boundary magnon correlations revealed in 2-magnon Raman spectra [93]. The PP signal is generated by two field interactions during pulse A that create the magnon population, and, after delay τ, one interaction with pulse B that generates a third-order 1QC that radiates the signal.

Relevant pathways of the various $\chi^{(3)}$ THz field-spin interactions described here are further elaborated by the double-sided Feynman diagrams shown in Fig. 31a. In the 2D spectrum of the AF mode, type-I $\chi^{(2)}$ signals due to SHG and THz rectification are also present, which are described by the Feynman diagrams shown in Fig. 31b. These signals are emitted by a $\chi^{(2)}$ magnetization due to the sum- and difference-frequency mixing of the magnon 1QCs generated by each THz pulse, which are given by

$$M^{(2)}(2\omega_{AF}) = \chi^{(2)}(2\omega_{AF};\omega_{AF},\omega_{AF})B_A(\omega_{AF})B_B(\omega_{AF}), \tag{10}$$

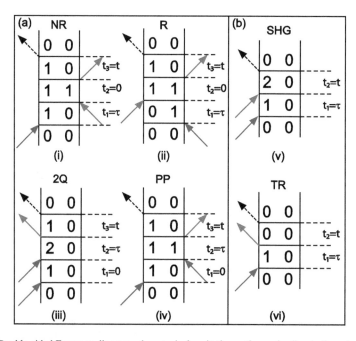

Fig. 31 Double-sided Feynman diagrams show typical excitation pathways leading to the coherent emission of third-order [**a**, (i)–(iv)] and second-order [**b**, (v)–(vi)] nonlinear signals. From [36]

$$M^{(2)}(0) = \chi^{(2)}(0; -\omega_{AF}, \omega_{AF})B_A^*(\omega_{AF})B_B(\omega_{AF}). \qquad (11)$$

These signals are observed because of the anharmonic precessions of the sublattice spins S_1 and S_2 whose excursions away from the equilibrium are about $\pm 0.5°$ [88]. $\chi^{(2)}$ signals are in principle also observable in the 2D spectrum of the F mode. But for the F mode, the $\chi^{(2)}$ signals that are expected to be parallel to the direction of the net magnetization are perpendicular to the fundamental field polarization. Phase-matching for the type-II $\chi^{(2)}$ process is not satisfied in the thick sample used in this experiment, as the birefringent YFO crystal has a large index difference between the two crystal axes along which the fundamental and $\chi^{(2)}$ magnetic fields are polarized.

5.3 Extensions of 2D THz EPR Spectroscopy

In this section, we have discussed an example of 2D THz magnetic resonance spectroscopy that has directly revealed the full set of $\chi^{(3)}$ nonlinear signals originating from the magnons in a magnetic material. The methodology is expected to be applicable to the study of many chemical and biologic systems with spin resonances at THz frequencies. To date, linear THz-frequency EPR spectroscopy has been conducted based on THz time-domain spectroscopy [94, 95], coherent synchrotron radiation [96], or blackbody radiation [97, 98]. These studies have measured the ZFSs in single-molecule magnets [95, 96] and other molecular and biologic systems [82, 95, 99] at zero and nonzero external magnetic field. We anticipate that 2D EPR spectroscopy of molecular complexes and biomolecules in the THz frequency range will be demonstrated and will provide wide-ranging new insights just as it has in lower frequencies.

6 Conclusion and Outlook

Utilizing existing THz spectroscopic techniques, it is now possible to record 2D THz spectra originating from various material degrees of freedom and phases, including the rotations of gas-phase molecules, spin precessions in a magnetic crystal, lattice vibrations in solids, and intra- and intermolecular dynamics in liquids. We note again that we have not reviewed 2D THz or multi-THz spectroscopy of electronic responses [27, 29, 30, 100]. Even including those examples, it is evident that 2D THz spectroscopy is still in a very nascent stage. The use of recently developed nonlinear optical crystals [22–24, 101] for generation of strong THz fields that span a wide frequency range, extending all the way to the long-wavelength IR region will substantially increase the scope of 2D THz studies. Further use of frequency-selected and otherwise tailored THz fields [102–104], a third THz pulse [33, 34] to establish an additional variable time period, multiple THz sources or beam paths to permit non-collinear THz field polarizations [37], and combined THz and optical fields for signal generation and detection [37, 39, 60] will extend the information content that can be extracted in a wide range of samples. Incorporating single-shot

THz detection methods [77] could drastically reduce the long data acquisition times that presently hinder virtually all 2D THz measurements. As in other frequency ranges including radio, microwave, infrared, and visible, multidimensional spectroscopy in the THz regime can reveal otherwise elusive structure, interactions, and dynamics. With further advances in the relevant technologies and refinements of the measurement methodology, 2D THz spectroscopy can be expected to take its place among the multidimensional spectroscopies that have produced broad and deep advances in all areas of chemistry.

Acknowledgements We thank Sharly Fleischer, Takayuki Kurihara, and Tohru Suemoto for contributions to this work. This work was supported, in part, by Office of Naval Research Grant N00014-13-1-0509 and Defense University Research Instrumentation Program Grant N00014-15-1-2879, National Science Foundation Grants CHE-1111557 and CHE-1665383, and the Samsung Global Research Outreach program.

References

1. Hamm P, Zanni MT (2011) Concepts and methods of 2D infrared spectroscopy. Cambridge University Press, Cambridge
2. Mukamel S (2000) Multidimensional femtosecond correlation spectroscopies of electronic and vibrational excitations. Annu Rev Phys Chem 51:691–729. https://doi.org/10.1146/annurev.phys chem.51.1.691
3. Jonas DM (2003) Two-dimensional femtosecond spectroscopy. Annu Rev Phys Chem 54:425–463. https://doi.org/10.1146/annurev.physchem.54.011002.103907
4. Zheng J, Kwak K, Fayer MD (2007) Ultrafast 2D IR vibrational echo spectroscopy. Acc Chem Res 40:75–83. https://doi.org/10.1021/ar068010d
5. Ramasesha K, De Marco L, Mandal A, Tokmakoff A (2013) Water vibrations have strongly mixed intra- and intermolecular character. Nat Chem 5:935–940. https://doi.org/10.1038/nchem.1757
6. Baiz CR, Reppert M, Tokmakoff A (2013) Amide I two-dimensional infrared spectroscopy: methods for visualizing the vibrational structure of large proteins. J Phys Chem A 117:5955–5961. http s://doi.org/10.1021/jp310689a
7. Krummel AT, Mukherjee P, Zanni MT (2003) Inter and intrastrand vibrational coupling in DNA studied with heterodyned 2D-IR spectroscopy. J Phys Chem B 107:9165–9169. https://doi.org/10.1021/JP035473H
8. Fuller FD, Ogilvie JP (2015) Experimental implementations of two-dimensional Fourier transform electronic spectroscopy. Annu Rev Phys Chem 66:667–690. https://doi.org/10.1146/annurev-phys chem-040513-103623
9. Stone KW, Gundogdu K, Turner DB et al (2009) Two-quantum 2D FT electronic spectroscopy of biexcitons in GaAs quantum wells. Science (80-) 324:1169–1173. https://doi.org/10.1126/scie nce.1170274
10. Turner DB, Nelson KA (2010) Coherent measurements of high-order electronic correlations in quantum wells. Nature 466:1089–1092. https://doi.org/10.1038/nature09286
11. Brixner T, Stenger J, Vaswani HM et al (2005) Two-dimensional spectroscopy of electronic couplings in photosynthesis. Nature 434:625–628. https://doi.org/10.1038/nature03429
12. Cheng Y-C, Fleming GR (2009) Dynamics of light harvesting in photosynthesis. Annu Rev Phys Chem 60:241–262. https://doi.org/10.1146/annurev.physchem.040808.090259
13. Eisele DM, Arias DH, Fu X et al (2014) Robust excitons inhabit soft supramolecular nanotubes. Proc Natl Acad Sci 111:E3367–E3375. https://doi.org/10.1073/pnas.1408342111
14. Krebs N, Pugliesi I, Hauer J, Riedle E (2013) Two-dimensional Fourier transform spectroscopy in the ultraviolet with sub-20 fs pump pulses and 250–720 nm supercontinuum probe. New J Phys 15:85016. https://doi.org/10.1088/1367-2630/15/8/085016
15. Lee Y-S (2009) Principles of terahertz science and technology. Springer, Berlin

16. Blanchard F, Razzari L, Bandulet HC et al (2007) Generation of 1.5 μJ single-cycle terahertz pulses by optical rectification from a large aperture ZnTe crystal. Opt Express 15:13212. https://doi.org/10.1364/OE.15.013212

17. Ferguson B, Zhang X-C (2002) Materials for terahertz science and technology. Nat Mater 1:26–33. https://doi.org/10.1038/nmat708

18. Schmuttenmaer CA (2004) Exploring dynamics in the far-infrared with terahertz spectroscopy. Chem Rev 104:1759–1779. https://doi.org/10.1021/cr020685g

19. Tonouchi M (2007) Cutting-edge terahertz technology. Nat Photonics 1:97–105. https://doi.org/10.1038/nphoton.2007.3

20. Yeh KL, Hoffmann MC, Hebling J, Nelson KA (2007) Generation of 10 μJ ultrashort terahertz pulses by optical rectification. Appl Phys Lett 90:171121. https://doi.org/10.1063/1.2734374

21. Hirori H, Doi A, Blanchard F, Tanaka K (2011) Single-cycle terahertz pulses with amplitudes exceeding 1 MV/cm generated by optical rectification in $LiNbO_3$. Appl Phys Lett 98:91106. https://doi.org/10.1063/1.3560062

22. Shalaby M, Hauri CP (2015) Demonstration of a low-frequency three-dimensional terahertz bullet with extreme brightness. Nat Commun 6:5976. https://doi.org/10.1038/ncomms6976

23. Vicario C, Ruchert C, Hauri CP (2015) High field broadband THz generation in organic materials. J Mod Opt 62:1480–1485. https://doi.org/10.1080/09500340.2013.800242

24. Lee S-H, Jazbinsek M, Hauri CP, Kwon O-P (2016) Recent progress in acentric core structures for highly efficient nonlinear optical crystals and their supramolecular interactions and terahertz applications. CrystEngComm 18:7180–7203. https://doi.org/10.1039/C6CE00707D

25. Kampfrath T, Tanaka K, Nelson KA (2013) Resonant and nonresonant control over matter and light by intense terahertz transients. Nat Photonics 7:680–690. https://doi.org/10.1038/nphoton.2013.184

26. Hwang HY, Fleischer S, Brandt NC et al (2015) A review of non-linear terahertz spectroscopy with ultrashort tabletop-laser pulses. J Mod Opt 62:1447–1479. https://doi.org/10.1080/09500340.2014.918200

27. Woerner M, Kuehn W, Bowlan P et al (2013) Ultrafast two-dimensional terahertz spectroscopy of elementary excitations in solids. New J Phys 15:25039. https://doi.org/10.1088/1367-2630/15/2/025039

28. Kuehn W, Reimann K, Woerner M et al (2011) Strong correlation of electronic and lattice excitations in GaAs/AlGaAs semiconductor quantum wells revealed by two-dimensional terahertz spectroscopy. Phys Rev Lett 107:67401. https://doi.org/10.1103/PhysRevLett.107.067401

29. Somma C, Reimann K, Flytzanis C et al (2014) High-field terahertz bulk photovoltaic effect in lithium niobate. Phys Rev Lett 112:146602. https://doi.org/10.1103/PhysRevLett.112.146602

30. Maag T, Bayer A, Baierl S et al (2015) Coherent cyclotron motion beyond Kohn's theorem. Nat Phys 11:1–6. https://doi.org/10.1038/NPHYS3559

31. Lu J, Zhang Y, Hwang HY et al (2016) Nonlinear two-dimensional terahertz photon echo and rotational spectroscopy in the gas phase. Proc Natl Acad Sci USA 113:11800–11805. https://doi.org/10.1073/pnas.1609558113

32. Lu J, Zhang Y, Hwang HY et al (2016) Two-dimensional terahertz photon echo and rotational spectroscopy in the gas phase. In: International conference on ultrafast phenomena. OSA Technical Digest (online), Optical Society of America, Paper UTu1A.6. https://doi.org/10.1364/UP.2016.UTu1A.6

33. Somma C, Folpini G, Reimann K et al (2016) Two-phonon quantum coherences in indium antimonide studied by nonlinear two-dimensional terahertz spectroscopy. Phys Rev Lett 116:1–6. https://doi.org/10.1103/PhysRevLett.116.177401

34. Somma C, Folpini G, Reimann K et al (2016) Phase-resolved two-dimensional terahertz spectroscopy including off-resonant interactions beyond the $\chi^{(3)}$ limit. J Chem Phys 144:184202. https://doi.org/10.1063/1.4948639

35. Lu J, Li X, Hwang HY et al (2016) 2D nonlinear terahertz magnetic resonance spectroscopy of magnons in a canted antiferromagnet. In: International conference on ultrafast phenomena. OSA Technical Digest (online), Optical Society of America, Paper UTh3A.2. https://doi.org/10.1364/UP.2016.UTh3A.2

36. Lu J, Li X, Hwang HY et al (2017) Coherent two-dimensional terahertz magnetic resonance spectroscopy of collective spin waves. Phys Rev Lett 118:207204. https://doi.org/10.1103/PhysRevLett.118.207204

 Springer

37. Finneran IA, Welsch R, Allodi MA et al (2016) Coherent two-dimensional terahertz-terahertz-Raman spectroscopy. Proc Natl Acad Sci USA 113:6857–6861. https://doi.org/10.1073/pnas.1605 631113

38. Finneran IA, Welsch R, Allodi MA et al (2017) 2D THz-THz-Raman photon-echo spectroscopy of molecular vibrations in liquid bromoform. J Phys Chem Lett 8:4640–4644. https://doi.org/10.1021/acs.jpclett.7b02106

39. Savolainen J, Ahmed S, Hamm P (2013) Two-dimensional Raman-terahertz spectroscopy of water. Proc Natl Acad Sci USA 110:20402–20407. https://doi.org/10.1073/pnas.1317459110

40. Shalit A, Ahmed S, Savolainen J, Hamm P (2016) Terahertz echoes reveal the inhomogeneity of aqueous salt solutions. Nat Chem 9:273–278. https://doi.org/10.1038/nchem.2642

41. Boyd R (2007) Nonlinear optics. Academic Press, Boston

42. Hebling J, Almasi G, Kozma I, Kuhl J (2002) Velocity matching by pulse front tilting for large area THz-pulse generation. Opt Express 10:1161. https://doi.org/10.1364/OE.10.001161

43. Hebling J, Yeh K-L, Hoffmann MC et al (2008) Generation of high-power terahertz pulses by tilted-pulse-front excitation and their application possibilities. J Opt Soc Am B 25:B6–B19. https://doi.org/10.1364/JOSAB.25.0000B6

44. Vicario C, Jazbinsek M, Ovchinnikov AV et al (2015) High efficiency THz generation in DSTMS, DAST and OH1 pumped by Cr:forsterite laser. Opt Express 23:4573–4580. https://doi.org/10.1364/OE.23.004573

45. Dai J, Liu J, Zhang XC (2011) Terahertz wave air photonics: Terahertz wave generation and detection with laser-induced gas plasma. IEEE J Sel Top Quantum Electron 17:183–190. https://doi.org/10.1109/JSTQE.2010.2047007

46. Clough B, Dai J, Zhang XC (2012) Laser air photonics: beyond the terahertz gap. Mater Today 15:50–58

47. Carr GL, Martin MC, McKinney WR et al (2002) High-power terahertz radiation from relativistic electrons. Nature 420:153–156. https://doi.org/10.1038/nature01175

48. Wu Z, Fisher AS, Goodfellow J et al (2013) Intense terahertz pulses from SLAC electron beams using coherent transition radiation. Rev Sci Instrum 84:22701. https://doi.org/10.1063/1.4790427

49. Wu Q, Zhang XC (1995) Free-space electro-optic sampling of terahertz beams. Appl Phys Lett 67:3523. https://doi.org/10.1063/1.114909

50. Nahata A, Auston DH, Heinz TF, Wu C (1996) Coherent detection of freely propagating terahertz radiation by electro-optic sampling. Appl Phys Lett 68:150. https://doi.org/10.1063/1.116130

51. Novelli F, Fausti D, Giusti F et al (2013) Mixed regime of light-matter interaction revealed by phase sensitive measurements of the dynamical Franz-Keldysh effect. Sci Rep 3:1227. https://doi.org/10.1038/srep01227

52. Pein BC, Chang W, Hwang HY et al (2017) Terahertz-driven luminescence and colossal Stark effect in CdSe-CdS colloidal quantum dots. Nano Lett 17:5375–5380. https://doi.org/10.1021/acs.nanolett.7b01837

53. Cook DJ, Chen JX, Morlino EA, Hochstrasser RM (1999) Terahertz-field-induced second-harmonic generation measurements of liquid dynamics. Chem Phys Lett 309:221–228. https://doi.org/10.1016/S0009-2614(99)00668-5

54. Hoffmann MC, Brandt NC, Hwang HY et al (2009) Terahertz Kerr effect. Appl Phys Lett 95:231105. https://doi.org/10.1063/1.3271520

55. Kampfrath T, Sell A, Klatt G et al (2011) Coherent terahertz control of antiferromagnetic spin waves. Nat Photonics 5:31–34. https://doi.org/10.1038/nphoton.2010.259

56. Allodi MA, Finneran IA, Blake GA (2015) Nonlinear terahertz coherent excitation of vibrational modes of liquids. J Chem Phys 143:234204. https://doi.org/10.1063/1.4938165

57. Hwang HY, Hoffmann MC, Brandt NC, Nelson KA (2010) THz Kerr effect in relaxor ferroelectrics. In: International conference on ultrafast phenomena. OSA Technical Digest (CD), Optical Society of America, Paper ThE41. https://doi.org/10.1364/UP.2010.ThE41

58. Lu J, Li X, Hwang HY et al (2016) Terahertz Kerr effect in an organic ferroelectric. In: International conference on ultrafast phenomena. OSA Technical Digest (online), Optical Society of America, Paper UW4A.14. https://doi.org/10.1364/UP.2016.UW4A.14

59. Fleischer S, Zhou Y, Field RW, Nelson KA (2011) Molecular orientation and alignment by intense single-cycle THz pulses. Phys Rev Lett 107:163603. https://doi.org/10.1103/PhysRevLett.107.1636 03

60. Fleischer S, Field RW, Nelson KA (2012) Commensurate two-quantum coherences induced by time-delayed THz fields. Phys Rev Lett 109:123603. https://doi.org/10.1103/PhysRevLett.109.1236 03

61. Hamm P (2005) Principles of nonlinear optical spectroscopy : a practical approach or : Mukamel for dummies. University of Zurich. http://www.mitr.p.lodz.pl/evu/lectures/Hamm.pdf

62. Tokmakoff A (2014) Time-dependent quantum mechanics and spectroscopy. University of Chicago. http://tdqms.uchicago.edu/

63. Hamm P, Shalit A (2017) Perspective: echoes in 2D-Raman–THz spectroscopy. J Chem Phys 146:130901. https://doi.org/10.1063/1.4979288

64. Tokmakoff A, Lang M, Larsen D et al (1997) Two-dimensional Raman spectroscopy of vibrational interactions in liquids. Phys Rev Lett 79:2702–2705. https://doi.org/10.1103/PhysRevLett.79.2702

65. Frostig H, Bayer T, Dudovich N et al (2015) Single-beam spectrally controlled two-dimensional Raman spectroscopy. Nat Photonics 9:339–343. https://doi.org/10.1038/nphoton.2015.64

66. Harde H, Keiding S, Grischkowsky D (1991) THz commensurate echoes: periodic rephasing of molecular transitions in free-induction decay. Phys Rev Lett 66:1834–1837. https://doi.org/10.1103/PhysRevLett.66.1834

67. Fleischer S, Averbukh IS, Prior Y (2007) Selective alignment of molecular spin isomers. Phys Rev Lett 99:93002. https://doi.org/10.1103/PhysRevLett.99.093002

68. Karras G, Hertz E, Billard F et al (2015) Orientation and alignment echoes. Phys Rev Lett 114:153601. https://doi.org/10.1103/PhysRevLett.114.153601

69. Engel GS, Calhoun TR, Read EL et al (2007) Evidence for wavelike energy transfer through quantum coherence in photosynthetic systems. Nature 446:782–786. https://doi.org/10.1038/nature05 678

70. Roberts ST, Loparo JJ, Tokmakoff A (2006) Characterization of spectral diffusion from two-dimensional line shapes. J Chem Phys 125:84502. https://doi.org/10.1063/1.2232271

71. Damari R, Kallush S, Fleischer S (2016) Rotational control of asymmetric molecules: dipole- versus polarizability-driven rotational dynamics. Phys Rev Lett 117:103001. https://doi.org/10.1103/PhysRevLett.117.103001

72. Damari R, Rosenberg D, Fleischer S (2017) Coherent radiative decay of molecular rotations: a comparative study of terahertz-oriented versus optically aligned molecular ensembles. Phys Rev Lett 119:33002. https://doi.org/10.1103/PhysRevLett.119.033002

73. Fecko CJ, Eaves JD, Loparo JJ, Tokmakoff A, Geissler PL (2003) Ultrafast hydrogen-bond dynamics in the infrared spectroscopy of water. Science (80-) 301:1698–1702. https://doi.org/10.1126/science.1087251

74. Asbury JB, Steinel T, Kwak K et al (2004) Dynamics of water probed with vibrational echo correlation spectroscopy. J Chem Phys 121:12431–12446. https://doi.org/10.1063/1.1818107

75. Hamm P, Savolainen J (2012) Two-dimensional-Raman–terahertz spectroscopy of water: theory. J Chem Phys 136:94516. https://doi.org/10.1063/1.3691601

76. Hamm P, Savolainen J, Ono J, Tanimura Y (2012) Note: Inverted time-ordering in two-dimensional-Raman–terahertz spectroscopy of water. J Chem Phys 136:236101

77. Teo SM, Ofori-Okai BK, Werley CA, Nelson KA (2015) Invited article: Single-shot THz detection techniques optimized for multidimensional THz spectroscopy. Rev Sci Instrum 86:51301. https://doi.org/10.1063/1.4921389

78. Bajaj VS, Mak-Jurkauskas ML, Belenky M et al (2009) Functional and shunt states of bacteriorhodopsin resolved by 250 GHz dynamic nuclear polarization-enhanced solid-state NMR. Proc Natl Acad Sci USA 106:9244–9249. https://doi.org/10.1073/pnas.0900908106

79. Borbat PP, Costa-Filho AJ, Earle KA et al (2001) Electron spin resonance in studies of membranes and proteins. Science (80-) 291:266–269. https://doi.org/10.1126/science.291.5502.266

80. Koppens FHL, Buizert C, Tielrooij KJ et al (2006) Driven coherent oscillations of a single electron spin in a quantum dot. Supplementary notes. Nature 442:766–771. https://doi.org/10.1038/natu re05065

81. Craig GA, Murrie M (2015) 3d single-ion magnets. Chem Soc Rev 44:2135–2147. https://doi.org/10.1039/C4CS00439F

82. Nehrkorn J, Martins BM, Holldack K et al (2013) Zero-field splittings in metHb and metMb with aquo and fluoro ligands: a FD-FT THz-EPR study. Mol Phys 111:2696–2707. https://doi.org/10.1080/00268976.2013.809806

83. Andersson KK, Schmidt PP, Katterle B et al (2003) Examples of high-frequency EPR studies in bioinorganic chemistry. J Biol Inorg Chem 8:235–247. https://doi.org/10.1007/s00775-002-0429-0

84. Möbius K, Savitsky A, Wegener C et al (2005) Combining high-field EPR with site-directed spin labeling reveals unique information on proteins in action. Magn Reson Chem 43:S4–S19. https://doi.org/10.1002/mrc.1690

85. Yamaguchi K, Nakajima M, Suemoto T (2010) Coherent control of spin precession motion with impulsive magnetic fields of half-cycle terahertz radiation. Phys Rev Lett 105:237201. https://doi.org/10.1103/PhysRevLett.105.237201

86. Kozlov GV, Lebedev SP, Mukhin AA et al (1993) Submillimeter backward-wave oscillator spectroscopy of the rare-earth orthoferrites. IEEE Trans Magn 29:3443–3445. https://doi.org/10.1109/20.281190

87. Baierl S, Hohenleutner M, Kampfrath T et al (2016) Nonlinear spin control by terahertz-driven anisotropy fields. Nat Photonics 10:715–718. https://doi.org/10.1038/nphoton.2016.181

88. Baierl S, Mentink JH, Hohenleutner M et al (2016) Terahertz-driven nonlinear spin response of antiferromagnetic nickel oxide. Phys Rev Lett 117:197201. https://doi.org/10.1103/PhysRevLett.117.197201

89. Mukai Y, Hirori H, Yamamoto T et al (2016) Nonlinear magnetization dynamics of antiferromagnetic spin resonance induced by intense terahertz magnetic field. New J Phys 18:13045. https://doi.org/10.1088/1367-2630/18/1/013045

90. Herrmann GF (1963) Resonance and high frequency susceptibility in canted antiferromagnetic substances. J Phys Chem Solids 24:597–606. https://doi.org/10.1016/S0022-3697(63)80001-3

91. Herrmann GF (1964) Magnetic resonances and susceptibility in orthoferrites. Phys Rev 133:A1334–A1344. https://doi.org/10.1103/PhysRev.133.A1334

92. Morello A, Stamp PCE, Tupitsyn IS (2006) Pairwise decoherence in coupled spin qubit networks. Phys Rev Lett 97:207206. https://doi.org/10.1103/PhysRevLett.97.207206

93. Fleury PA, Loudon R (1968) Scattering of light by one- and two-magnon excitations. Phys Rev 166:514–530. https://doi.org/10.1103/PhysRev.166.514

94. Kozuki K, Nagashima T, Hangyo M (2011) Measurement of electron paramagnetic resonance using terahertz time-domain spectroscopy. Opt Express 19:24950. https://doi.org/10.1364/OE.19.024950

95. Lu J, Li X, Skorupskii G et al (2017) Rapid and precise determination of zero-field splittings by terahertz time-domain electron paramagnetic resonance spectroscopy. Chem Sci 8:7312–7323. https://doi.org/10.1039/c7sc00830a

96. Schnegg A, Behrends J, Lips K et al (2009) Frequency domain Fourier transform THz-EPR on single molecule magnets using coherent synchrotron radiation. Phys Chem Chem Phys 11:6820–6825. https://doi.org/10.1039/b905709a

97. Champion PM, Sievers AJ (1977) Far infrared magnetic resonance in FeSiF$_6$·6H$_2$O and Fe(SPh)$_4^{2-}$. J Chem Phys 66:1819–1825. https://doi.org/10.1063/1.434200

98. Brackett GC (1971) Far-infrared magnetic resonance in Fe(III) and Mn(III) porphyrins, myoglobin, hemoglobin, ferrichrome A, and Fe(III) dithiocarbamates. J Chem Phys 54:4383. https://doi.org/10.1063/1.1674688

99. Nehrkorn J, Telser J, Holldack K et al (2015) Simulating frequency-domain electron paramagnetic resonance: bridging the gap between experiment and magnetic parameters for high-spin transition-metal ion complexes. J Phys Chem B 119:13816–13824. https://doi.org/10.1021/acs.jpcb.5b04156

100. Kuehn W, Reimann K, Woerner M et al (2011) Two-dimensional terahertz correlation spectra of electronic excitations in semiconductor quantum wells. J Phys Chem B 115:5448–5455. https://doi.org/10.1021/jp1099046

101. Lee S-H, Lu J, Lee S-J et al (2017) Benzothiazolium single crystals: a new class of nonlinear optical crystals with efficient THz wave generation. Adv Mater 29:1701748. https://doi.org/10.1002/adma.201701748

102. Chen Z, Zhou X, Werley CA, Nelson KA (2011) Generation of high power tunable multicycle terahertz pulses. Appl Phys Lett 99:71102. https://doi.org/10.1063/1.3624919

103. Lu J, Hwang HY, Li X et al (2015) Tunable multi-cycle THz generation in organic crystal HMQ-TMS. Opt Express 23:22723–22729. https://doi.org/10.1364/OE.23.022723

104. Liu B, Bromberger H, Cartella A et al (2017) Generation of narrowband, high-intensity, carrier-envelope phase-stable pulses tunable between 4 and 18 THz. Opt Lett 42:129. https://doi.org/10.1364/OL.42.000129

CPSIA information can be obtained
at www.ICGtesting.com
Printed in the USA
LVHW082052190519
618387LV00001B/130/P

9 783030 024772